Handbook of Air Pollution Analysis

Handbook of
Air Pollution Analysis

Edited by Roger Perry
and Robert J. Young

LONDON
CHAPMAN AND HALL

A Halsted Press Book
John Wiley & Sons, New York

First published 1977
by Chapman and Hall Ltd
11 New Fetter Lane, London EC4P 4EE
© 1977 Chapman and Hall Ltd
Typeset by Alden Press, Oxford, London and Northampton
and printed in Great Britain at the
University Printing House, Cambridge

ISBN 0 412 12660 5

 MAR. 5 1979

Distribution in the U.S.A. by Halsted Press,
a Division of John Wiley & Sons, Inc., New York

Library of Congress Cataloging in Publication Data
Main entry under title:

Handbook of air pollution analysis.

  "A Halsted Press book."
  Includes bibliographical references.
  1. Air-Pollution-Measurement. I. Perry, Roger,
1940-   II. Young, Robert, 1949-
TD890.H36   628.5'3   77-12646
ISBN 0-470-99316-2

Contents

Preface

This book is intended to provide a comprehensive working knowledge of the theory and practice of air pollution analysis. Its contributors are drawn from a wide range of backgrounds and have written from extensive personal experience in the field. An attempt has been made both to review recently reported developments in air pollution analysis and to present detailed descriptions of established analytical procedures which have been used in the evaluation of air pollution problems of various types.

In preparation of the text, considerable emphasis has been placed upon a uniformity of approach by the contributing authors thus enabling the book to be used as a comprehensive reference and working manual. In order to achieve this objective and, in addition, to draw to the full upon the specific expertise of the contributors, several of the chapters although written principally by one or two authors have been extended or modified in some form.

The contributing authors include Robert L. Byer, Richard A. Cox, Roy M. Harrison, Stephen E. Hrudey, Alun E. McIntyre, David J. Moore, Ed. Murray, Don Ratnayaka and we are extremely grateful to them for their considerable effort in the preparation of this book. It was with deep regret that we learned of the death of James Parker shortly after he had submitted his valuable contribution.

Our thanks are also due to John Cima for preparation of the figures used and to Ian Davies for help with the manuscript preparation.

Roger Perry
Robert J. Young

Contributors

R.L. Byer (Chapter 8) Department of Physics, Stanford University, California, U.S.A.

R.A. Cox (Chapters 4 and 5) Environmental and Medical Sciences Division, Atomic Energy Research Establishment, Harwell, Oxfordshire, U.K.

R.M. Harrison (Chapters 2 and 3) Department of Environment Sciences, University of Lancaster, U.K.

S.E. Hrudey (Chapter 1) Department of Civil Engineering, University of Alberta, Edmonton, Canada

A.E. McIntyre (Chapter 7) Public Health and Water Resource Engineering, Imperial College, London, U.K.

D.J. Moore (Chapter 9) Central Electricity Generating Board, Research Laboratories, Leatherhead, Surrey, U.K.

E. Murray (Chapter 8) Department of Physics, Stanford University, California, U.S.A.

J. Parker (Chapter 2) Air Pollution Section, Warren Spring Laboratory, Stevenage, Herts., U.K.

R. Perry (Chapter 6) Public Health and Water Resource Engineering, Imperial College, London, U.K.

D. Ratnayaka (Chapter 7) Binnie and Partners, Consulting Engineers, London, U.K.

R.J. Young (Chapter 6) Public Health and Water Resource Engineering, Imperial College, London, U.K.

CHAPTER ONE

General sampling techniques

1.1 Sampling goals and requirements

1.1.1 Ambient sampling

1.1.1.1 General objectives

Ambient air sampling may be considered the collection of air samples in any unconfined location exposed to the atmosphere. Within this broad classification an infinite variety of air sampling locations and air mass quality will exist, from samples collected 0.5 m from the ground in a busy parking lot to samples collected from a free floating balloon over the Atlantic Ocean. Because of the variety of sampling schemes which may be used to obtain 'ambient' air samples, the objectives to be achieved by the sampling programme and subsequent analysis must be thoroughly and clearly defined before proceeding further.

Among the objectives which may commonly form the basis for ambient air sampling programmes are [1–3]:

The determination of community air quality as related to local health, social and environmental effects.

The determination of the influence of specific emission sources or groups of sources on local air quality.

The generation of information to aid in planning overall pollution control and industrial and municipal zoning strategies.

Research into topics such as procedures for identification of emission source contributions or mechanisms of air pollutant reaction and dispersion.

In order to achieve one or more of the stated objectives, the sampling system must be considered as the 'receptor' for the influence or effect to be measured. All subsequent planning of the sampling system to be used must be performed with constant referral, at each decision step, back to the consequence that the decision will have on the success of the sampling system as a valid 'receptor'.

It is often a useful exercise to seek data from past or current air monitoring systems [4–7] and determine the advantages and shortcomings of such data when manipulated in the manner necessary to achieve the study objectives. The shortcomings of a current data source may often suggest shortcomings in the proposed sampling system which can be rectified.

In conjunction with satisfying the study objectives, the definitive planning of ambient air sampling systems requires consideration of meteorological factors, sampling site criteria and sample scheduling. These considerations will be dealt with in turn.

1.1.1.2 Meteorological considerations

Meteorology is the study of the physics and geography of atmospheric phenomena. The nature and changes of the atmosphere constitute what we perceive as weather [8].

Atmospheric phenomena play an important role in the determination of ambient air quality. Diurnal and seasonal fluctuations in source emissions tend to be reflected in ambient air pollution levels but the diurnal and seasonal variations in meteorological conditions superimpose an effect on those due to emission variations. Clearly, the objectives of a sampling programme will not normally be attained if due consideration is not given to the assessment of meteorological information.

Meteorological parameters have varying degrees of influence on ambient air quality. The parameters of major significance are atmospheric stability, wind speed and direction and precipitation. Other influences may be attributed to temperature, humidity and solar radiation.

The understanding of atmospheric stability is based on the knowledge that air, being a compressible gas, will cool upon expansion and will heat upon compression. Because of the downward pressure of the air mass above the earth's surface, a parcel of air at the earth's surface experiences a greater atmospheric pressure than a parcel of air experiences at some height above that point. Thus, if a parcel of air at a relatively high pressure at the earth's surface were moved to a higher altitude, it would experience a lower pressure which would allow it to expand. As the parcel of air expands, it cools. As a result of this condition and the fact that the atmosphere is heated from the surface of the earth outwards, the air temperature should be expected to decrease with increasing altitude. The rate of decrease of temperature with altitude is called the lapse rate. The dry adiabatic lapse rate is the rate of decrease in temperature of a parcel of unsaturated air as it moves upward without exchanging heat with the surrounding air and is equal to 9.8° C per 1000 m. If the air parcel is saturated, water will tend to condense as cooler temperatures are encountered. Since the condensation will release latent heat, the effective lapse rate due to expansion cooling will be lessened. The saturated adiabatic lapse rate is variable but is consistently less than 9.8° C per 1000 m.

Atmospheric stability depends on the lapse rate in that under clear sunny conditions the air near the earth's surface is heated, as the surface is warmed by solar radiation, whereupon air expands, becomes buoyant and rises. The air will continue to rise until it reaches air at the same density, which for a given altitude, will be at the same temperature. Under these conditions, the lapse rate would normally be positive so that the rising air parcel will continue to encounter cooler air and will continue to rise until it reaches equilibrium with its surroundings. If, however, the lapse rate is less than the adiabatic rate or is inverted (temperature increase with altitude), the air parcel will not rise above its current level because it would tend to lose temperature with altitude faster than the temperature change with altitude in the surrounding air. As a result, there would be no buoyant force to encourage further rising. This condition is stable, in that the air mass at the surface is prevented from mixing vertically.

The implications of stability conditions to the quality of surface air receiving consistent emission sources are clearly important. The lapse rate itself is very dependent upon the thermal capacity and physical properties of the ground [3]. Lapse rates will be found to vary from urban to suburban to rural environments. Details of the causes and air pollution effects of atmospheric stability have been discussed by Scorer [9].

Unfortunately, the continuous reliable determination of stability conditions is difficult for many ambient sampling applications. For weather purposes the lapse rate is determined by means of a radiosonde, a balloon equipped to measure temperature and other meteorological parameters as it rises and telemeter them back to a receiving station. By measuring the temperature profile with altitude, the lapse rate and, therefore, the stability can be determined. Regular determination of the lapse rate by this method is beyond the means of most ambient sampling systems. However, estimates of the lapse rate may be obtained by continuous measuring of the temperature at the level of the sampler and at the top of a pole several meters above. This method requires care to ensure that temperature anomalies at the two measurement points are not introduced by variable solar or wind exposure. In some cases, stability information may be obtained from the local weather office. In all cases, consideration must be given to the possible effects that stability changes may have upon the air samples taken.

Wind speed and direction will directly affect the movement and dispersion of pollutants from emission sources within a given study area. Ambient pollution levels have been found to be inversely related to wind speed [10]. Wind speed can

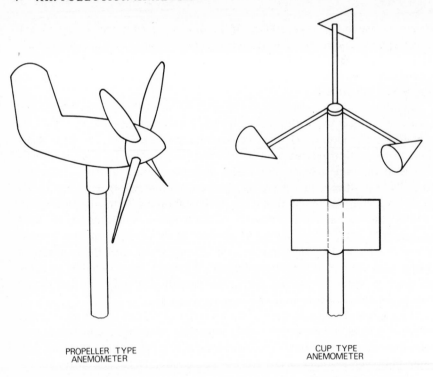

PROPELLER TYPE
ANEMOMETER

CUP TYPE
ANEMOMETER

Fig. 1.1 Cup type anemometer (right); propeller type anemometer (left).

be measured by various instruments [11]. Rotation anemometers of the cup or
propeller type are commonly used (Fig. 1.1). Other devices include pressure
tube anemometers which operate on the principle of the pitot tube and hot wire
anemometers which utilize the relative cooling effect of air movement past a
heated element to measure wind velocity.

Wind direction is normally detected by various types of wind vane, which
rotate to face into the wind. The axial rotation of the wind vane is transmitted
to a recording device by one of two basic methods. A potentiometer may be
used in which the angular position of the wind vane corresponds to a contact on
a variable resistance. As the vane turns, the contact moves and varies the resistance to the recording device. The other system uses position motors. These are
two or more small motors electrically interconnected so that the rotation imposed
by the vane on one motor activates another motor to drive the recorder.

Directional devices are useful in concert with monitoring instruments in that
a control system can be designed such that the sampling system samples only
when the wind is coming from a specified direction. In practice, an intake sector
must be defined, allowing sampling only when the wind direction is within a

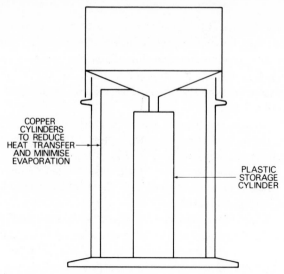

Fig. 1.2 Canadian standard rain gauge.

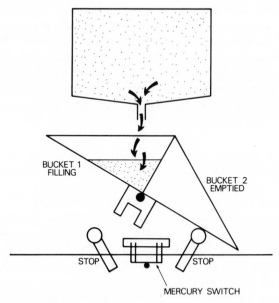

Fig. 1.3 'Tipping bucket rain gauge'.

range of so many degrees either side of the direction of interest. Also, some damping of time delay must usually be built into the system to avoid frequent starts and stops due to wind direction fluctuation. An alternative approach to directional sampling involves the positioning of fixed and continuous collection openings facing four or more directions. This method is discussed for dustfall collection in Section 1.2.2.

Precipitation can take the form of rain, sleet, snow, hail and various combinations thereof. The general effect of precipitation is the scavenging of particulates and gases from the atmosphere [12—15]. The net result of precipitation may be the removal of pollutants from the atmosphere before dispersion takes place. In such cases, a high concentration of the pollutant might be measured during the initial stages of a rainfall.

Precipitation is commonly measured [11] by a standard rain gauge which consists of a funnel of specified area running into a measuring cylinder, which is periodically checked for collected volume (Fig. 1.2). Various types of recording rain gauges are available. One device operates with a balanced bucket with two equal sized compartments. Rain falls into one compartment until enough weight has been collected to tip the bucket. This trips a recording switch and puts the other compartment in position to collect rainwater, A continuous record is maintained based on the number of times the switch has been tripped and on the calibrated volume required to tip the bucket (Fig. 1.3).

Other meteorological factors which should be considered are temperature, humidity and sunlight. Temperature tends to have its main effect due to the resulting changes in domestic heating requirements during colder weather. Shorter term effects might be determined by continuous temperature monitoring. For example, [3] turbulence within an inversion layer might bring higher level pollutants down to ground level overnight. A corresponding rise of temperature due to the higher temperature of the inverted air mass along with a rise in pollutant concentration would indicate such an effect.

Temperature may be measured by mercury-in-glass thermometers, thermal resistors or thermocouples. The latter two can provide continuous temperature recording.

Humidity can effect air quality in a variety of ways [3]. Low humidity can result in increased suspended particulate concentrations due to suspension of surface dust. High humidity, as exhibited in fog conditions, can block solar heating of the ground surface and thereby prolong the life of inversion layers. Air pollution incidents with increased morbidity and mortality were generally associated with low temperature and fog [16].

Humidity can be measured by several devices [11]. The commonly used psychrometer consists of two mercury-in-glass thermometers, one with a dry bulb, the other with the bulb covered by a thin moist wick. The wet bulb thermometer will record a lower temperature because of the evaporation of water from the bulb. The resulting different temperatures can be used to enter a psychometric chart which will provide the specific and relative humidity. Other devices for humidity measurement include the hair hygrometer which measures the length of a piece of human hair as a function of humidity: the dewpoint hygrometer which measures the temperature of a metal plate which is cooled to the current dewpoint and the electrical hygrometer which measures the electrical resistance of a lithium chloride film as a function of humidity.

Photochemical reactions and generation of secondary pollutants are generally dependent on the solar radiation exposure available [17]. The degree of cloud cover, which will affect surface solar radiation is normally recorded based on observation and may often be obtained from local weather offices on a daily basis.

1.1.1.3 Sampling site criteria

Sampling site criteria generally fall into two classes. Firstly, there are criteria necessary for the proper siting of individual sampling systems in order that each site should provide a true representation of the receptor defined in the study objectives. The other class of criteria refer to the location of sites relative to one another to form sampling networks which will provide the area-wide data required to achieve the study objectives.

Many sources have discussed proper siting criteria for individual sampling systems [1–3, 18–20]. However, no comprehensive set of rules has been adopted as standard procedure, because of the infinite diversity that will be encountered in choosing new sampling sites for a specific purpose.

Some general rules of thumb include the following points.

Sampling inlets should generally be more than 2 m above ground level. The maximum height is determined by the objective of the survey but should be consistent from one site to another within a given network designed for a common objective.

Sampling sites should not be located in the lee of major obstructions such as tall buildings. A general rule is that the top of obstructions should subtend less than a 30° angle with the horizontal at the sampling point.

The sample intake should not be exposed to contamination from specific localized sources (i.e., a chimney on a roof top). For general area air quality monitoring, the site should not be directly downwind from major emission sources, such as motorways, parking lots or industrial stacks.

The site must be accessible and yet secure from tampering.

The site must provide an adequate, reliable, power supply to run the sampling equipment as necessary.

Within these general rules, wide variations in sampling system siting and set-up are possible [19, 21]. Therefore, once the general siting criteria are satisfied, consistent with the objectives of the study, all sites should be made to conform with one another in as much detail as possible, if valid comparisons are to be made using the collected data.

Location criteria for spacing network sites are very dependent on the specific objectives of the study. Various network configurations have been used including [13, 20, 22]:

location of sites on concentric circular lines centered on the area of interest;

location of sites on typical trajectories of surface winds;

location of a random heavy density of sites in the core of interest with random open spacing further out;

location of sites on an equally spaced grid pattern.

The actual choice of a sampling network pattern is often dictated by financial constraints and local conditions. Where flexibility exists and there is scope for putting the data into an air quality simulation model, preliminary work with the model using artificial data from possible sampling locations will often indicate efficient choices.

Meteorological and topographical factors may play an important role in defining the optimum sampling network. An emission source inventory, even in rough form, will often be required to locate sampling sites so that they will generate the information necessary to satisfy the study objectives.

Most networks require background locations for comparative purposes. These sites should normally be free from the influence of any major sources. Enough sites should be provided so that when the wind blows from the central core to a background site, there are still other background sites located upwind of the central core, so that upwind–downwind comparisons can be made.

1.1.1.4 Sample scheduling

Upon locating ambient air sampling sites which will provide representative samples, free from undesirable interference, the choice remains of how frequently and for what duration to obtain samples. The frequency and duration of sampling are very much dependent upon the objectives of the study and the nature of the pollutants and their emission sources.

The variation in requirements for sampling duration is evidenced by the variety of time periods specified for various pollutant criteria in the U.S. National Air Quality Objectives (Table 1.1). The variability is due in part to the nature of a pollutant's effect upon the receptor. For example, semi-annual averages provide useful information for lead concentrations because the effect of lead on receptor organisms is cumulative and long term. Sulphur dioxide (SO_2), on the other

Table 1.1   National air quality standards reference methods

Pollutant	Average time	Reference method	Principle of detection
SO_2	3 h, 24 h annual	Pararosaniline	Colorimetric
Particulate matter	24 h, annual	Hi-vol sampler	Gravimetric
CO	1 h, 8 h	Non-dispersive infrared spectrometry	Infrared
Photochemical oxidants (ozone)	1 h	Gas phase O_3−ethylene reaction (calibrated against neutral buffered Kl method)	Chemiluminescence
Hydrocarbons (nonmethane)	3 h	Gas chromatographic	Flame ionization
NO_2	Annual	24 h integrated samples collected in alkaline solution	Colorimetric

hand, can wreak its harmful effects upon receptor organisms over short term fumigation periods and so mainly short period, high frequency information is relevant.

Saltzman [23] considered a model for air pollutant concentration fluctuations in an attempt to rationalize sampling time considerations. The model is based on a series of sine functions of varying periodicity superimposed upon an arithmetic mean concentration. Varying periodicity for pollutant fluctuations would represent factors such as hourly fluctuations due to traffic emissions, daily variations due to industrial emissions and seasonal fluctuations due to domestic heating sources. For any sampling period other than continuous sampling, the values obtained from individual samples will not fully reflect the variance of the true concentration fluctuations about the long term mean value. Application of Saltzman's model indicated that concentration fluctuations with a cycle period shorter than the sampling period were almost completely attenuated (averaged out) by looking strictly at the periodic sample results. However, when the concentration fluctuation cycle period becomes 2.25 times the sample period, 50% of the true concentration variance is observed and when the concentration fluctuation cycle period becomes 5.6 times the sample period, 90% of the true concentration variance is observed. The important application of the Saltzman analysis is in providing an estimate of the degree to which a particular sampling programme will reflect the actual pollutant concentration variations.

One further step in this analysis is the application of the pollutant fluctuation model to a biological receptor model. Since a biological receptor will tend to eliminate a given pollutant from its system at the same time it is ingesting the pollutant, the fluctuations in the effective pollutant concentration within the receptor organ will be attenuated in comparison with the ambient concentration fluctuations. Saltzman proposes that if the required information could be gathered to define the factors needed for a valid biological model, the model could be used to choose rationally the sampling period. Provided that the chosen sampling period consistently provided less attenuation of the true pollutant concentration than would be provided by the biological receptor, then the maximum useful biological information will be obtained using the chosen sampling period. As an example, based on a biological half life of 20 min for SO_2 and a pollutant retention function value of 0.5, a sampling period of 20 min would be expected from the model, to give the maximum useful biological information.

Sampling schedule possibilities have been discussed by Akland [24]. The modified random system, commonly used in air sampling networks, calls for sampling intervals of fixed length (i.e. weekly). During the sampling interval one day is randomly chosen for sampling. The random choice may be restricted to provide for approximately the same number of sampling occurrences on each day of the week over the year. The systematic system calls for starting the sampling programme on a day picked at random, followed by sampling at fixed intervals, other than 7 days, from that day onward. Comparison of the relative precision of the two methods for hypothetical sampling programmes drawn from an existing data base which had daily samples, Akland found the systematic approach to be consistently more precise.

The precision of air pollutant sampling that may be expected, given that these usually follow a log normal distribution, can be calculated. Hunt [25] provides the formulae to calculate the confidence interval about the geometric mean.

Confidence interval bounds at $1 - \alpha$ confidence level

Lower bound $= \bar{X}_{geo} - m_1 \bar{X}$

Upper bound $= \bar{X}_{geo} + m_2 \bar{X}$

where

$$m_1 = 1 - \exp\left[-t_{1-\alpha/2} \frac{S_{log}}{n^{1/2}} \left(1 - \frac{n}{N}\right)^{1/2} \right]$$

$$m_2 = \exp\left[t_{1-\alpha/2} \frac{S_{log}}{n^{1/2}} \left(1 - \frac{n}{N}\right)^{1/2} \right]$$

$\bar{X}_{geo}$ = geometric mean

and

$1 - \alpha$ = level of confidence (i.e. 95%, 99%);

n = number of samples taken during given interval;

N = maximum possible number of samples during given interval (i.e.

$N = 30$ daily samples in average month);

$S_{\log}$ = Standard deviation (S.D.) of the logarithms of air pollutant measurements (given by the log of the geometric S.D.)

$t_{1-\alpha/2}$ = the t statistic with $n - 1$ degrees of freedom.

Thus if the overall study objectives require a certain level of precision to ensure the detection of subtle effects, provided an estimate of the geometric S.D. of the pollutant concentrations can be obtained, it is possible to calculate the number of samples that are necessary, within a given time interval, to provide the required precision.

1.1.2 Source sampling

1.1.2.1 General objectives

Source sampling in air pollution work may be considered as the collection of airborne pollutants before emission to and dilution by the atmosphere. Emission sources are usually categorized as stationary or mobile sources. Stationary sources include various emissions from industrial plants such as steel mills, pulp mills, chemical plants and oil refineries, from municipal sources such as electricity generating plants and refuse incinerators and from domestic sources such as house chimneys. Mobile sources include emissions from petrol and diesel engined vehicles and from aircraft.

Emission sources are studied for several reasons, including:

determination of the mass emission rate of particular pollutants from a particular source and how it is affected by process variations;

evaluation of control devices for the reduction of pollutant emissions;

data gathering of emissions from several sources for input to air quality management models.

The approaches and requirements for the two source categories are basically different and will be discussed in turn.

1.1.2.2 Stationary source sampling

Planning and preparation

The requirements for source sampling often necessitate long sampling periods. As a result, source sampling requirements can often be met only by careful advance planning so that the time spent at the sampling location is efficiently used.

Proper planning requires a thorough knowledge of the process to be sampled.

The nature of process conditions and their effect upon emission parameters should be determined by discussion with a knowledgeable person in charge of the facility. The frequency of any process cycles should be considered and the reported level of emissions from similar processes studied [26].

The nature of the emission source for the purposes of planning source testing can be considered by applying the classification scheme of Achinger and Shigehara [27]. These authors specify two requirements for valid source testing. Firstly, the sample should accurately reflect the true magnitude of the pollutant emission at a specific point in a stack at a specific instant of time. This requirement is determined by the design of the sampling instrument and is considered in Section 1.2. The second requirement is to obtain enough measurements varied in space and in time such that their combined results will accurately represent the entire source emission. Satisfying the latter requires consideration of the fluctuations of the source emission both in space, across the stack diameter, and in time. Achinger and Shigehara classify sources into categories with the four possible combinations of steady conditions (no variation with time) and uniform conditions (no variations in space).

CLASS I. Sources in this classification are both steady and uniform. Theoretically, only one measurement need be taken as this will represent the whole stack cross-section for the whole period of the time of steady operation. The example specified for this condition is sampling for a gaseous pollutant from a turbulent gas stream.

CLASS II. Sources in this classification are steady but non-uniform. For this type of operation the collection of a composite sample at several locations on the stack cross-section will produce results representing the whole stack for the period of steady operation. The example specified for this condition is sampling for particulates at a large continuous feed coal fired power station.

CLASS III. Sources in this classification are unsteady but uniform. In this case, only one sample need be used, but sampling must take place for the entire cycle of a cyclic operation or for as long as possible in a non-cyclic operation. The example specified for this condition is sampling a gaseous pollutant from a cyclic operation with turbulent stack flow.

CLASS IV. Sources in this classification are both unsteady and non-uniform. The sampling approach to be taken depends upon the nature of the source variability with time. All measured variables may vary proportionally or non-proportionally with time. Furthermore, the non-proportional variations may be reproducible or non-reproducible.

If all the parameters vary proportionally with time, individual rather than composite samples must be taken. Simultaneous sampling at a reference location and at the specified sampling points is recommended. The results for the individual sample points can then be adjusted back to the original point in time by applying a factor determined by comparing the simultaneously measured reference point value with the initial reference point value.

If stack parameters vary non-proportionally with time, but vary over reproducible short cycles, then complete cycles may be sampled at each sample location. Further, composite samples are valid provided the same number of cycles are sampled at each location.

If stack parameters vary non-proportionally with time and over long or erratic cycles (non-cyclic), then all measurements must be made simultaneously over the entire cycle. This requirement becomes impossible for an adequate number of sample points in most cases. A recommended compromise is the use of two to four statistically random sites for simultaneous measurement and corresponding caution in determining statistical confidence in the results.

Sampling site selection is the next major step in conjunction with classifying the nature of the source emission. The criteria used in site selection must include:

safety of the location for the test personnel;

relationship to the points of particular interest (i.e. at the pollution control device, for device efficiency testing);

availability of a platform for men and equipment;

accessibility of the platform to the men and equipment;

access to the stack interior from a suitable port;

provision of power supply for sampling equipment; and

satisfaction of flow disturbance criteria.

Many of the above criteria will have to be compromised as the ideal sampling location seldom exists. However, safety considerations cannot be compromised as stack sampling is a hazardous undertaking even under good conditions. Inability to satisfy safety criteria is sufficient reason to abandon a given site until modifications to allow safe operations can be made.

Various criteria have been specified for avoiding flow disturbances [28–32]. Generally, a sampling location in a vertical flue sited eight flue equivalent diameters* downstream and two flue equivalent diameters* upstream from a flow disturbance such as a bend, inlet or outlet, is considered good. Since many emission sources will not provide a site meeting this criteria, Fig. 1.4 [28, 30] has been developed to compensate for non-ideal sampling locations with increased sample points across the stack cross-section.

The location of sampling points at the centroids of equal areas across the stack cross-section is recommended. For a rectangle, sample points are located at the centroids of smaller equal area rectangles (Fig. 1.5) and for a circular cross-section, at the centroids of equal area annular segments (Fig. 1.6). The location of the centroids for the annular segments of Fig. 1.6 are summarized in Table 1.2 for various numbers of sample points on a stack diameter.

---

* Flue (stack) equivalent diameter = 4 [area of flue cross-section/perimeter of flue cross-section] = actual diameter for circular flue.

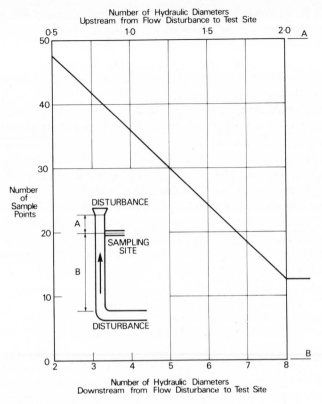

Fig. 1.4 Number of sample points. (After [28].)

Data requirements

Upon locating a suitable sampling site and classifying the source as to the sampling approach required, certain basic physical measurements must be performed on the stack gas [28, 30]. These are:

a stack gas composition analysis by Orsat apparatus (primarily for combustion sources);

a stack gas moisture determination;

a stack gas temperature determination;

a stack gas pressure determination;

a stack gas velocity determination.

The stack gas Orsat analysis [28–30, 34–37] may be obtained by drawing a small gas sample through a sample probe into the Orsat analyser using a hand operated or hydraulic aspirator, or a small electric pump. The stack gas is

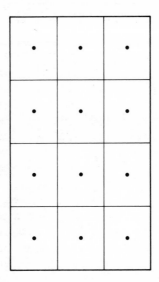

Fig. 1.5 Cross-section of rectangular stack divided into twelve equal areas, showing location of traverse points at centroid of each area. (After [28].)

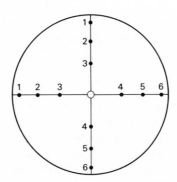

Fig. 1.6 Cross-section of circular stack divided into twelve equal areas, showing location of traverse points at centroid of each area. (After [28].)

Table 1.2   % of circular stack diameter from inside wall to traverse point (After [28]).

Traverse point number along diameter[†]	Number of traverse points on a single diameter[‡]											
	2	4	6	8	10	12	14	16	18	20	22	24
1	14.6	6.7	4.4	3.3	2.5	2.1	1.8	1.6	1.4	1.3	1.1	1.1
2	85.4	25.0	14.7	10.5	8.2	6.7	5.7	4.9	4.4	3.9	3.5	3.2
3	–	75.0	29.5	19.4	14.6	11.8	9.9	8.5	7.5	6.7	6.0	5.5
4	–	93.3	70.5	32.3	22.6	17.7	14.6	12.5	10.9	9.7	8.7	7.9
5	–	–	85.3	67.7	34.2	25.0	20.1	16.9	14.6	12.9	11.6	10.5
6	–	–	95.6	80.6	65.8	35.5	26.9	22.0	18.8	16.5	14.6	13.2
7	–	–	–	89.5	77.4	64.5	36.6	28.3	23.6	20.4	18.0	16.1
8	–	–	–	96.7	85.4	75.0	63.4	37.5	29.6	25.0	21.8	19.4
9	–	–	–	–	91.8	82.3	73.1	62.5	38.2	30.6	26.1	23.0
10	–	–	–	–	97.5	88.2	79.9	71.7	61.8	38.8	31.5	27.2
11	–	–	–	–	–	93.3	85.4	78.0	70.4	61.2	39.3	32.3
12	–	–	–	–	–	97.9	90.1	83.1	76.4	69.4	60.7	39.8
13	–	–	–	–	–	–	94.3	87.5	81.2	75.0	68.5	60.2
14	–	–	–	–	–	–	98.2	91.5	85.4	79.6	73.9	67.7
15	–	–	–	–	–	–	–	95.1	89.1	83.5	78.2	72.8
16	–	–	–	–	–	–	–	98.4	92.5	87.1	82.0	77.0
17	–	–	–	–	–	–	–	–	95.6	90.3	85.4	80.6
18	–	–	–	–	–	–	–	–	98.6	93.3	88.4	83.9
19	–	–	–	–	–	–	–	–	–	96.1	91.3	86.8
20	–	–	–	–	–	–	–	–	–	98.7	94.0	89.5
21	–	–	–	–	–	–	–	–	–	–	96.5	92.1
22	–	–	–	–	–	–	–	–	–	–	98.9	94.5
23	–	–	–	–	–	–	–	–	–	–	–	96.8
24	–	–	–	–	–	–	–	–	–	–	–	98.8

† Points numbered from inside wall toward opposite wall.
‡ The total number of points along two diameters would be twice the number of points along a single diameter.

bubbled through the distilled water in the Orsat levelling bottle (Fig. 1.7) for 10 min. The stack gas sample is then taken with the Orsat apparatus and analysed for CO_2, O_2, CO, and N_2 by difference. The dry molecular weight of the stack gas is calculated from the Orsat volumetric analysis with the formula

$$M_D = (0.44)(\%CO_2) + (0.28)(\%CO) + (0.32)(\%O_2) + (0.28)(\%N_2)$$

where M_D = dry molecular weight of stack gas.

The stack gas moisture determination may be obtained by sampling a known volume of air and condensing the water vapour in ice cooled condensers or by measurement of the wet and dry bulb temperatures of the stack gas [28–30, 34–37]. In practice, the former method is performed in conjunction with the actual sampling for many of the available source sampling procedures, while the

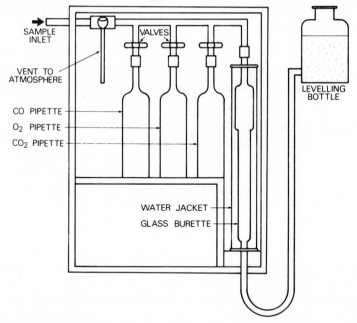

Fig. 1.7 Orsat apparatus.

latter method may be performed before testing to provide data for necessary calculations to control the sampling rate.

The wet and dry bulb method requires withdrawing the stack gas through a sample system (Fig. 1.8) provided with a psychrometer. By determining the wet and dry bulb temperatures (the wet bulb must be lower), it is possible to use a hygrometric (psychrometric) chart [38] to obtain the specific humidity ω in kilograms of H_2O per kilogram of dry air. The volumetric humidity θ in m^3 of water vapour per m^3 of dry air can be obtained from the formula

$$\theta = \frac{28.8\,\text{kg (mol dry air)}^{-1}}{18.0\,\text{kg (mol H}_2\text{O)}^{-1}}\,\omega$$

$$= (1.6)\omega\,\frac{\text{mol H}_2\text{O}}{\text{mol dry air}}$$

and because a molar ratio is equivalent to a volumetric ratio using Amagat's Law

$$= (1.6)\omega\,\frac{\text{m}^3\,\text{H}_2\text{O}}{\text{m}^3\,\text{dry air}}$$

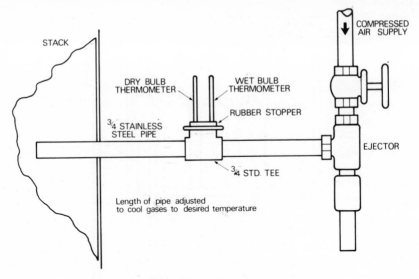

Fig. 1.8   Moisture measurement. (After [28].)

Therefore, the moisture content of the stack gas as a volume fraction is given by

$$B_{w_0} = \frac{1.6\omega}{1 + 1.6\omega}$$

For the condensation method, the volume of liquid collected is measured and converted into the equivalent volume of water vapour at standard conditions using

$$V_w = V_c \frac{\rho_{H_2O}}{M_{H_2O}} \frac{RT}{P}$$

where

V = volume of liquid collected, litre;

ρ_{H_2O} = density of water, 1 kg litre^{-1};

M_{H_2O} = molecular weight of water, 18 kg mol^{-1};

R = ideal gas constant, $8.31 \text{ kJ (mol}^\circ\text{K)}^{-1}$;

T = standard temperature, 293°K;

and P = standard pressure, 1.01 b.

$$V_w = (1.35 \text{ m}^3 \text{ litre}^{-1})V_c$$

The moisture content of the stack gas as a volume fraction is then given by

$$B_{w_0} = \frac{(1.35 \text{ m}^3 \text{ litre}^{-1})V_c}{(1.35 \text{ m}^3 \text{ litre}^{-1})V_c + V_m}$$

where V_m = volume of dry gas through the meter, at standard conditions.

The stack gas temperature determination can be relatively easy depending upon temperature range encountered. For stack temperatures up to $400°$ C mercury-in-glass thermometers may be used directly. Maximum reading thermometers which retain the stack temperature while the thermometer is pulled from the stack offer an advantage. For higher temperatures or for conducting temperature traverses across the stack to assure uniformity, direct reading thermocouples or thermistors may be used.

The relative stack gas pressure determination can be included in the determination of stack gas velocity by determining the static pressure in the stack with the standard pitot tube. A barometer should be used to establish the atmospheric pressure at the stack sampling location if conditions are likely to vary markedly from standard conditions.

The average stack gas velocity determination is obtained by performing a pitot tube traverse [28, 30, 34–36, 39]. The velocity head is measured at several points in the stack cross-section, chosen according to the discussion on location of sampling points (Figs. 1.5, 1.6). Either a standard pitot tube (Fig. 1.9) or a Staubscheibe, or type S pitot tube (Fig. 1.10) may be used, but the latter must be calibrated against a standard pitot tube.

The velocity head at each point must be determined by directing the pitot tube opening along the axis of the stack into the oncoming flow. The velocity head is read from an inclined gauge manometer indicating the difference between the dynamic pressure of the flow and the static stack gas pressure. If the velocity head is found to be negative at one or more sampling points, the pitot tube should be checked. If it is found to be functioning reliably and the negative reading is correct, the site is unsuitable for further sampling [31]. Likewise, if the direction of flow is found to be more than $30°$ from the axis of the stack, as determined by rotating the standard pitot tube in $5°$ increments to determine the maximum and minimum velocity heads, the site is unsuitable and another site consistent with the other specified criteria should be chosen.

The average velocity is obtained by calculating the square roots of the velocity pressures at the various sample points and applying the formula

$$(V_s)_{av} = K_p C_p (\sqrt{\Delta p})_{av} \sqrt{\left(\frac{(T_s)_{av}}{P_s M_s}\right)}$$

where $(V_s)_{av}$ = average stack gas velocity, m s^{-1};

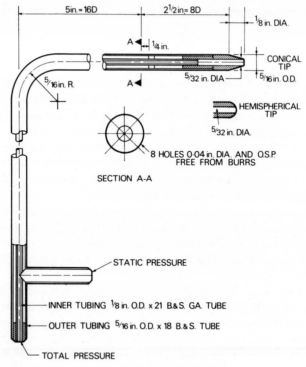

Fig. 1.9 Standard pitot tube details. (After [29].)

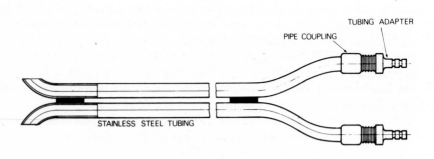

Fig. 1.10 Type S pitot tube (special). (After [29].)

C_p = pitot tube coefficient, determined by calibration for an S type or
= 1 for a standard pitot tube;

$K_p = 4.05\,\mathrm{m\,s^{-1}(kg\,mol^{-1}\,{}^\circ K^{-1})}$, a unit conversion factor for the units specified herein;

$(\sqrt{\Delta p})_{av}$ = the average of the roots of the velocity heads in mb (not equal to the root of the average velocity head);

$(T_s)_{av}$ = average absolute stack gas temperature, ${}^\circ K$;

P_s = absolute stack gas pressure, b;

and M_s = molecular weight of the stack gas on a wet basis, $\mathrm{kg\,mol^{-1}}$.

The molecular weight of the stack gas on a wet basis is determined from

$$M_s = M_D(1 - B_{w_0}) + 18\,B_{w_0}$$

where M_D = dry molecular weight of the stack gas $(\mathrm{kg\,mol^{-1}})$ calculated from the Orsat analysis;

and B_{w_0} = volumetric moisture fraction calculated from the moisture determination.

Finally, the volumetric emission rate for the stack may be calculated by applying the formula

$$Q_s = 3600\,(V_s)_{av}A_s$$

where Q_s = volumetric emission rate at wet stack conditions in $\mathrm{m^3\,h^{-1}}$;

$(V_s)_{av}$ = average stack gas velocity in $\mathrm{m\,s^{-1}}$;

A_s = stack area at the location of the velocity traverse, in $\mathrm{m^2}$.

The volumetric emission rate corrected to standard conditions on a dry basis may be calculated from

$$Q_{std} = 3600\,(1 - B_{w_0})\,(V_s)_{av}\,A_s\left[\left(\frac{293^\circ\,K}{(T_s)_{av}}\right)\left(\frac{P_s}{1.01\,b}\right)\right]$$

with the symbols as defined previously.

After obtaining the required physical information, several sampling procedures may be used for aerosols and gases. Some of these techniques are discussed in Sections 1.2.2.4 and 1.2.3.4.

When sampling has been successfully completed, the analysis will provide the information to calculate the concentration of a given pollutant in the stack emission. Given the pollutant concentration and the volume emission rate determined during sampling, the total mass emission rate can be calculated as follows

$$M_s = Q_{std}C_s$$

where M_s = pollutant mass emission rate $g\,h^{-1}$;

$\quad Q_{std}$ = volumetric emission rate corrected to dry standard conditions, $m^3\,h^{-1}$;

$\quad C_s$ = concentration of pollutant, measured at dry standard conditions, $g\,m^{-3}$.

Finally, it is often useful in assessing industrial sources to normalize the pollutant emission rates with production units so that comparison may be maintained between various plants. Thus, for example, where a refuse incinerator burns $5\,t\,h^{-1}$, and emits particulates at a rate of $20 \times 10^3\,g\,h^{-1}$, the normalized emission rate would be

$$\frac{20 \times 10^3\,g\,h^{-1}}{5\,t\,h^{-1}} = 4.0\,kg \text{ particulates (t of refuse burned)}^{-1}.$$

The value of this procedure for comparing different operations with similar processes is self evident.

1.1.2.3 Mobile source sampling

Testing conditions and requirements

Exhaust gas sampling from vehicles and aircraft is an involved process requiring sophisticated equipment. Vehicle and aircraft emissions are heavily dependent upon the engine operating mode and therefore the operating cycle used during exhaust sampling is fundamental to the interpretation of the results.

Soltau and Larbey [40] have reviewed driving cycles which are available for vehicle testing. Exhaust emission tests are usually performed with the vehicle on a dynamometer equipped with inertia flywheels to represent the vehicle weight and brake loading to reproduce the level road load at a given speed.

The California 7 - mode driving cycle is illustrated in Fig. 1.11. The emissions during the shaded portion of the curve are not used in the calculation of the total vehicle emission. Tests with the California 7 - mode cycle require the use of seven consecutive complete cycles with only the first four and last two monitored and emissions recorded.

The US Federal cycle [41] for the 1972 model year is shown in Fig 1.12. The vehicle is set up on a dynamometer as for the 7 - mode California cycle and is operated from a cold start through the entire non-repetitive cycle. The total

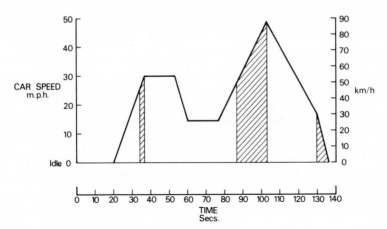

Fig. 1.11 California driving cycle. (After [40].)

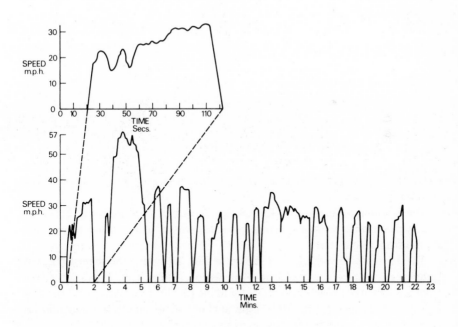

Fig. 1.12 U.S. Federal driving schedule (II). (After [40].)

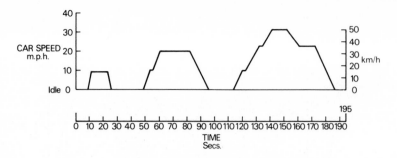

Fig. 1.13   European E.C.E. driving cycle. (After [40].)

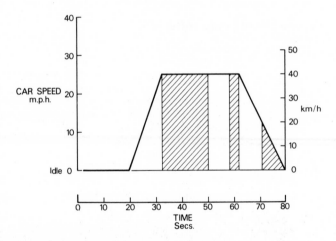

Fig. 1.14   Japanese driving cycle. (After [40].)

emissions throughout the whole cycle are collected for the calculation of the total vehicle emission.

The Economic Commission for Europe driving cycle is shown in Fig. 1.13. This cycle is similar to the California 7 - mode cycle, in that four repetitive cycles are run from a cold start.

The Japanese driving cycle is shown in Fig. 1.14 with the shaded portions being neglected from the emission calculation. Three such cycles are driven from a warm start. Weighting factors for idling (0.11), accelerating (0.35), cruising (0.52) and decelerating (0.02) are applied to the emission calculation for each segment by averaging the results of three successive test cycles.

The test procedures for aircraft engines [42] tend to follow the approach of the California and Japanese operating cycles in that emissions are measured during each phase of the cycle with the results averaged and weighted to produce a calculated total emission. Aircraft emission cycles tend to be based on steady-state operating conditions only. The approach of the 1972 US Federal cycle and the E.C.E. cycle where sampling is composited through a representative complete operating cycle has not been applied to aircraft engine testing [42].

Some actual sampling system applications to mobile source exhaust particulate and gaseous emissions are considered in Sections 1.2.2.4 and 1.2.3.4.

1.2 Sampling methods

1.2.1 General sampling system considerations

Sampling systems for airborne pollutants generally consist of four component subsystems:

intake and transfer component;
collection component;
flow measuring component;
air moving component.

Malfunction by any one component will hinder the successful performance of the whole system. Therefore, the performance of each component must be evaluated when synthesizing the overall sampling system.

1.2.1.1 Intake and transfer component

The nature of the intake is basically determined by the objective of the sampling programme, varying from thin-walled probes used for aerosol source sampling to free vertical access for dustfall deposit gauges. Specific types of intakes will be discussed, as necessary, with the various types of sampling, but certain general considerations apply to the intake and transfer function.

The primary consideration in evaluating the intake for a given sampling system is the ability of the device to inhale faithfully the total quantity or a reproducible representative portion of the airborne constituent being studied from a given volume of air sampled. Although fundamental to the interpretation of air pollution measurements, the difficulties encountered in verifying intake performance for a given constituent have resulted in limited evaluation of many devices commonly used in air sampling. Therefore, serious consideration should be given to a preliminary evaluation of intake performance, particularly when embarking on a sampling programme for airborne constituents not commonly reported.

Upon satisfaction of the above constraint on the intake, the transfer system must then transport the constituent of interest to the collection device without modifying the quantity or any properties of the constituent. Phenomena such as adhesion of aerosols to tube walls, condensation of volatile components within the transfer lines, reaction of gaseous components with transfer system materials and adsorption and reaction with collected particulates are some of those which have been reported [1, 18, 43–46]. Many specific problems have been recognized for given sampling systems and these will be included in the discussion of specific procedures.

1.2.1.2 Collection component

The collection component chosen for the sampling system is determined by the airborne constituent or constituents which are sought. Aerosol sampling may incorporate filtration, impingement, thermal or electrostatic precipitation, and gravity or centrifugal collection. Gaseous sampling may incorporate adsorption, absorption, condensation or grab sampling. Descriptions of these components are provided in Sections 1.2.2.2 and 1.2.3.2. However, the need to determine the collection efficiency of the device for the specific airborne constituent or constituents sought is common to all devices. Although 100% collection efficiency would be desirable, in practice, lower efficiencies can be used, provided the efficiency can be precisely and reproducibly measured for the specific constituent being sampled. The lowest collection efficiency that can be tolerated is determined by the importance attached to quantitative results, but collection efficiencies of 90% or better are generally acceptable [46] for quantitative analysis. Verification of collection efficiency has received attention with most commonly used sampling systems but specific verification of the collection component selected for a given constituent is a recommended practice.

Evaluation of the collection component requires the preparation of known test atmospheres [46]. For gaseous constituents this may be performed by adding measured quantities of the test gas into a container providing a known dilution volume. Collapsible containers provide the advantage of preventing dilution or pressure changes within the test atmosphere as the sample is drawn from the container.

For aerosol testing, the preparation of an aerosol dispersion of known size distribution is required [47, 48]. Aerosol standard generators usually function by dispersing a material dissolved in a liquid solvent in a manner to create aerosol droplets. The solvent subsequently evaporates from the small particles leaving relatively non-volatile particles. Examples of such materials include polystyrene latex, dioctyl phthalate and methylene blue.

For either gaseous or aerosol constituents, losses within the test generating equipment must be considered as well as the possible inefficiency of a given collector. Alternately, the use of two collectors in series has been used in the past

to evaluate the collection efficiency of a given collector [46]. This method, however, will only highlight overloading or poor efficiency on the part of the first
collector, but it cannot, by itself, guarantee that both collectors are not, in fact,
inefficient, if no collection is seen on the second collector [46]. A second collector may be of more value where it is proved to be efficient for the constituent
sought while the first collector is an unknown which requires testing.

When collection components are used in series for regular sampling, specific
evaluation of the effect of upstream components on the constituents reaching
downstream components is usually necessary. For example, particulates collected
on an upstream filter may adsorb a gaseous constituent intended for collection
in a downstream collector. In some cases, it is necessary to place intentionally a
screening collection component upstream of the primary collector, in order to
remove a constituent that will interfere with the collection or analysis of the
constituent of interest.

While some of these problems may be unavoidable for the system required, a
knowledge of such interferences will allow modifications to the analytical
schemes chosen or the quantitative interpretation placed on the results.

1.2.1.3 Flow measurement component

Any attempt to measure accurately the concentration of a given airborne constituent is fundamentally dependent upon the accurate knowledge of the original
air volume relationship with the sample being analysed. Unfortunately, the flow
measurement component of an air sampling system often receives less attention
than other components.

Flow measuring components generally fall into two classes: volume meters
and rate meters [18, 49]. Volume meters measure the total integrated volume
which has passed through them for a given period of time. As such, they have
the advantage of providing a direct record of the volume of air sampled. The
common types of volume meters that are available are the dry test gas meter, the
wet test gas meter and the cycloid gas meter. Of these, the dry test meter is most
commonly used because of its relative sturdiness, low cost and weight, in comparison with the latter two meters.

The dry test gas meter measures volume flow by mechanical displacement of
internal bellows by the air flow. The displacement is recorded on a mechanical
counter via a series of levers. A properly maintained and calibrated unit is considered accurate to ± 1% [18].

Rate meters measure the instantaneous volume flow rate through the sampling system and, therefore, have the disadvantage that frequent checks are
required to ensure accurate calculation of the total volume sampled. In some
cases, permanent flow rate recording can be provided, which will solve this

problem. The advantage of rate meters is their relatively small size in comparison to the bulky dry test gas meter. The common types of rate meters that are available include: venturi meters, orifice meters, flow nozzle meters, rotameters, pitot tubes, turbine and hot wire anemometers.

Venturi, orifice and flow nozzle meters all depend on the air stream flowing through a constriction. The measured flow through the constriction is a function of the static pressure before and after (or at) the constriction. Thus, the measurement of static pressure at two selected points allows the calculation of flow, given the appropriate constants for the constriction. The venturi meter provides the highest accuracy with the smallest pressure loss, but is more expensive and requires more installation space than either the orifice or flow nozzle meters. These devices may all be adapted to provide continuous flow recording.

The rotameter consists of an expanding conical flow section which is mounted vertically and contains a pointed float which rides the upward flow in the flow section and effectively provides a variable annular flow area between itself and the flow section walls. The downward gravity force on the float is counteracted by the upward pressure, buoyant and drag forces due to the upward air flow. The flow section is graduated in arbitrary units. By means of calibration, the flow value of the arbitrary scale is established and the flow is then indicated by noting the level of the float on the scale.

Pitot tubes may be used to establish flow by determining the velocity profile of a flow section as discussed in Section 1.1.2.2. Turbine and hotwire anemometers are rarely used in current air sampling equipment, but these devices may become popular with the development of more automated sampling equipment. The former utilizes the rate of rotation of various types of propeller devices within the flow stream to determine the volume flow rate. The latter utilizes the dependence of the convective cooling rate of a hot wire on the velocity of air flow past the wire to determine flow velocity and thereby flow rate.

The flow devices discussed tend to be inaccurate at low flow rates, i.e. less than $10\,\mathrm{ml\,min^{-1}}$ [1], but sampling rates for most sampling systems are usually adequate to maintain accurate flow measurements. Frequent calibration of all measuring devices is recommended and can be performed according to the manufacturer's recommendations or standard calibration procedures [1, 50]. Calibration should be performed at anticipated ambient temperature conditions with the intake and collection components in place so that the calibration will be valid for field conditions.

Flow measuring device maintenance must be aimed at possible fouling of meter flow restrictions with particulates or reactive gas corrosion products. Ideally, however, the meter should be protected from upstream contamination by an efficient particulate collector and an adsorber for reactive gases. Provided that suitable calibration, maintenance and protection of the flow measuring component are used, the primary remaining source of error will be leaks in the sampling system. The latter may be significant and must be evaluated by measuring inflow to the system when the intake is sealed. A leakage rate of less than 0.6 litre min^{-1} at 0.5 l b of vacuum is considered acceptable for source testing [30] while ambient gas sampling containers should maintain a vacuum of 0.98 ± 0.0013 b for 1 h without losing more than 0.0013 b [46].

Finally, because air volume is both pressure and temperature dependent, the measurement of these two parameters at the inlet to the volume or rate measuring device is an essential part of the flow measuring component. The determination of these parameters allows the conversion of the measured air volume to chosen standard conditions. Without this information, concentration determinations under different conditions could not be compared and would be relatively meaningless.

1.2.1.4 Air moving component

The final component of the sampling system is necessary to draw or force the air to be sampled through the overall system. In practice, a device creating a vacuum to draw the air through the collection component is preferable to devices which would have to intake the air and then push it through the collection component. The latter scheme obviously provides greater scope for sample contamination and change as well as greater likelihood of damage by the air sample to the air moving component itself.

Types of vacuum source include mechanical blowers, aspirators and hand operated pumps. The major determining factors in choosing the air moving component are the volumetric flow rate required and availability and type of power source.

For continuous operation at medium to high flow rates, mechanical blowers are usually chosen. Electrically powered pumps are preferable, but where a power supply is not available petrol driven pumps may be necessary. However, the possibility of sample contamination by the motor exhaust is a serious limitation to the use of petrol driven pumps. Under no circumstances should they be used without conducting the exhaust gases to a discharge point remote from the sampling intake and in all cases the degree of contamination possible must be evaluated by sampling in a clean laboratory atmosphere. The brushes on electric motors for vacuum blowers have been reported to produce aerosols [51, 52] which may contaminate aerosol sampling systems. Proper maintenance of air moving equipment is required to minimize such problems. Aspirators may be used where low to medium flow rates are required and a flow of water, air or steam under pressure is available. Head loss build up in the collection component must be avoided as water or steam may be drawn back into the sampling system from the aspirator if the water or steam flow through the aspirator should decrease because of a drop in supply line pressure.

Hand pumps and syringes may be used for grab sampling of gases, but these devices are of little use where continuous sampling is required.

1.2.2 Aerosols

1.2.2.1 Aerosol sampling considerations

Aerosol is defined [53] as a dispersion of solid or liquid particles in a gaseous media, where particle means a small discrete mass of solid or liquid matter. As a result of the discrete mass nature of the particles of an aerosol, the collection of a representative sample of the aerosol requires special consideration.

The flow pattern of a homogeneous gas is described by a pattern of streamlines (lines drawn tangentially to the direction of instantaneous velocity for all points in the gas flow). If the flow is steady, then the actual path followed by the flow will correspond to the streamline pattern.

If, for example, an uniform horizontal flow exists in the gas stream, (Fig. 1.15a) and a sampling intake probe, connected to a closed valve, is inserted in the flow, the streamlines will be distorted, as the flow diverts around the obstruction (Fig. 1.15b). If the valve to the sampling probe is opened and the vacuum source activated, then a certain flow into the probe will result (Fig. 1.15c). As the intake flow increases, until the velocity at the face of the probe matches the velocity immediately upstream from that point in the uniform flow, then the air will flow into the probe with minimal disturbance of the streamline pattern (Fig. 1.15d). Although some distortion of the streamline pattern is inevitable, even at this sampling condition, distortion can be minimized by the use of a sharp edged, thin-walled probe. Further increases in the intake flow rate will result in a distorted pattern with streamlines converging into the probe (Fig. 1.15e). The various sampling conditions which distort the streamline pattern do not interfere with obtaining a representative sample when sampling only gaseous constituents of the air mass, since gaseous constituents, provided they are thoroughly mixed with the air mass, will follow the streamlines.

However, because components of an aerosol are discrete mass particles, they will be subject to inertial effects relative to the gas stream. Particles will possess inertia determined by the product of their mass and velocity. Their inertia is a vector quantity and as such has a directional component, which coincides with the directional component to the particle's velocity. Therefore, the particles travelling in the uniform gas stream of Fig. 1.16a will have inertia directed along the streamlines. If as in Figs. 1.16b, c, the streamlines are forced to change direction abruptly in order to avoid the obstruction presented by the sampling probe,

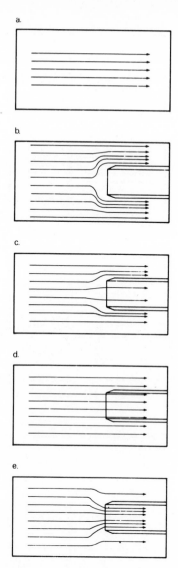

Fig. 1.15 Streamline pattern around a sampling probe in a uniform flow field.

the interactive forces of the molecular gas stream may be insufficient to deflect the particles to follow the streamline. As a result, the particle may depart from the streamline and continue its path into the sampling probe. The net result will be the collection in the probe of a greater proportion of particles per unit volume of gas than exists in the actual gas flow. Only when the velocity at

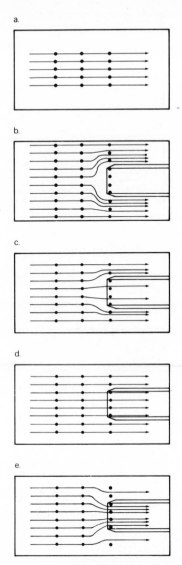

Fig. 1.16 Suspended particulate trajectories around a sampling probe in a uniform flow field.

the face of the probe is equal to the approach velocity of the gas stream, will the streamline pattern remain undisturbed, with the result being the correct proportion of particles per unit volume of the gas being sampled into the probe. This condition is referred to as isokinetic (equal velocity) sampling (Fig. 1.16d). If

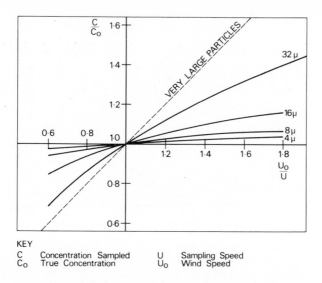

Fig. 1.17 Relationship of concentration sampled/true concentration to wind speed/sampling speed for spheres of unit density [56].

the intake flow rate is increased beyond isokinetic conditions, as in Fig. 16e, the inertia of particles originally in the outer streamlines which are now being drawn into the probe will cause these particles to continue on their original paths and miss the probe. The net result will be a lower concentration of particles per unit volume of gas sampled than exists in the external gas flow. Clearly the attainment of isokinetic sampling conditions is fundamental to the collection of quantitatively representative aerosol samples, for which particle inertia is significant.

The errors incurred by anisokinetic sampling have been studied by several workers including Zenker [54], Hemeon and Haines [55], Watson [56], Sehmel [57], Badzioch [58] and Vitols [59] and various correction formulae have been developed. A plot based on the work of Watson is presented in Fig. 1.17.

In addition to the collection of an inaccurate total mass concentration of particles when sampling anisokinetically, an inaccurate particle size distribution will be obtained because of the relation of particle size to inertial effects.

Particle inertial effects will also cause sampling errors if the probe is not properly aligned with the flow direction, as this will also result in a disturbance of the streamline pattern at the probe intake. The errors due to misalignment of the probe are demonstrated in Fig. 1.18 [56].

Inaccurate particle distributions in the flow will be caused by inertial effects when the flow is diverted around bends. Sansone [60] studied sampling for

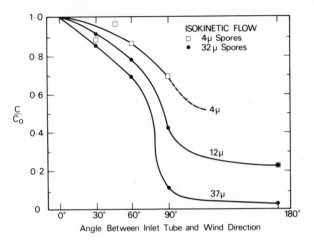

Fig. 1.18   Isokinetic sampling factors with the inlet tube aligned at various angles to the windstream, for 4, 12 and 37 MMD diethyl phthalate clouds (after Mayhood and Langstroth, from Watson [56]).

particulates following a 90° bend. As would be expected, the particulate concentration was found to be higher at the wall further from the centre of curvature of the bend. It was also found that under turbulent flow conditions, particle concentrations were higher around the periphery of the flow duct.

The requirements for isokinetic sampling can generally be relaxed when sampling particles exclusively less than 3 to 5 microns in diameter, as the inertial effect on these smaller particles is insufficient to cause significant sampling error. Likewise, the need for isokinetic sampling is lessened by sampling from stagnant or low velocity air masses. Thus, ambient air sampling does not generally require isokinetic conditions, which is fortunate, because isokinetic sampling of ambient air masses would be difficult to achieve consistently in most cases. However, gravitational and wind velocity effects can interfere with sampling of aerosols from ambient air.

The gravitational effect is simply described for the case where the collecting surface is pointed upwards as [1]

$$C_c = 1 + \frac{V_s}{V_c} C_a$$

where C_c = collected particulate concentration; V_s = settling velocity of particles; V_c = collecting velocity of sampler; C_a = ambient particulate concentration and where the collecting surface is pointed downwards

$$C_c = \left[1 - \frac{V_s}{V_c} \right] C_a$$

Since the particle settling velocity is directly dependent upon the size and density of the particle, the correction factor is small for smaller particles. The deviation from accurate sampling has been calculated for the standard high volume sampler [1] (Section 1.2.2.3) in Table 1.3.

Wind velocity effects on aerosol sample intake efficiencies were studied by Raynor [61]. He concluded that the entrance efficiencies for larger particles sizes ($6\,\mu$m and $20\,\mu$m) were quite sensitive to small changes in wind speed and approach angle. As would be expected, smaller particles were sampled more accurately than larger particles for various combinations of wind speed, flow rate and approach angle. Entrance efficiency generally decreased with increasing wind speed except at forward angles where inertial effects would enhance the collection of larger particles. However, the vertical position for the sampler intake ($90°$ to wind direction) was found to be the most inefficient for all but the smallest particles ($0.68\,\mu$m) which were insensitive to the variables tested anyway. The findings of Raynor [61] highlight the need to assess the collection efficiency of a chosen sampling system in relation to the anticipated size range of aerosols to be sampled and the ambient air flow conditions.

Table 1.3 Gravitational effect on collected particles* (after [13]).

Particle size (μm)	% reduction of collected particles
11	1
36	10
85	50
135	100

* Assumes $645\,\text{cm}^2$ of opening; $1.42\,\text{m}^3\,\text{min}^{-1}$ sampling rate.

Table 1.4 Permissible radii of tubes (cm) for sampling aerosols in calm conditions (after [62]).

Particle diameter (μm)	Rate of suction F (cm^3 s^{-1})					
	1	10	10^2	10^3	10^4	10^5
1	0.033–1.9	0.071–6.0	0.15–19	0.33–60	0.71–190	1.5–600
2	0.051–1.0	0.11–3.2	0.23–10	0.51–32	1.1–100	2.3–320
5	0.093–0.41	0.20–1.3	0.43–4.1	0.93–13	2.0–41	4.3–130
10	0.15–0.21	0.31–0.65	0.68–2.1	1.5–6.5	3.1–21	6.8–65
20	(0.23 ~ 0.10)	(0.50 ~ 0.33)	(1.1 ~ 1.0)	2.3–3.1	5.0–10.3	11.0–31
50	(0.42 ~ 0.042)	(0.90 ~ 0.13)	(1.9 ~ 0.42)	(4.2 ~ 1.33)	(9.0 ~ 4.2)	(19 ~ 13.3)
100	(0.63 ~ 0.023)	(1.4 ~ 0.071)	(2.9 ~ 0.23)	(6.3 ~ 0.71)	(14 ~ 2.3)	(29 ~ 7.1)
200	(0.89 ~ 0.014)	(1.9 ~ 0.037)	(4.1 ~ 0.14)	(8.9 ~ 0.37)	(19 ~ 1.4)	(41 ~ 3.7)
500	(1.26 ~ 0.008)	(2.7 ~ 0.025)	(5.8 ~ 0.08)	(12.6 ~ 0.25)	(27 ~ 0.80)	(58 ~ 2.5)

Table 1.5 Specified sampling rates and probe inlet dimensions of some commonly used sampling instruments (after [63]).

Sampler	Sampling Curve and Sampling Designation	Rate (litres min^{-1})	Actual Diameter of Sampler (cm)
DEL electrostatic precipitator	A	750	6.1
MSA electrostatic precipitator	B	85	3.6
Anderson cascade impactor	C	28	1.8
Standard G-S impinger	D	28	1.2
Half-inch cyclone	E	10	0.41
Half-inch cyclone	F	8	0.41
Midget impinger	G	2.8	0.36
MRE elutriator	H	2.4	0.10
10 mm cyclone	J	2.0	0.21
10 mm cyclone	K	1.7	0.21
10 mm cyclone	L	1.4	0.21
Micro impinger	M	0.56	0.36

Davies [62] has studied the theoretical limitations on the entrance efficiency due to inertia and gravitational effects. In order to avoid significant inertial effects, Davies suggests that the radius of the sampling intake must be much larger than the stopping distance* for the particle concerned. In turn, for gravitational effects, the inlet velocity at the sampling intake must be large in comparison to the settling velocity for the particle concerned. The inlet velocity for a given flow rate is a function of the inlet radius and, thus, Davies derived an expression for the upper and lower bounds of the sample inlet radius at a given flow rate for a given particle size. A tabulation of these bounds for sample inlets is presented in Table 1.4.

Bien and Corn [63] applied Davies' criteria to several commercially available samplers. Their findings, summarized in Table 1.5 and Fig. 1.19, indicate that many commercially available samplers fail to conform to the criteria of Davies' theoretical model for unit density spheres in calm ambient air.

Although the need for considering the inertial and gravitational effects on the collection of aerosols requires primary attention, several other sources of error in aerosol sampling must also be considered, such as:

---

* The stopping distance is the distance a particle would require to stop if injected into a stagnant air mass with the velocity it possessed in the air stream. Thus, the stopping distance is a function of the particle momentum.

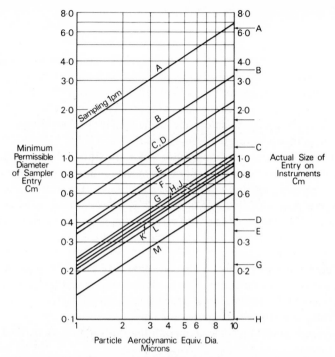

Fig. 1.19 Particle size efficiently captured by selected aerosol samplers, according to the criteria of Davies. The letter designations A to M refer to the instruments and conditions as listed in Table 1.5. (After [63].)

aerosols may collect on sample transfer tubing walls, particularly after bends or where condensation is allowed;

aerosols may be attracted to tubing surfaces by electrostatic forces particularly where non-conducting glass and plastic tubing is used;

liquid aerosols may evaporate after collection or react with solid particles or one another;

aerosols may be hygroscopic and adsorb water vapour.

Some changes in the aerosol may be tolerated depending on the analysis to be performed. For example, agglomeration of particles without weight change would hinder particle size analysis but would not hinder a mass concentration or chemical analysis. However, a complete and thorough evaluation of possible sources of loss or change to aerosols within a sampling system is necessary in order to ensure that the aerosol which reaches the collection component quantitatively and qualitatively resembles, for the purposes of the subsequent analysis, the original aerosol sampled.

1.2.2.2 Aerosol sampling collection components

Filtration

Filtration for the collection of aerosols depends upon drawing the air sample through a network of small openings which may be formed from:

the overlap of fibres as in fibreglass or cellulose filters;
random pores with a fixed media as in sintered glass filters;
controlled sized pores in an organic media as in membrane filters;
random pathways through a granular media.

In all cases, the small size of openings within the filtration media results in the presentation of a large surface area of solid media for direct impingement and collection of aerosol particles. For particles which follow the air flow into the filter the tortuous path with frequent directional changes encourages the deviation of the particles from the air stream with resulting impingement on the internal filter media. As the filter collects more and more particles from the air stream, the flow pathways become further restricted and particles become collected on the mat of collected material. Finally, some collection may accrue because of attractive forces such as electrostatic attraction between the filter and the particles.

The type of filter media chosen depends upon several factors:

The collection efficiency required for a given particle size must be established. Collection efficiencies for some common filter types are shown in Table 1.6.

The pressure drop developed across the filter as the filter mat builds up must be considered in assembling the overall sampling system. Head loss characteristics for some common filter types are shown in Fig. 1.20.

The background concentration of trace constituents within the filter media which might interfere with subsequent analysis must be determined. Direct analysis of the filter media under consideration is the safest method of determining possible interferences.

The chemical and physical resistance of the filter media to the particular air environment to be sampled must be considered. Apart from the obvious cases such as the attack of a solvent gas on an organic filter media, most problems of this nature will require field evaluation.

The hygroscopic properties of the filter media are important when the humidity of the air to be sampled cannot be controlled. Glass fibre and membrane filters are preferable in this regard to cellulose filters which tend to be hygroscopic.

In most cases the filter media is provided as a thin sheet which must be mounted in a filter housing to provide structural support. Several types of screen and mesh supports are commercially available. The support material must be consistent with the sampling task and should not react with any of the constituents which could conceivably be present in the air sample.

Table 1.6 Examples of high-efficiency filters (after [22]).

Filter	Compostion	Efficiency[a] (%)
W. & R. Balston, England		
Whatman GF/C	glass fibre	> 99.9
Whatman no. 32	cellulose	99.5
Whatman no. 42	cellulose	99.2
Whatman no. 44	cellulose	98.6
Whatman no. 50	cellulose	97.0
Gelman Instrument Co., U.S.A.		
type VM-4	polyvinyl chloride	~ 100
Acropor, type AN-3000	acrylonitrile/polyvinyl-chloride copolymer reinforced with nylon	~ 100
type A	glass fibre, no organic binder	> 99.99
type E	glass fibre with acrylic binder	> 99.99
Hollingsworth & Vose, U.S.A.		
HV-70, 18-mil[b]	cellulose & asbestos	99.3
Millipore Filter Corp., U.S.A.		
types HA & AA	cellulose esters	> 99.9
Mine Safety Appliance Co., U.S.A.		
type 1106-B	glass fibre	99.99
type 1106-B	flash-fired glass fibre	99.97
Scleicher & Schüll, Federal Republic of Germany		
Ultrafilter	cellulose ester	99.99[c]
Schneider-Poelman, France		
type 'rose'	esparto & asbestos fibres	99.98[d]
type 'jaune'	esparto & asbestos fibres	97.0[d]
U.S.S.R.		
FPP-15-4.5		99.995[e]
FPP-15-3.0		99.99[e]
BF		99.9[f]
AFA RMP		99.5[e]
AFA RMA-20		97.0[e]
U.S. Atomic Energy Commission		
AEC-1	asbestos & vegetable fibres	> 99.9
CWS-6	asbestos & cellulose fibres	> 99.9

[a] It should be noted that, owing to differences in methods of measurement, not all the figures for efficiency are comparable. Unless otherwise noted, the figures given are based on retention of di-(2-ethylhexyl)phthalate particles 0.3 μm in diameter, at a flow rate of 0.1 m min^{-1}.

[b] The figure given is for pore size. 1 mil = 1 milli-inch = 0.002 54 cm.

[c] Efficiency based on retention of di-(2-ethylhexyl)phthalate particles not exceeding 1 μm in diameter at flow rate of 0.1 m min^{-1}.

[d] Efficiency based on retention of Indigo aerosol particles 0.3 μm in diameter.

[e] Efficiency based on retention of decay particles of radon.

[f] Efficiency based on retention of oil aerosol droplets, diameter not stated.

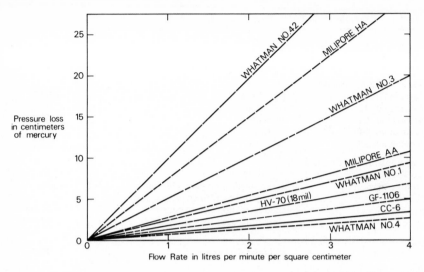

Fig. 1.20 Head-loss characteristics for various types of common filter media. (After [18].)

Impingement

Impingement for the collection of aerosols consists of forcing the air flow through a jet which increases the velocity of the air stream and thereby the momentum of associated particles, followed by collision of the airstream with an abrupt obstruction. Because of their high momentum and the abrupt change in flow direction necessitated by the obstruction, the particles will tend to collect on the obstruction.

Impingers have been developed to operate either wet or dry. Wet impingers are used with the jet and collecting obstruction below the surface of a liquid (in which the particles must be insoluble). Dry impingers, as the name implies, have the jet and collecting obstruction dry, although an adhesive may be utilized to prevent the particles from becoming resuspended.

Two common types of collection impingers, the midget impinger and the full sized Greenburg–Smith impinger are illustrated in Fig. 1.21. The midget impinger is commonly used wet, while the Greenburg–Smith may be used either wet or dry.

Because the impinger uses particle inertia for collection and that, in turn, depends upon the mass and velocity of the particles, the size collection efficiency of impingers depends on the flow rate. The midget impinger is commonly operated at 3 litres min^{-1} while the Greenburg–Smith impinger is operated at 30 litres min^{-1}.

The dependence of the particle size collection efficiency on jet velocity

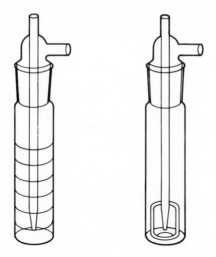

Fig. 1.21   Impingers. Left: Midget; right: Greenburg-Smith.

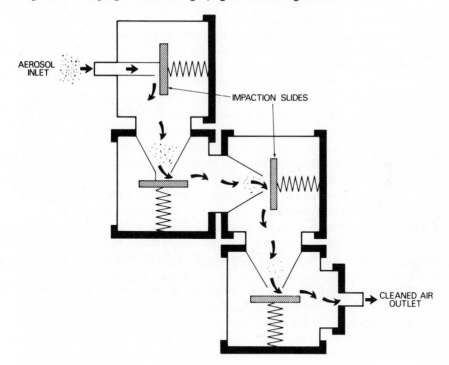

Fig. 1.22   Cascade impactor (after [3]).

has been used to advantage in the design of cascade impactors. These are devices which use a series of sequential jets and collection plates with increasing jet velocities and/or decreasing gaps between the jet and the collecting plate (Fig. 1.22). As the air stream progresses through the device smaller particles will be collected more efficiently. As a result, the cascade impactor fractionates the air sample to obtain a particle size distribution. Unfortunately, in come cases large agglomerated particles will shatter upon impinging on the collecting plate and may resuspend giving an inaccurate size distribution.

Precipitation

Precipitation for the collection of aerosols depends upon exposing the air stream to either a thermal or electrostatic gradient.

In the case of the former, a sharp temperature gradient is created by a high temperature wire suspended across a cylindrical collection chamber (Fig. 1.23) [64]. The temperature gradient causes the particles to migrate towards the cooler collecting surface, apparently because molecules of gas migrate around suspended particles from the cooler region to the warmer region, causing the particles to move toward the cooler region by reaction. The device is extremely efficient for particle sizes from $5\,\mu$m down to $0.01\,\mu$m and possibly smaller [22], but the basic instrument is limited to a relatively low flow rate (0.007 to $0.02\,$litre min^{-1}). Although higher flow rate samplers have been developed ($0.5\,$litres min^{-1}), the thermal precipitator is recommended primarily for specialized research studies where efficient collection of particles for microscopic investigations is required, but not for general aerosol sampling usage.

In the case of electrostatic precipitation, a high voltage potential (12 kV, d.c. or 20 kV, a.c.) is maintained across the plates of a capacitor with the air to be sampled flowing between the capacitor plates. The ionizing potential created by the electrostatic field produces ions in the gas stream which, in turn, tend to collide with particles and charge them. The charged particles then migrate under the influence of the electrostatic field to the collecting electrode. This device has been found to be virtually 100% efficient for a wide range of particle sizes at flow rates up to 85 litre min^{-1} [64]. Precautions must be observed to avoid excessive voltage, as direct arcing across the plates may occur and sampling efficiency will drop to nil. Also, the device is clearly unsuitable for use in explosive atmospheres [22]. However, other than possible agglomeration of particles due to charge effects, this device collects at a high efficiency with relatively little physical disruption of the particles.

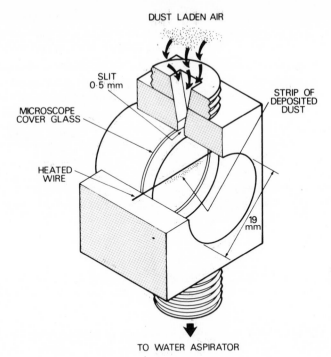

Fig. 1.23 Sampling head of standard thermal precipitator. Crown Copyright. Reproduced by permission of the Controller of Her Majesty's Stationery Office, London.

Gravity and centrifugal collection

Gravity and centrifugal collection depend upon the action of an external remote force on the aerosol particles to cause their movement toward a collecting surface. In the case of the former, the gravitational attraction of the particle mass towards the earth causes particles to settle at a rate dependent on the fluid drag forces counteracting gravity. For smaller particles the settling rate is small in relation to air movements with the effect that small (less than $7\,\mu$m) particles remain suspended.

In the case of centrifugal collection, the air stream is made to follow a spiral pathway in a cyclone, with the resulting centrifugal action on the particles causing them to move outward to a collecting surface. Both collection methods are limited to larger particles than are collected by other collection components, with their use aimed primarily at preliminary separation of larger particles before a secondary collector.

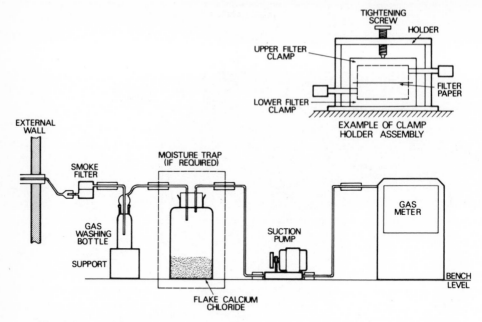

Fig. 1.24 Typical example of aerosol sampling apparatus assembly. (After [65].)

1.2.2.3 Ambient aerosol sampling applications

British suspended particulate sampler

The suspended particulate or smoke sampler is commonly used in the United Kingdom National Survey of Air Pollution. The method is the subject of a British Standard [65].

Although most commonly used as a relatively low volume sampler (1 to 1.6 litres min^{-1}) for estimating smoke density, a system is also described for sampling at 25 to 100 litres min^{-1}. This system is believed to sample only particles less than 20 to 30 μm.

The whole system is shown in Fig. 1.24. The inlet consists of an inverted glass funnel located more than 2.5 m above the ground and 1 m from the adjacent wall. The inlet may be protected by a mesh guard, but this must be constructed of a material which will not flake or corrode in the atmosphere.

A 100 mm diameter glass fibre filter is recommended and the method calls for the use of a control filter to determine weight changes due to handling. The system employs a volume meter, capable of being read to 0.025 m^3 or less. For the determination of suspended particulate mass concentration, a sample of at least 10 mg should be obtained, which will normally require a sampling period of 24 h.

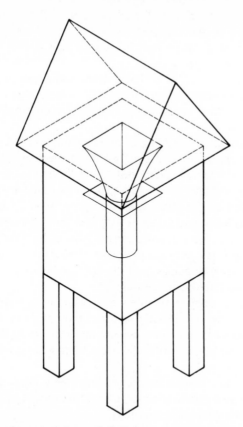

Fig. 1.25   High volume sampler and shelter.

For a mass concentration of 10 mg or greater, the precision of the procedure has been found to be within 10%.

American high volume sampler for suspended particulates

The 'high volume sampler' is commonly used for the United States National Air Sampling Network. Detailed standard procedures have been published for this method [1, 50].

A high flow rate blower (1.1 to 1.7 m³ min⁻¹) draws the air sample into a covered housing and through a 20 cm × 25 cm rectangular glass fibre filter (Fig. 1.25). The intake dimensions and normal flow rate are believed to limit aerosol collection to particles less than 100 μm in diameter [50]. The covered housing is necessary to protect the filter from precipitation and falling debris and should be constructed from non-flaking or corroding material, such as aluminium.

The unit is normally operated for 24 h although other sampling periods may be used depending on the expected concentrations of suspended particulates. Glass fibre filters are normally used because of their gradual headloss build up characteristics and non-hygroscopic properties.

Units may be purchased with timers which will automatically sample for a predetermined time during a seven day period. Likewise, flow recorders are available to measure the flow rate during the sampling period to allow calculation of the sampled volume.

Even with continuous recording of the sampling flow rate, it is desirable to maintain the sampling rate constant, because collection efficiency has been reported to vary with sampling rate [66]. The collection of oily particles or sampling during foggy conditions may lead to decreased sampling flow rates. Avera [67] has developed an electronic flow regulation system using a hot wire anemometer as the feedback sensing device to a flow controller which will adjust the blower to compensate for the increasing head loss at the filter. The anemometer is set for a null signal at a given sampling rate. Fluctuations from the set rate produce a feedback signal to compensate the flow produced by the blower to bring the anemometer back to the null setting.

Contamination or damage to filters is possible when inserting or removing them from the sampler, particularly on a windy or rainy day. King and Fordyce [68] have developed a filter-holding cassette which allows the filter to be inserted and sealed in the laboratory with subsequent rapid and trouble-free installation in the field.

Operating the high volume sampler at lower flow rates, $0.85\,m^3\,min^{-1}$ has been reported to prolong electric motor brush life as well as maintain less flow drop off over a 24 h sampling period [69].

The sampler must be calibrated regularly at least every six months. Standard calibration procedures are available [1, 50]. Lynam *et al.* [70] have criticized calibration procedures which are based on varying the flow rate only with the use of a variac transformer on the vacuum blower power supply. They recommend calibration by modifying the flow resistance by placing extra filters on the sampler. Tebbens [71] explains the discrepancies noted by Lynam *et al.* as due to inward leakage between the filter and the flow recorder, but he does recommend calibration with the normal flow resistance in place.

Clements *et al.* [72] compared 239 pairs of particulate mass concentration results obtained by high volume samplers and found that the average deviation from the mean of the paired values was less than 4.5% and 90% of the values deviated by less than 14%.

Tape filter samplers

Filter tape samplers have been developed to take a series of sequential samples for predetermined sampling periods. A common sampler consists of an intake nozzle, a sampling head to hold and seal the filter tape, spools to collect the filter tape, an interval timer and clock, a flow meter and a vacuum blower.

The instrument is normally operated at 7 litres min^{-1} with cellulose filter tape for sequential sampling periods of 1 to 4 h to collect a series of smoke spots which are subsequently analysed for their relative light transmittance. However, the unit can be used to collect sequential samples for chemical analysis or, with adaptation, for particle size determination. Heard and Wiffen [73] developed a tape sampler using AA Millipore membrane filter tape to collect aerosols for electron microscopy. The device uses a 5 cm wide strip of membrane filter to collect 4 samples side by side over a 3 h sampling period. The sampling rate corresponds to 0.25 litre min^{-1} per sample spot and particulates from 0.01 to 5 μm are efficiently collected. The sampled tape is wound on an open spiral spool to prevent damage to collected particles. Sufficient tape is provided to sample for 8 consecutive weeks.

High volume cascade impactors

Cascade impactors for particle size fractionation of high volume air samples are relatively recent innovations. Gussman *et al.* [74] have developed and evaluated a 5-stage, slit type high volume cascade impactor, while Burton *et al.* [75] have evaluated a commercially available high volume, multi-orifice cascade impactor.

The unit developed by Gussman *et al.* [74] was developed on a single slit principle, with 4 fractionating stages and a final collection filter (Fig. 1.26). The device was designed to fit over a standard high volume air sampler and to operate at a flow rate of 0.85 m^3 min^{-1}. The cutoff diameters for 50% collection of unit density particles were found to be 9.4, 2.5, 1.8 and 1.5 μm for the successive stages.

The cascade impactor evaluated by Burton *et al.* [75] consisted of a stack of five 30 cm diameter plates, each with 300 orifice jets, mounted over a standard high volume glass fibre filter. The collection media consisted of perforated sheets of aluminium foil or glass fibre which are mounted on successive aluminium plates. The device provides approximate particle size fractions of greater than 7; 3.3 to 7; 2.0 to 3.3; 1.1 to 2.2; and less than 1.1 μm for a flow rate of 0.56 m^3 min^{-1}.

Field testing of the instrument using duplicate sampling provided reproducible results. The overall mass collected also compared favourably to a regular filter high volume sampler operated at the same sampling rate, provided that a neutral pH filter paper, which would not preferentially absorb acidic vapours, was used.

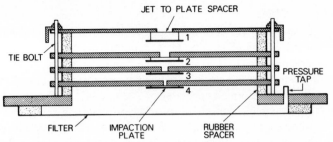

Fig. 1.26   Schematic cutaway view of high volume cascade impactor. (After [74].)

Centrifugal impinger for micro-organism sampling

Buchanan *et al.* [76] have developed a liquid scrubber which uses centrifugal action to impinge aerosol micro-organisms on to a thin liquid film. Collecting fluid is aspirated into the sampled air stream and the combined flow enters a cylindrical tube tangentially, creating a spiral flow pattern. The aspirated mist is forced out to the tube walls by centrifugal effect and forms a continuous thin film which then moves in a spiral pattern along the tube to a collection flask under slight vacuum. The micro-organisms are collected on the liquid film as a result of centrifugal impingement and subsequently flow with the collection liquid into the collection flask.

The system operates at flow rates up to 950 litres min^{-1} and is found to be effective for sampling viable micro-organisms intact. The relatively gentle collection mechanism may have application to the collection of other aerosols.

Respirable particulate sampling

Aerosol sampling aimed at measuring respirable particulate concentration has been receiving increasing attention in recent years. Lippman [77] has discussed the rationale applied by the Aerosol Technology Committee [78] of the American Industrial Hygiene Association in evaluating several sampling systems for their performance in respirable particulate sampling. Criteria which were applied (Table 1.7) require the sampling systems to sample discriminately various ranges of the particle size spectrum with different efficiencies. The aim of this type of respirable sampling is to reproduce the particle intake and retention characteristics of the human respiratory system and thereby allow sampling of particulates directly relevant to the assessment of particulate inhalation health hazards. In practice, the particle size discrimination characteristics for respirable particulate sampling systems require the use of two or more collection components in series. Most devices use a first stage component with a very high efficiency for 10 μm or

Table 1.7   Respirable particulate criteria (after [78]).

Aerodynamic diameter (unit density sphere) (μm)	% Passing
American Conference of Government Industrial Hygienists criteria	
$\leqslant 2$	90
2.5	75
3.5	50
5.0	25
10	0
British Medical Research Council	
1	100
5	50
7	0

larger particulates, but which drop off quickly to low efficiency at $2\,\mu$m. The second stage then collects those particulates which have passed the first stage screening.

The Aerosol Technology Committee have reviewed the calibration specifications of several commercially available sampling devices and report that five devices appear capable of meeting one of the respirable particulate criteria [78].

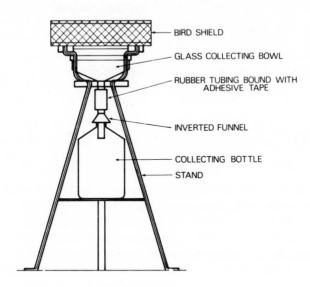

Fig. 1.27   British standard deposit gauge.

British deposit gauge

The deposit gauge for the collection of particulates which settle is described in detail by a British Standard [79]. Particulates which will be measured with this method are those which will fall into the gauge as a result of their own settling velocity or particulates which are carried down with precipitation. This gauge tends to underestimate deposition of particulates during high winds.

The device consists of a collection bowl connected to a bottle and supported by a galvanized steel stand (Fig. 1.27). The stand may be provided with a bird guard to protect the sample from avian interference.

The deposit gauge should be sited at a distance from any object of twice the height of any object in the vicinity. The gauge may be sited above ground level, but never more than 5 m above the ground. The sample is collected over a prolonged period ranging up to a month. During spring and summer it may be necessary to add an algicide, such as copper sulphate, to the collected water to prevent algal blooms which would modify the sample character.

Directional dust gauges

A recent modification to the standard deposit gauge provides for wind directional dust collection. The standard deposit gauge may suffer losses of settled particulates during high winds and fails to collect particles which would be deposited by wind driven impingement.

The directional device, covered by British Standard [80] consists of four cylinders mounted on a common post facing the four quadrants. Each cylinder has an open face directed outward and each is connected at the bottom to a collection bottle (Fig. 1.28). The effect of this arrangement is to collect wind driven and direct settling dust from four directions simultaneously.

In addition to the advantages of collecting wind driven particulates, the gauge is useful for pinpointing the effects of source emissions of particles which settle. The siting requirements are similar to those applied to the standard deposit gauge, with the added condition that the top of the gauge should be 1.5 ± 0.1 m above ground.

The dustfall canister

Particulates which settle, defined as solid or liquid particulates small enough to pass a 1 mm screen while large enough to settle out from the atmosphere, are commonly collected in dustfall canisters. The sampling system [1, 81, 82] involves the placing of an open top cylinder, greater than 15 cm in diameter (height should be two to three times the diameter), in a support stand (Fig. 1.29). The cylinder and support are constructed from non-corrosive and non-flaking materials.

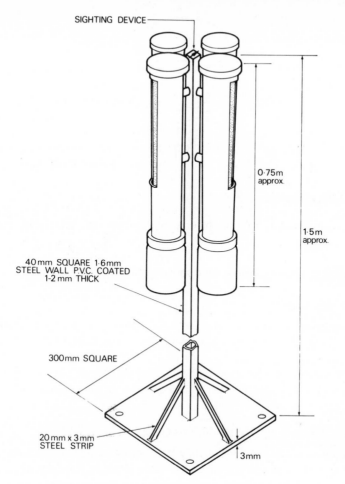

SIGHTING DEVICE

0·75m
approx.

1·5m
approx.

40mm SQUARE 1·6mm
STEEL WALL P.V.C. COATED
1-2 mm THICK

300mm SQUARE

20 mm x 3 mm
STEEL STRIP

3mm

Fig. 1.28   Directional dust gauge, typical assembly. (After [80].)

The cylinder is filled less than half full with deionized water and mounted in the holder, more than 0.9 m above the supports. In winter an antifreeze and in summer an algicide may be added to the collecting liquid. The canister should be sited more than 2.4 m but less than 15 m above ground and in all cases more than 1.2 m above any other adjacent surface. The tops of buildings and large objects should be less than 30° above the horizontal at the sampling location.

Although the results from this and other settled particulate collectors can be extrapolated over a small representative area, the common practice of expressing dustfall rates measured in this way as tons square mile^{-1} month^{-1} or

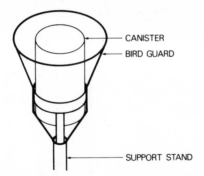

CANISTER

BIRD GUARD

SUPPORT STAND

Fig. 1.29   Dustfall canister.

tonnes km^{-1} month^{-1} is misleading and results are preferably expressed as g m^{-2} month^{-1}.

1.2.2.4 Emission source aerosol sampling applications

Stationary sources

The sampling of aerosol emissions from stationary sources has received a great deal of attention over the last fifteen years. As a result, there exists a wealth of detailed literature on overall source sampling procedures [18, 27, 34, 37, 83, 84] and standardized source sampling techniques [28–33, 36, 85–87]. Some of these will be discussed in order to illustrate the techniques available.

British Standard, B.S. 3405: 1971 [31, 32]

This standard provides for sampling from stacks for particulates larger than 1 μm with an expected accuracy of ± 25%. The equipment for the method includes: a pitot static tube, inclined gauge manometer, thermometer, probe tube, flow meter, flow rate control valve, connecting tubing, vacuum blower, and a collection component. The latter should be capable of filtering 98% of particulates over 1 μm. Some different collection component assemblies have been approved as complying with the requirements of the standard.

Testing is performed at a minimum of 4 sampling points at the centroids of equal area sectors of the stack cross-section for a minimum of 2 min at each location as illustrated in Fig. 1.30. For stacks in excess of 2.5 m^2 cross-sectional area or where velocity pressures at sampling points vary from one another by more than 4 to 1, 8 sample locations are used.

Following a gas velocity traverse, as described in Section 1.1.2.1, the sample probe is positioned at the sampling location. The vacuum blower is started with

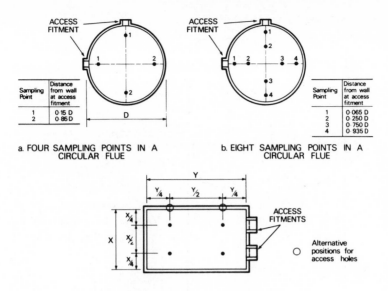

Sampling Point	Distance from wall at access fitment
1	0.15 D
2	0.85 D

a. FOUR SAMPLING POINTS IN A CIRCULAR FLUE

Sampling Point	Distance from wall at access fitment
1	0.065 D
2	0.250 D
3	0.750 D
4	0.935 D

b. EIGHT SAMPLING POINTS IN A CIRCULAR FLUE

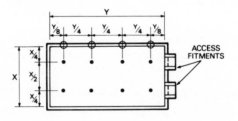

c. FOUR SAMPLING POINTS IN A RECTANGULAR FLUE

d. EIGHT SAMPLING POINTS IN A RECTANGULAR FLUE

Fig. 1.30   Sampling points in circular and rectangular flues. (After [32].)

the control valve closed. The control is opened until a precalculated flow rate for isokinetic conditions is reached. The probe is moved from location to location, providing the same sampling period at each. After completion of sampling at all locations, the probe may be withdrawn, or improved accuracy may be achieved by repeating sampling at each location.

After removal of the probe from the stack, the collection device is detached, deposits from the probe and sample lines are added and the collected particulate matter is sent to the laboratory for gravimetric and/or chemical analysis.

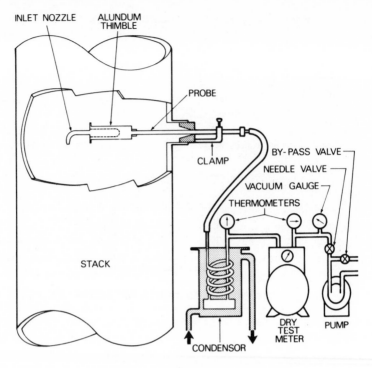

Fig. 1.31   Schematic diagram of sampling train for dry particulate matter using an in-stack sampling arrangement. (After [29].)

American Society for Testing and Materials (A.S.T.M.) [29]

This standard provides a procedure for sampling stacks utilizing an in-stack particulate collector. Equipment for this method is similar to that required for B.S. 3405: 1971 with the addition of a condenser. An example of one assembled sampling train which is consistent with this standard is shown in Fig. 1.31. Three types of filter media are specified for the in-stack filter holder. Alundum thimbles are useful for applications at temperatures up to 550° C. Filter holders providing wire mesh or sintered stainless steel support for various round sheet filter media are included, but these are generally not suitable for high temperature operation. Filter tubes, packed with glass wool provide a good media for sampling wet gas streams which would clog the alundum thimble. The glass wool packed tubes may generally be used up to 480° C.

The condenser in the system must provide for the anticipated volume of water condensate, as well as allowing easy measurement of the collected condensate volume. Flow metering is generally provided by a dry test meter. Although a rate meter is not included in the basic sampling train, a device such as an orifice meter is useful for monitoring isokinetic sampling flow conditions. The vacuum source may be a blower, pump or aspirator provided it can maintain sufficient flow through the sampling train resistance to provide for isokinetic sampling conditions.

The method calls for a minimum of 4 sample locations for rectangular stacks and a minimum of 3 for circular stacks. These may increase to 24 or more for larger stack cross-sections. Furthermore, the number of sample points should be doubled when only 4 to 6 diameters of straight stack are available from the closest upstream disturbance rather than the required 8 stack diameters. When less than 4 stack diameters are available, isokinetic sampling is considered to be difficult to achieve.

Following a gas velocity traverse and prior to sampling, the equipment is leak tested by plugging the probe and drawing the maximum vacuum the pump can produce. If no flow is produced, the test is started with the probe in position and nozzle facing downstream. When the test is started the vacuum source is actuated, the control valve opened and the nozzle turned to face directly upstream. The sampling time at each location should be at least 5 min under steady conditions with a total sampling time of at least 1 h when possible. Sampling is maintained isokinetically during the sampling period by adjusting the flow through the sample train to precalculated values determined from the velocity traverse and stack gas conditions.

Upon completion of the test, the control valve is closed, the vacuum blower shut off, the nozzle turned downstream and the probe removed from the stack. The whole apparatus, or the collection component only (where it is feasible to change the component in the field) is removed and sent to the laboratory for analysis.

United States Environmental Protection Agency (E.P.A.) procedure [30]

This standard provides a procedure for sampling stacks utilizing an out-of-stack collection component. The relative accuracy of out-of-stack versus in-stack collectors has been discussed by Hemeon and Black [88] who maintain that out-of-stack collectors are subject to serious error.

The E.P.A. procedure described is currently the most restrictive and specific with regard to optional equipment as it is used for regulatory purposes. Details of construction [89] and operation, maintenance and calibration [39] have been published for this sampling method.

The sampling equipment required includes the same general components as the A.S.T.M. procedure except that an orifice flow meter is mandatory. The assembled E.P.A. sampling train is illustrated in Fig. 1.32. A sharp edged stainless

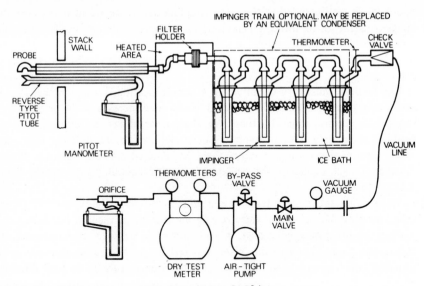

Fig. 1.32   Particulate-sampling train.(After [30].)

steel nozzle is used with a heated pyrex or corrosion resistant alloy sample probe. A type S pitot tube is attached directly to the sample probe to allow simultaneous sample collection and velocity measurement.

The particulate collection medium is specified as a glass fibre filter MSA 1106BH or equivalent filter paper. The filter is mounted against a sintered glass plate which is mounted in a heated pyrex filter holder. This is followed by an impinger train or a measuring condenser, then an air tight pump, dry test meter and orifice flow meter.

The number and positions of sample locations are determined from Figs. 1.4–1.6. Following a gas velocity traverse and before sampling, the equipment is leak tested by plugging the probe. A leakage rate of less than 0.6 litre min^{-1} at 0.51 b vacuum is considered acceptable.

The sampling procedure is similar to the A.S.T.M. and B.S. procedures. A minimum of 5 min is required at each sample location. During sampling, isokinetic flow rates are maintained by adjusting the flow control valve to achieve calculated isokinetic flow rates through the orifice, using the simultaneous stack gas velocity measurement from the pitot tube. Field nomographs are available and necessary to perform the flow rate calculations, since adjustment must be made for the different gas conditions in the stack and at the orifice.

At the completion of the test, the blower is turned off and the probe and nozzle removed from the stack. The equipment is removed to a clean location and disassembled. Particulate matter collected in the nozzle or probe is washed out with acetone and collected in a sealed container. The water volume collected in the condenser is measured for the moisture determination and the filter holder with filter in place and probe washings are returned to the laboratory for analysis.

The high volume stack sampler [83]

The stack sampling devices described in the foregoing procedures are suitable for high velocity, confined stack emissions. Boubel [83] has developed a stack sampling system for use with low flow rate and unconfined emission sources.

Boubel considered the disadvantages of the common standard sampling techniques. Among those listed are: expensive equipment, complicated use and analysis, inability to use collected sample for particle size analysis, and inability to sample at flow rates greater than 30 litres min^{-1}.

The high volume stack sampler developed, is constructed entirely from aluminium, to provide the convenience of light weight. The device incorporates a nozzle connected to a sample probe which in turn is connected to a filter holder designed to hold a standard 20×25 cm high volume glass fibre filter. From the filter holder, the flow runs through a valve and a sharp edged orifice, connected to a dial reading manometer (magnehelic gauge). The dial is located in a control panel positioned on the sampler body with a magnehelic gauge reading from a pitot tube attached to the sampling probe. The unit is then connected to a high volume blower by means of flexible tubing.

After a preliminary velocity traverse with the pitot tube and a preliminary sampling run to determine sample temperature through the filter and orifice, the isokinetic sampling rate is determined by reference to precalculated operating curves. Sampling may then proceed at selected sampling locations as long as necessary to collect a measurable sample. For particulate emissions of $0.25 \, \mathrm{g\,m^{-3}}$, a 1 min sample period would produce an adequate sample. At the completion of testing, the particulates collected in the nozzle and probe are collected by rinsing with methyl alcohol. The filter and probe rinsings are then returned to the laboratory for gravimetric, chemical and/or particle size analysis.

Mobile sources

Aerosol sampling from mobile sources provides a difficult challenge. Vehicle aerosol emissions cover a wide spectrum from particles several millimeters in diameter down to submicron particles. The wide range in particle sizes, high temperature and humidity of exhaust gases and the unsteady and non-uniform vehicle exhaust flow all contribute to make exhaust aerosol sampling difficult.

Hirschler et al. [90] pioneered vehicle exhaust aerosol sampling in a study aimed at determining vehicle lead emissions. The entire exhaust stream was passed through an electrostatic precipitator after approximately a 10 to 1 dilution with filtered air. The system efficiency was determined to be 90 to 95% for lead particulates based on collection with a backup membrane filter. For this study, the precipitator was lined with a polyvinyl acetate liner which was dissolved in trichloroethylene after sampling. The inorganic particulates were then collected from the solvent by filtration.

McKee and McMahon [91] collected particulate samples on 20 cm × 25 cm sheets of glass fibre filter, using a vacuum blower to draw the exhaust sample through the filter. Samples for electron microscopy were collected from the exhaust line, downstream from the exhaust manifold, and diluted with 2 to 3 volumes of dry air to prevent condensation. The diluted air was sampled with a thermal precipitator which collected particulates directly onto a perforated brass screen mounted on a glass slide for electron microscopy.

Mueller *et al.* [92] studied isokinetic probe sampling for particulates directly from the vehicle tailpipe. The collected exhaust was diluted with 4 volumes of filtered, dry air. The flow stream passed through a baffle into a sampling chamber from which aerosol samples were drawn for particle size fractionation.

Habibi [93, 94] studied the requirements for vehicle particulate sampling and developed criteria for a suitable partial (proportional) collection system. Included in the criteria were a suitable vehicle operating cycle, prevention of condensation by dilution or heating, equal proportion collection of all particle sizes, isokinetic sampling, no loss or change of particles in the sampling system and capability of long duration sampling. Habibi developed the system illustrated in Fig. 1.33. With this system, the exhaust stream is led into a large tunnel and diluted with filtered ambient air. A dew point calculation showed that a 4 to 1 dilution was

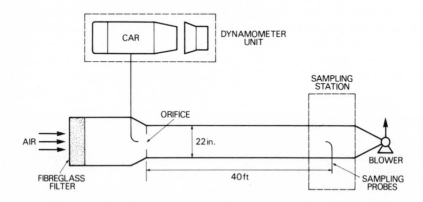

Fig. 1.33   Details of proportional vehicle exhaust sampling system. (After [93].)

required to avoid condensation, but the system provided much higher dilutions (23 to 1 for $70\,km\,h^{-1}$ road load). The tunnel was operated on the variable dilution principle whereby a fixed outflow from the tunnel is maintained by the exhaust blower while the total fixed inflow is provided by the variable engine exhaust flow and a compensating dilution air inflow. A tunnel mixing length of 12.2 m was provided along with an inlet orifice to promote turbulent mixing. The combination of tunnel length and inlet orifice was a compromise to produce good mixing along with a steady velocity profile at the sampling point. Isokinetic sampling was undertaken with 0.95 and 1.9 cm diameter sampling probes. The system performed satisfactorily except for slightly anisokinetic sampling conditions due to temperature fluctuations and the inevitable loss of particulates which settled to the bottom of the dilution tunnel.

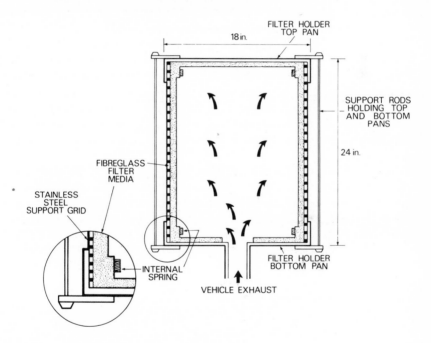

Fig. 1.34   Total exhaust filter. (After [93].)

Habibi [93] also developed a total flow filter which mounts directly on the exhaust pipe and provides less than 5 mb pressure drop at $110\,km\,h^{-1}$, making the system suitable for road testing. The device (Fig. 1.34) consists of a cylinder, packed with a glass fibre filter medium. The exhaust flows into the cylinder and passes radially out through the media which is supported on a stainless steel grid.

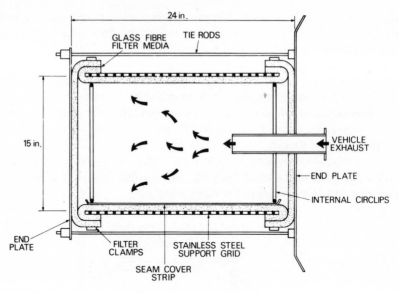

Fig. 1.35 Alternate total exhaust filter. (After [95].)

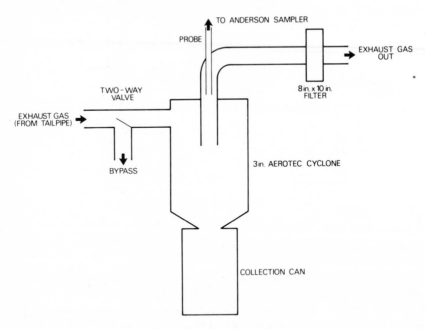

Fig. 1.36 Cyclone total exhaust filter. (After [96].)

The unit is sealed by internal springs and stainless steel strip over the seams. When used for extraction and recovery of exhausted lead particulates, the total filter was found to be more than 99% efficient.

Campbell and Dartnell [95] worked with a total filter (Fig. 1.35) and found difficulties with condensation and temperature effects when trying to determine total emitted particulate concentration. However, they found the system was useful for the determination of metallic particulates. Exhaust gas flow would collect on the glass filter media which was removed, solvent extracted to remove hydrocarbons and then macerated prior to metals analysis.

Ter Haar *et al.* [96] developed a system to run the total exhaust into a $70\,m^3$ black polyethylene bag along with a minimum of 8 to 1 filtered, dry (less than 10% relative humidity) dilution air, to minimize the effects of humidity on particle agglomeration and fallout. Upon completion of a driving cycle, sampling commenced within the bag using a 0.6 cm diameter stainless steel probe inserted 1.8 m into the bag. The sample was drawn through a 47 mm Millipore type AA, $84\,\mu m$ membrane filter at a flow rate of 7 litres min^{-1}. Sampling was also performed with the bag using an open faced filter holder with a 4.25 cm glass fibre filter.

Ter Haar *et al.* [96] also developed a total filter for connection directly to the exhaust pipe (Fig. 1.36). The unit uses a cyclone having a 50% cutoff for 0.5 to

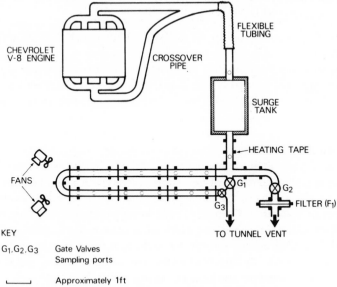

Fig. 1.37 Schematic diagram of exhaust system.(After [97].)

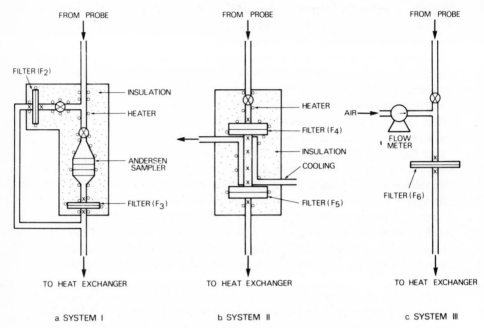

Fig. 1.38 Collection components for particulate sampling of automobile exhaust (after Sampson and Springer [97]).

$5\,\mu$m particulates backed up by a glass fibre filter with a 99.9% efficiency for $0.3\,\mu$m particulates.

Sampson and Springer [97] conducted particulate sampling at several points along a simulated exhaust system shown in Fig. 1.37, using three separate collection components (Fig. 1.38). Each component was used in turn with a sharp nosed probe for isokinetic sampling. The first collection component was a high temperature cascade impactor with backup and bypass filters. This component provided particle size fractionation at exhaust gas temperatures up to $450°$ C or total particulate mass counts depending on which valve was open. The second collection system consisted of two filters in series with a heat exchanger between them, to effect exhaust gas cooling. The third system provided metered ambient dilution air flow into the system to cool the exhaust before the filter. In all cases, glass fibre filters were used and mounted in filter holders capable of withstanding $450°$ C. After passing through the collection component in use, the sampled air flow was passed through heat exchangers to bring exhaust gas temperatures to $21°$ C. Total gas flow was measured with a wet test gas meter.

Sampson and Springer [97] evaluated several of the procedures for their system. They found that 250 cm downstream, from the exhaust port, the unsteady flow had evened out sufficiently that relatively small sampling errors were incurred while attempting isokinetic sampling. However, they added a surge tank to smooth further the unsteady exhaust flow before the sampling points. Variations in sampling flow rate around the isokinetic rate were found to have limited effect on measured particle concentrations, but isokinetic sampling was used anyway. The effect of sampling probe diameter was evaluated and found to have no consistent effect on measured particle concentrations. The particle distribution across the exhaust pipe was evaluated and found to vary by less than 4% so that sampling in the centre of the pipe was adopted, along with a 10 min sampling period.

1.2.3 Gases

1.2.3.1 Gas sampling considerations

Gaseous sampling from the atmosphere requires careful attention to avoid the presentation of a modified or diminished sample to the analysis step. Although gaseous constituents of the atmosphere are not subject to the inertial and electrostatic effects which were discussed for aerosols, special precautions are necessary for accurate sampling. Several general considerations in gaseous sampling will be mentioned.

Where heavy concentrations of a gas may be present in a stagnant or slow moving air mass, a density differential between the gas and air may result in an uneven distribution of the gas within the air mass. This possibility is generally overcome by sampling from a turbulent flow or in a well mixed air mass.

Gaseous constituents of the atmosphere are in a relatively high energy state and may be reactive with components of the sampling system or other gaseous constituents. It is essential to evaluate all components of the sampling system carefully for their possible reactivity with sampled gaseous constituents. In many cases, it may be necessary to pretreat the sampled air stream before the gas reaches the collection component of the sampling system. In such cases, for example, prefiltration of particulates or moisture removal, the pretreating units may remove significant quantities of the gaseous constituent to be sampled. Where pretreatment is used, the pretreatment units should be analysed for their content of the relevant gaseous constituents.

All possible precautions should be taken against inward or outward leakage, particularly for stored samples. However, sealing lubricants may collect gaseous constituents and their use should be avoided or carefully evaluated. Self lubricating PTFE joints and stopcocks are preferable and these can generally provide relatively air tight seals. Diffusion of gases through apparently impermeable materials such as rubber, neoprene and plasticized PVC is possible [1] and should be evaluated where samples will be stored for periods in excess of a few hours.

Condensation of atmospheric moisture within sample lines can create reactive conditions for some combinations of gaseous constituents, or the moisture may simply collect gaseous constituents before they reach the collection component. Many gaseous constituents are photochemically reactive and may undergo reactive changes within the sampling system if they remain exposed to light energy. Light sealed containers for sampling systems would simply eliminate this problem.

The sampling system must be designed to draw continually a fresh sample of the atmosphere being studied and should not allow the resampling of air which has passed through the system. An extended exhaust line to a location remote from the sample intake will normally prevent resampling from occurring.

1.2.3.2 Gas sampling collection components

Adsorption

Adsorption of gases from the atmosphere is a surface phenomen whereby gas molecules are concentrated and bound by intermolecular attraction to the surface layer of a collection phase. Under equilibrium conditions at constant temperature the volume of gas adsorbed on the collection phase is proportional to a positive power of the partial pressure of the gas (Freundich's adsorption isotherm). As well, adsorption, being a surface phenomenon, is dependent upon the relative surface area of adsorbent material. Various gaseous constituents of a mixture will be adsorbed in amounts approximately inversely proportional to their volatility.

Adsorption is useful in a collection component for gaseous constituents of the atmosphere because of its ability to concentrate trace concentrations by sampling large volumes of air. Materials which are commonly used as adsorbents include activated carbon, silica gel, alumina and various gas chromatographic support phases.

Several factors need to be considered in selecting an adsorbent. A high relative surface area is important to maintain a large contact area for adsorption, while retaining the maximum space between adsorbent granules for maximum air flow rates. Silica gel and alumina offer surface areas of $200-600 \, m^2 \, g^{-1}$, activated carbon offers $500-2000 \, m^2 \, g^{-1}$, whereas GC phases can offer considerably higher surface areas per g.

Relative affinity for polar or non-polar compounds is important for selecting the adsorbent to sample a given gaseous constituent. Activated carbon is non-polar and therefore, has affinity for organic compounds to the exclusion of polar gases, including water vapour. Silica gel and alumina are polar and have increasing affinity for higher polarity gases. This may cause desorption of lower polarity gases in the presence of higher polarity gases. In addition, water vapour does collect on these adsorbents and may lower adsorption efficiency as well as clogging the granular bed, resulting in reduced flow rates.

The adsorbents must not be chemically reactive with the gases to be collected, unless chemisorption is used intentionally. In this instance, adsorbents are specially treated to provide a reactive surface for chemical interaction between a molecular layer of the gas and the chemisorption media. Where this technique is used, the quantitative and qualitative results will be evaluated with full awareness of the reactions which have occurred.

The adsorbent material should not be prone to fracturing, crushing or flaking which may result in carrying over fine particulates with the sampled air stream. Most adsorbents are sieved to remove fines before use.

The retention capacity of the adsorbent should be predictable and high in order to avoid non-quantitative recovery of gaseous constituents. For activated carbon, the adsorption of gaseous constituents is complete initially, but after the retention capacity has been reached, the gases will be incompletely adsorbed until the carbon becomes saturated and no further adsorption takes place.

The desorption properties of adsorbents must be amenable to quantitative recovery of the collected sample, preferably with regeneration of the adsorbent for subsequent use. Activated carbon, although a very efficient adsorber is very difficult to desorb quantitatively. Steam stripping, with the attendant risks of hydrolysis reactions with collected constituents, is often necessary for assurance of quantitative recovery of adsorbed materials. Alternately, heating under vacuum and distilling components to another collection medium is often possible. In contrast, support bonded GC phases have the advantage of allowing desorption by solvent extraction or by heating and flushing with an inert carrier gas.

Relative selectivity for atmospheric gases will be important depending upon the objectives of the sampling programme. Activated carbon is relatively unselective and as such is useful for screening atmospheric constituents while silica gel is somewhat more selective and particular GC phases may be very selective.

Adsorption is also a temperature dependent phenomenon and efficiency will be improved at lower temperatures, particularly for gases which have lower boiling points. Gases such as hydrogen, nitrogen, oxygen, carbon monoxide and methane are not normally adsorbable without resorting to chemisorption. Ammonia, ethylene, formaldehyde and hydrogen sulphide. are not completely adsorbed at normal ambient temperatures, but adsorption can be increased by cooling the adsorption device.

Adsorption systems provide a flexible means of collecting diverse gaseous constituents from the atmosphere and concentrating them prior to analysis, but the system chosen must be carefully evaluated to avoid non-quantitative adsorption, breakthrough effects due to exceeding the retention capacity or non-quantitative desorption of gaseous constituents.

Absorption

Absorption of gases from the atmosphere is a solubility phenomenon whereby gas molecules are preferentially dissolved in a liquid collecting phase. The degree of absorption of a given gaseous constituent in a particular solvent is limited by the equilibrium partial pressure of the vapour over its liquid solution. This limitation on absorption of gases into solution can be overcome where the gaseous constituent undergoes a relatively irreversible chemical reaction with the absorbent.

The efficiency of absorption of gases into a liquid phase is also very dependent on contact surface area and absorbing devices are designed to achieve the maximum area contact between gas bubbles and the absorbent. This is normally achieved by transforming the air stream to small, finely dispersed bubbles with a relatively long travel time through the absorbent.

As with adsorption, absorption is temperature dependent with lower temperatures improving efficiency for particularly volatile constituents. The efficiency of absorption on various gaseous constituents was studied by Sexton [98] and is summarized in Table 1.8. Various absorption devices have been devised to provide efficient absorption. Several absorber designs are illustrated in Fig. 1.39.

Table 1.8   Gas washing bottle efficiency tests (after [98].)

Chemical	Solution in gas washing bottle	Sampling rate (ft^3 min^{-1})	Gas washing bottle efficiency – %	Stack concentration ppm (by weight)
Sulphuric acid	Water	0.12	99.8	72
Hydrochloric acid	Water	1.00	97.1	47
Hydrochloric acid	Water	0.12	100	47
Perchloric acid	Water	0.12	99.8	41
Acetic acid	Water	0.12	99.5	1065
Nitric acid	Water	0.12	100	182
Ammonia	Water	0.12	84.3	103
Chlorine	Water	0.12	Very low[*]	103
Chlorine	2% Na$_2$CO$_3$	0.12	77.0	103
Ethylenediamine	Water	0.12	99.6	211

[*] Chlorine is only 1% soluble in water

The simple Dreschel bottle may be used in some applications. It has the advantage of high flow rate capacity, up to 30 litres min^{-1}, but is relatively

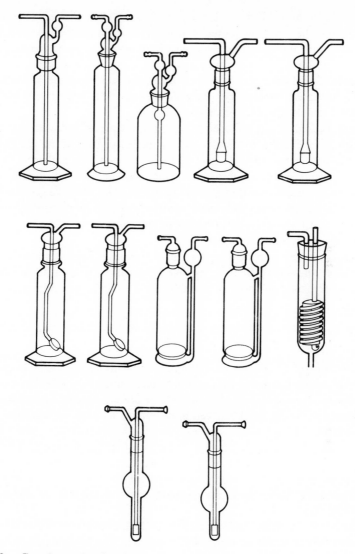

Fig. 1.39   Gas absorption bottles.

inefficient in comparison to other designs. Standard impingers, as used for aerosol sampling have been used as absorbing devices, but they are not markedly more efficient in many cases, than Dreschel bottles.

Absorbers with fritted glass diffusers have been constructed in various configurations. These have the advantage of creating a small bubble size to enhance efficient absorption. Fritted glass absorbers are the best choice in many applications, but they must be protected from blockage by particulates and adequate freeboard with the collecting fluid must be provided to avoid foaming over. As well, adsorption of gaseous constituent on the glass frit is possible and must be evaluated. Spiral absorbers have been designed to achieve longer detention times for the rising air bubbles, but these devices are flow rate limited and may be unable to sample adequate air volumes.

Bead packed absorbers operate on the principle of buffeting the gas flow through bead packing to provide mixing and dispersion of the gas throughout the absorbent. This principle may be used to create long pathways for the gas flow to ensure complete absorption. However, the beads themselves must be evaluated to ensure that they do not adsorb or react with the gaseous constituents. It is advantageous with most absorber configurations to concentrate the gaseous constituent into a small volume of absorbing solution in order to enhance subsequent analytical procedures.

Absorbers may provide effective concentration of gaseous constituents, but, in all cases, the actual efficiency of collection must be established and saturation values determined to avoid oversampling and subsequent misinterpretation of results.

Condensation

Condensation of gases from the atmosphere depends upon cooling the gas stream to temperatures below the boiling or freezing points of the gases to be collected. The temperatures used in a condensation device are selected to ensure adequate condensation of the gas by lowering its vapour pressure to less than 1.33 mb generally [1].

Various refrigerants are possible and a list of refrigerant temperatures is provided in Table 1.9. Because of the low temperatures involved, good insulation must be provided and, therefore, the equipment involved in condensation devices is rather bulky. Dewar flasks are commonly used to contain the refrigerant and collecting trap.

The primary problem encountered in condensation devices is the large volume of water vapour in the air which will freeze out. This is often overcome by providing sequential traps at progressively lower temperatures, with the first trap designed to collect primarily water vapour. The lowest temperature system which is recommended is liquid oxygen, because lower temperature systems will condense atmospheric oxygen into the collecting device which will present a serious combustion hazard.

Relatively low flow rates (less than 1 litre min^{-1}) are generally applicable with condensation systems, but the concentration efficiency of these devices

Table 1.9 Summary of cold bath solutions (after [18])

Coolant	Temperature (° C)
Ice and water[a]	0
Ice and NaCl	− 21
Carbon tetrachloride slush[a,b]	− 22.9
Chlorobenzene slush[a,b]	− 45.2
Chloroform slush[a,b]	− 63.5
Dry ice and acetone[a]	− 78.5
Dry ice and cellosolve[a]	− 78.5
Dry ice and isopropanol[a]	− 78.5
Ethyl acetate slush[a,b]	− 83.6
Toluene slush[b]	− 95
Carbon disulphide slush[a,b]	− 111.6
Methyl cyclohexane slush[a,b]	− 126.3
N-Pentane slush[b]	− 130
Liquid air	− 147
Isopentane slush[b]	− 160.5
Liquid oxygen	− 183
Liquid nitrogen	− 196

[a] Adequate for secondary temperature standard.
[b] The slushes may be prepared by placing solvent in a Dewar vessel and adding small increments of liquid nitrogen with rapid stirring until the consistency of a thick milkshake is obtained.

somewhat compensates. Higher flow rates would provide less refrigerant contact time and would tend to enhance the loss of condensation aerosols from the system. The latter should be recovered by filtration, in any case, to ensure quantitative sampling. Although condensation systems are not suitable for unattended operation, they have the advantage of providing preserved bulk atmospheric samples which are readily amenable to further analysis.

Grab sampling

Grab sampling of gases from the atmosphere involves the direct collection and isolation of the test atmosphere in an impermeable container. The sample may be returned to the laboratory for direct analysis. Clearly, this sampling technique is limited to those gaseous constituents for which sensitive analytical techniques are available, or for which there is a high concentration of a given constituent in the test atmosphere. Due to the fact that the sample is not concentrated, it is much more sensitive to contamination by leaks.

Many devices including evacuated bottles, syringes and various types of plastic bags have been used for this type of sampling. In all cases, the containers must be evaluated for their reactivity with or adsorption of, gaseous constituents. In some cases, where adsorption is reproducible, preconditioning of the container with large concentrations of the gaseous constituent to be sampled may be used. However, care must be exercised to avoid exposing conditioned containers to temperature and pressure conditions different to those at the time of conditioning, because desorption of the gaseous constituents may occur.

An advantage of most grab sampling techniques is that the volume of air sampled need not be accurately determined, if a small sample of air is to be withdrawn from the grab for direct analysis. However, the temperature and pressure conditions of the subsample taken must be known in order to relate measured quantities of gaseous constituents back to reference conditions. Grab sampling techniques generally require relatively simple equipment and offer variety and flexibility to many air sampling applications.

1.2.3.3 Ambient gas sampling applications

Charcoal adsorption tubes

Charcoal adsorption tubes are commonly used to collect many gaseous organic constituents. Reid and Halpin [99] report the successful use of activated charcoal for sampling halogenated and aromatic hydrocarbons while a standard method [100] uses charcoal for the collection of odorous vapours. Sampling efficiencies for adsorption of aliphatic hydrocarbons were reported by Fraust and Hermann [101] to be independent of mass sampling rate and to have an efficiency greater than 95% until breakthrough occurred.

Specifications for activated carbon used in adsorption sampling have been reported by the American Society for Testing and Materials [100] and are summarized in Table 1.10. The carbon fines may be removed by sieving or washing with distilled water and drying above $100°$ C.

The carbon granules are usually packed in a glass U tube, providing 5 to 10 cm length of carbon. The carbon packing is capped at either end by glass wool plugs. Reid and Halpin [99] recommend adjusting the plugs to provide a 46–53 mb head loss through the adsorption tube at a flow rate of 2 litres min^{-1}. Sampling rates of 1 to 2 litres min^{-1} can normally be maintained with carbon adsorption tubes. The required sampling time will normally be determined by experimentation in order to provide adequate sample quantity for subsequent analysis.

Activated carbon desorption is generally done by steam stripping or vacuum distillation [100]. For the former, superheated steam at $300°$ C or higher is flushed through the carbon bed and the steam effluent is condensed for subsequent analysis. With vacuum distillation the carbon is heated to 200 to $250°$ C and distilled into a condensation train maintained under vacuum.

Table 1.10 Requirements for activated carbon for air puri-
fication (after [100])

Property	Requirement
Activity for CCl_4[a]	at least 50%
Retentivity for CCl_4[b]	at least 30%
Apparent density	at least $0.42\,g\,ml^{-1}$
Hardness (ball abrasion)[c]	at least 80%
Particle size range[d]	
Passing sieve	6
Retained on sieve	16

[a] Maximum saturation of carbon, at 20°C and 760 mm Hg, in air stream equilibrated with CCl_4 at 0°C.
[b] Maximum weight of adsorbed CCl_4 retained by carbon on exposure to pure air at 20°C and 760 mm Hg.
[c] % of 6 to 8-mesh carbon that remains on a 14-mesh screen after shaking with 30 steel balls of 0.25 to 0.37 in. diameter/50 g carbon, for 30 min on a mechanical sieve-shaking device such as the 'Ro-Tap' machine.
[d] Particle size in terms of sieves conforming to ASTM Specification E 11, Wire-Cloth Sieves for Testing Purposes, *Annual Book of ASTM Standards*, Part 30.

Gas chromatographic phase adsorption media

A wide variety of gas chromatographic phases have been reported in use for gaseous sampling [102–113]. Convential GC support materials coated with a liquid phase may be used for sampling, but these materials are limited by their relatively low volumetric retention capacity. Cantuti and Cartoni [104] utilized the relationship of the gas phase concentration to the partition coefficient between the gas and liquid phases and the relationship for specific retention volume to the partition coefficient to calculate sampled concentrations of lead alkyls. The GC phase material was placed in a syringe within an upper leg of a U tube located in the sampling line. Sampling was either done at ambient conditions or at 0°C with the excess moisture collecting at the bottom of the U tube. Flow rates of $1.5\,litres\,min^{-1}$ for 10 to 15 min were used. Desorption was performed by flushing the syringe with the GC carrier gas.

Porous polymer GC phases have shown utility for sampling organic gases. Jeltes [102] found virtually 100% recovery of trichloroethylene at sampling rates of $0.05\,litre\,min^{-1}$. Mann *et al.* [103] used 10 cm lengths of Chromosorb 101 in 10 mm internal diameter glass tubing to achieve efficient recoveries of hexachlorobenzene at sampling rates of $2\,litres\,min^{-1}$ for 3 h. Desorption of the sample from the Chromosorb was achieved by shaking the sample with hexane.

Aue and Teli [105] have evaluated the utility of support bonded Chromosorb GC phases for air sampling and found them to be convenient for the collection of high molecular weight organic gases. If the phase material can undergo exhaustive solvent extraction without loss of polymer materials, then solvent extraction of adsorbed material can be used in place of temperature programmed desorption.

Perry and Twibell [106] used Chromosorb 102 to collect atmospheric hydrocarbons, after preparation of the media by extensive solvent extraction clean up. The use of gas chromatographic phases for adsorption of gaseous samples is a versatile, rapidly evolving sampling technique for a variety of gaseous constituents of the atmosphere.

Condensation traps

A condensation trap designed to collect gaseous pollutants was designed by Shepherd *et al.* [114]. The system takes account of the formation of frozen aerosols from water vapour by providing a glass wool filter within the condensation tube to prevent loss of the aerosol from the system. The sampler is immersed in liquid oxygen contained in a Dewar flask. The sampling system also includes a wet test meter, a control needle valve and mechanical pump capable of drawing several litres min^{-1} through the sytem. One variation to the basic condensation trap is the preconditioning for moisture removal, although this runs the risk of causing gas losses by adsorption to the preconditioner. Another variation is the provision of sequential condensation traps starting with one trap just cold enough to freeze the majority of the incoming water vapour, followed by sequentially colder traps to condense sequentially gaseous constituents. This latter modification has the advantage of some preliminary fractionation of gaseous components collected. However, sequential condensation traps are involved and cumbersome, making their routine usage often impractical.

Flexible container sampling

Various flexible containers have been evaluated and used for grab sampling of gaseous constituents. Bags constructed of polyester film, aluminium foil lined polyester, polyvinyl chloride film and various fluorinated plastics have been used successfully for particular applications.

Foil lined polyester bags are recommended in a standard procedure for low molecular weight atmospheric hydrocarbons [115]. These bags are reported as unsuitable for the collection of highly polar gases like SO_2, NO_2 or O_3 [1]. Desbaumes and Imhoff [116] evaluated saran bags for solvent collection by determining the diffusion curves of the solvents escaping the container. They concluded that saran bags were suited for all the common solvents tested except styrene which exhibited too high a diffusion rate through saran.

Schuette [117] reviewed the available sources and applications for several plastic films. Although the references, provided, reported various satisfactory applications, Schuette recommends specific testing of the selected materials to investigate their effectiveness in sampling and storing the specific pollutants to be collected.

For all bag sampling techniques, some means of filling is required. Conner and Nader [118] developed a bag sampling method in which a deflated bag is contained in a rigid box. The sample is taken by applying suction from a vacuum blower to the air surrounding the deflated bag in the box. The net result is the inflation of the bag and sampling rates of up to 10 litres min^{-1} were achieved using a 75 litre bag. Performance comparisons were done between mylar and teflon bags and the former were found to be superior. Predictable decay of NO_2, SO_2, O_3 and hydrocarbon samples occurred in the mylar bags, but the decay rate was low enough to permit several days storage.

Oord [119] reports equipping bags with hardboard sides and handles to allow manual flushing and filling of the bag by compression and expansion of the bag. Curtis and Hendricks [120] developed a gravity operated sampling system with a collapsed bag with cardboard sides supported on a wooden frame. Sampling was activated by opening the valve to the bag which concurrently allowed one of the side plates to drop, expanding the bag and, thereby, collecting the air sample. A sampling rate of 50 litres min^{-1} for 5 min was achieved by this method. Bag sampling methods using small hand held blowers, and by holding the bag in the wind have been suggested [46].

In all cases, the bags used must be carefully checked for leaks, particularly at valves and seals. They must be flushed and completely evacuated before sampling by collapsed bag techniques or thoroughly flushed where manual sampling is done. The bags should not be filled to capacity in order to allow for temperature and pressure variations which may cause gas expansion and encourage leakage. Finally, the bags must be protected in transit against rupture or association with contaminated atmospheres and samples should be analysed as soon as possible after collection.

Syringe sampling

Syringe sampling techniques are useful for rapid and convenient sampling of atmospheres for gases with sensitive analytical techniques or high concentrations of the gas. Lang and Freedman [121] report the use of disposable 10 ml glass barrel syringes equipped with butyl rubber plungers for sampling low molecular weight gases such as CO_2, CO, O_2, N_2, and CH_4. They evaluated several glass and plastic barrel syringes with the criteria of one week's storage without significant gas loss before choosing the glass barrel—butyl rubber combination. Losses from

syringes were attributed to leakage through pinholes or around seals and permeation losses by passage of the gas through interstices on the sealing surfaces.

Meader and Bethea [122] developed the use of polypropylene syringes in
preference to glass syringes for sampling reactive gases such as NO_2, Cl_2, HCl and
HF. However, the syringes required preconditioning with high concentrations of
the gas to be tested in order to control adsorption and reaction effects. Different
syringe materials may be expected to react differently depending upon the gases
to be sampled so that direct evaluation of the syringes to be used is necessary.

Syringe sampling offers a relatively simple sampling technique but applications
are limited to specific small volume sampling requirements.

Rigid container sampling

Rigid containers may be used for gaseous sampling, either by pre-evacuation of
the container and filling in the test atmosphere or flushing the test atmosphere
through the container several times before sealing the container.

Glass tubes may be evacuated and provided with a breakable seal to allow
the inward flow of the test atmosphere. After collection, the tube may be
capped with a wax plug. Tubes evacuated for sampling must be structurally
sound and they must be used with care after evacuation to avoid accidental
implosion. Evacuation to less than 1.3 mb should be achieved. If this is not possible, the bottle should be evacuated in a clean atmosphere and the result corrected to the volume sampled according to the formula [46]

$$V_s = V_b \times \frac{P_1 - P_2}{P_1}$$

where V_s = volume sampled; V_b = volume of bottle; P_1 = pressure after sampling;
P_2 = pressure after evacuation of tube.

Evacuated containers are not recommended for reactive gas sampling because
of possible reactions of the gas with the sealing wax plug [46]. This can be overcome by filling the container with an absorber with specific chemical reactivity
for the gaseous constituent sought.

Many types of rigid containers are possible for use in displacement sampling.
Van Houten and Lee [123] report the use of 120 ml glass bottles for sampling
solvents with a 50 ml hand operated aspirator bulb. The aspirator was squeezed
20 times to achieve 8 volume changes for the bottle resulting in over 99% air
change. The bottles were capped and sealed with three layers of saran and a
rubber gasket. Recoveries of 88 to 100% were found after 30 days storage for a
variety of organic solvents. Rigid container sampling may be adapted to provide
for sampling kits which are suitable for convenient sample collection and ship-
ment.

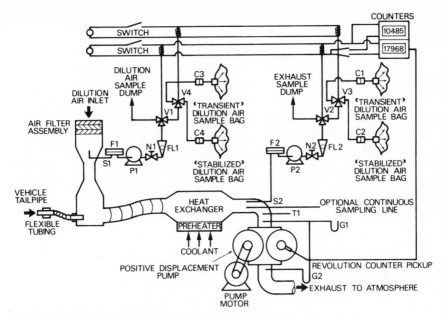

Fig. 1.40 Exhaust gas sampling system. (After [41].)

1.2.3.4 Emission source gas sampling applications

Mobile sources

A detailed standardized sampling apparatus has been developed for regulatory vehicle emission testing in the U.S. [41]. The system (Fig. 1.40) operates on the principle of variable dilution similar to that described for Habibi [93] in Section 1.2.2.4 on aerosol sampling from vehicle emissions. The entire exhaust cycle is sampled and diluted to prevent condensation.

The system provides for continuous constant flow sampling of filtered inlet dilution air and filtered, diluted and cooled exhaust into sample bags for either transient or stabilized vehicle operation modes. The minimum sampling flow rate required is 5 litres min^{-1}. The collected samples can then be run through analysing instruments for gaseous constituents including hydrocarbons, carbon monoxide and oxides of nitrogen. The total diluted exhaust volume is measured by recording revolution counts and pressure drop across a positive displacement blower. Total mass emissions can be calculated using the total diluted exhaust volume and the measured concentrations in the sample bags. The measured concentrations for the diluted exhaust are corrected for background concentrations measured on the inlet dilution air.

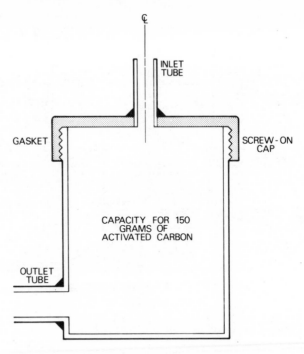

Fig. 1.41   Typical activated carbon trap (schematic diagram). (After [41].)

Soltau and Larbey [40] describe the advantages of variable dilution constant volume sampling in terms of reduction of reaction rates between hydrocarbons and oxides of nitrogen and prevention of water vapour and heavy hydrocarbon condensation.

A sampling system is also described [41] for measurement of evaporative losses from the carburettor and fuel tank. The collection device consists of a canister with a sealing screw cap, inlet and outlet tubes and capacity for 150 g of activated carbon (Fig. 1.41). The canister is connected to evaporative emission sources, as shown in Fig. 1.42. Connections must be sealed and the canister must be protected from heating by radiant or conductive heat from the engine during vehicle testing.

Siegel [42] reports that procedures for aircraft emissions must account for variations due to engine power level: time and spatial variations of exhaust composition; sampling line diameter, length, material and temperature; ambient temperature and humidity and ambient pollution levels. Sampling procedures for aircraft jet engines are currently being standardized by a committee of the Society of Automotive Engineers.

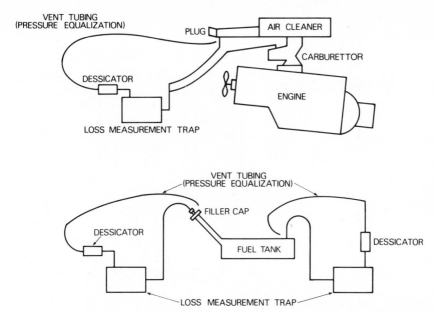

Fig. 1.42   Typical evaporative loss ⸻llection arrangements (schematic diagrams). (After [41].)

Probes constructed of stainless steel are currently being used for sample intake. These need not be constructed for isokinetic sampling since particulates in jet exhausts are generally less than $1 \mu m$. No variation in smoke density was found between facing the probe into the exhaust stream or facing $180°$ in the opposite direction.

Multipoint sampling is necessary and more than 12 locations (3 radial positions in each of the 4 quadrants) are recommended. The vertical sampling plane should be located no more than one nozzle diameter downstream of the exit nozzle. The sampling lines should be constructed of stainless steel and maintained at greater than $175°$ C to prevent condensation of water vapour and high boiling point organics.

Stationary sources

Sampling systems for the collection of gaseous constituents from emission sources are generally tailored to the specific constituent sought. Since there is no requirement for isokinetic conditions when sampling for gaseous constituents, operating procedures are somewhat simplified. However, where prolonged sampling is undertaken in variable flow emission sources, it is desirable to obtain a flow proportional sample. Alternatively, where variable flow conditions are

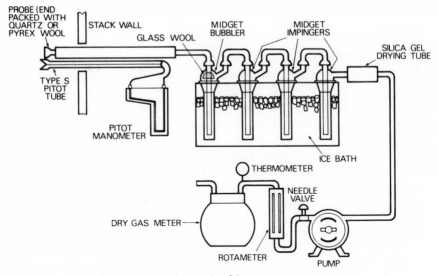

Fig. 1.43   SO$_2$ sampling train. (After [30].)

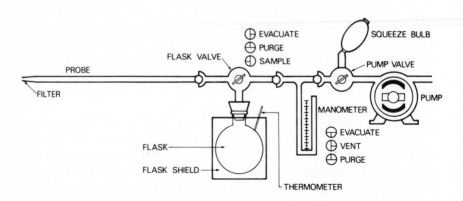

Fig. 1.44   Sampling train, flask valve and flask. (After [30].)

encountered, a series of grab samples may be taken with the flow velocity recorded. The results for the various samples may then be proportioned to flow.

Specific standard procedures for sampling for oxides of nitrogen, SO$_2$ and SO$_3$ are available [30, 124]. The sampling trains of the United States Environmental Protection Agency methods for SO$_2$ and NO$_x$ are shown in Figs. 1.43–44. These procedures use sampling trains similar to those employed for particulate sampling except that in addition to a filter to collect aerosols, absorbers or grab vessels are provided to collect the gaseous sample for subsequent analysis. Sampling equipment for gaseous emissions other than oxides of sulphur and nitrogen may be developed by using specific absorbers for other gases. Likewise, low temperature condenser or adsorber collection components may be used, depending on the properties of the gaseous constituents to be measured. Sampling train possibilities are very flexible provided that provision is made for accurate flow measurement, adequate sampling rate and protection of the sample from loss or contamination in sampling lines.

The actual sampling procedures generally follow the same pattern used for aerosol sampling with the exception of the acceptability of proportional sampling in place of isokinetic sampling.

Berger *et al.* [125] reviewed the precision of common stack gas sampling procedures for oxides of sulphur and nitrogen and found the results exhibited relatively good precision. In general, accurate results may be obtained, as well, provided that care is taken to avoid leaks in the actual sampling system.

The sampling points should be selected to take advantage of flow turbulence to provide a homogenous gas mixture rather than a stratified flow. Flow restrictions that provide for increased gas flow and turbulence, provided that a consistent velocity profile is available across the stack cross-section, are useful. Sample points should also be evaluated to avoid sampling in the proximity of flow stagnation or significant inward leakage of dilution air.

References

[1]   Intersociety Committee (1972). Methods of Air Sampling and Analysis. American Public Health Association, Washington.

[2]   ASTM Annual Book of Standards (1973). Standard Recommended Practice for Planning the Sampling of the Atmosphere, D 1357-57, Part 23.

[3]   American Industrial Hygiene Association (1972). Air Pollution Manual Part I, Evaluation, 2nd edn., Detroit.

[4]   National Air Sampling Network, U.S. Environmental Protection Agency, Washington D.C.

[5]   National Air Pollution Surveillence, Annual Summaries, Air Pollution Control Directorate, Environmental Protection Service Environment Canada, Ottawa, Canada.

[6]   Continuous Air Monitoring Program, U.S. Environmental Protection Agency, Washington D.C.

[7]   National Survey of Air Pollution in the United Kingdom, Warren Springs Laboratory, Stevenage.

[8]   Weather Ways (1964). Meterological Branch, Department of Transport, Ottawa, Canada.

[9]   Scorer R. (1968). *Air Pollution,* Pergamon Press, Oxford.

[10]   Brief, R.S. and Confer, R.G. (1972). Air Quality Monitoring: Procedures; Data analysis, *Heating, Piping, Air Conditioning*, pp. 103–10.
[11]   Gow, M.C. (1972). Meteorological Instruments – Atmospheric Stability, Pollution Control, Government of British Columbia, Victoria, B.C.
[12]   Georgii, H.W. (1963). *Int. J. Air Water Pollut.* **7**(11/12) 1057–9.
[13]   Bielbe S. and Georgii, H.W. (1968). *Tellus* **20**(3), 435–48.
[14]   Makhonko, K.P. (1967). *Tellus* **19**(3) 467–76.
[15]   Andersson, T. (1969). *Tellus* **21**(5), 685–92.
[16]   First, M.W. (1969). *Environmental Research* **2**(2) 88–92.
[17]   Schneider, F.A. (1970). *Environmental Research* **3**(5/6) 452–9.
[18]   Stern, A.C. (1968). *Air Pollution*, 2nd edn, Academic Press, New York.
[19]   Charlson, R.J. (1969). *J. Air Pollut. Control Ass.* **9**(10) 802.
[20]   Pooler, F. (1974). *J. Air Pollut. Control Ass.* **24**(3) 228–31.
[21]   Yamada, V.M. (1970). *J. Air Pollut. Control Ass.* **20**(4) 209–13.
[22]   Katz, M. (1969). Measurement of Air Pollutants, Guide to the Selection of Methods, WHO, Geneva.
[23]   Saltzman, B.E. (1970). *J. Air Pollut. Control Ass.* **20**(10) 660–5.
[24]   Akland, G.G. (1972). *J. Air Pollut. Control Ass.* **22**(4) 264–6.
[25]   Hunt, W.F. (1972). *J. Air Pollut. Control Ass.* **22**(9) 687–91.
[26]   Duprey, R.L. (1968). Compilation of Air Pollutant Emission Factors, National Air Pollution Control Administration, PHS Pub. No. 999-AP42, Washington, D.C.
[27]   Achinger, W.C. and Shigehara, R.T. (1968). *J. Air Pollut. Control Ass.* **18**(9) 605–9.
[28]   Pollution Control Branch, B.C. Water Resources Service (1974). Source Testing Manual for the Determination of Discharges to the Atmosphere. 3rd Edn, Victoria, B.C.
[29]   ASTM Annual Book of Standards (1973). Standard Method of Test for Sampling Stacks for Particulate Matter, D 2928-71, Part 23.
[30]   Environmental Protection Agency (Dec 23, 1971). Standards for Performance for New Stationary Sources, *U.S. Federal Register*, Part II, Vol. 36, No. 247, 24876–95.
[31]   British Standard Simplified Methods for Measurement of Grit and Dust Emissions from Chimneys (Metric Units), B.S. 3405: 1961.
[32]   British Standard Simplified Methods for Measurement of Grit and Dust Emissions (Metric Units), B.S. 3405: 1971.
[33]   American Society of Mechanical Engineers (1957). Test Code for Determining Dust Concentrations in a Gas Stream, Power Test Code 27.
[34]   Brenchley, D.L., Turley, C.D. and Yarmac, R.F. (1973). *Industrial-source Sampling*, Ann Arbor Science, Ann Arbor.
[35]   ASTM Annual Book of Standards (1973). Standard Method of Test for Average Velocity in a Duct, (Pitot Tube Method), D3154-72., Part 23.
[36]   Harlerd, H.H. (ed.) *Methods for Determination of Velocity, Volume, Dust and Mist Content of Gases*, Bulletin WP-50, 7th edn, Joy mfg. Co., Los Angeles.
[37]   Cooper, H.P.H. and Rossano, A.T. (1971). *Source Testing for Air Pollution Control*, Environmental Science Services.
[38]   Rogers, G.F.C. and Mayhew, Y.R. (1967). *Engineering Thermodynamics Work and Heat Transfer* (S.I. Units), Longmans, London.

[39] Rom, J.J. (1972). Maintenance, Calibrations and Operation of Isokinetic Source Sampling Equipment, Environmental Protection Agency, Office of Air Programs, Publications number APTD-0567.

[40] Soltau, J.P. and Larbey, R.J. (1971). *Symp. of Institution of Mechanical Engineers*, 218, London.

[41] Environmental Protection Agency, Protection of Environment, *Title 40*, Chap. 1, Part 85 Control of Air Pollution from New Motor Vehicles and New Motor Vehicle Engines, U.S. Federal Register, 606.

[42] Siegel, R.D. (1972). *J. Air Pollut. Control. Ass.* 22(11) 845–53.

[43] Slowik, A.A. and Sansome, E.B. (1974). *J. Air Pollut. Control Ass.* 24(3) 245–7.

[44] Wohlers, H.C., Newstein, H. and Daunis, D. (1967). *J. Air Pollut. Ass.* 17(11) 753.

[45] Byers, R.J. and Davis, J.W. (1970). *J. Air Pollut. Control Ass.* 20(4) 236.

[46] ASTM Annual Book of Standards (1973). Standard Recommended Practises for Sampling Atmospheres for Analysis of Gases and Vapours, D-1605-60. Part 23.

[47] Mercer, T.T. (1973). *Aerosol Technology in Hazard Evaluation*, Academic Press, London.

[48] Bergland, R.N. and Liu, B.Y.H. (1973). *Environ. Sci. Technol.* 7(2) 147–53.

[49] Beckwith, T.G. and Buck, N.L. (1961). *Mechanical Measurements*, Addison Wesley, London.

[50] Environmental Protection Agency (1971). Reference Method for the Determination of Suspended Particulates in the Atmosphere (High Volume Method), U.S. Federal Register 36, No. 84.

[51] Countess, R.J. (1974). *J. Air Pollut. Control Ass.* 24(6) 605.

[52] Hoffman, G.L., and Duce, R.A. (1971). *Environ. Sci. Technol.* 5(11) 1135.

[53] ASTM Annual Book of Standards (1973). Standard Definitions of Terms Relating to Atmospheric Sampling and Analysis, D 1356-67a, Part 23.

[54] Zenker, P. (1971). *Staub.* 31(6) 30–5.

[55] Hemeon, W.C.L. and Haines, G.F. (1954). *Air Repair (J. Air Pollut. Control Ass.)* 4, 159.

[56] Watson, H.H. (1954). *Am. Ind. Hyg. Ass. Quarterly*, 15, 21–5.

[57] Sehmel, G.A. (1970). *Am. Ind. Hyg. Ass. J.* 31(6) 758–71.

[58] Badzioch, S. (1960). *J. Inst. Fuel*, 33, 106–10.

[59] Vitols, V. (1966). *J. Air Pollut. Control Ass.* 16(2) 79–83.

[60] Sansone, E.B. (1969). *Am. Ind. Hyg. Ass. J.* 30(5) 487–93.

[61] Raynor, G.S. (1968). *Am. Ind. Hyg. Ass. J.* 29(4) 397–404.

[62] Davies, C.N. (1968). *Br. J. Appl. Phys.* Ser. 2, 1, 921–32.

[63] Bien, C.T. and Corn, M.T. (1971). *Am. Ind. Hyg. Ass. J.* 32(7) 453–6.

[64] Hodkinson, J.R. (1972). *Air Sampling Instruments for Evaluation of Atmospheric Contaminants*. 4th edn. American Conference of Government Industrial Hygienists, Cincinatti.

[65] British Standard Method for the Measurement of Air Pollution, Part II – Determination of Concentration of Suspended Matter, B.S. 1747: Part II, 1969.

[66]   Cohen, A.L. (1973). *Environ. Sci. Technol.* **7**(1) 60—1.
[67]   Avera, C.B. (1968). *Am. Ind. Hyg. Ass. J.* **29**(4) 397—404.
[68]   King, R.B. and Fordyce, J.S. (1971). *J. Air Pollut. Control Ass.* **21**(11) 720.
[69]   Giever, P.M. and Ruch, W.E. (1971). *Am. Ind. Hyg. Ass. J.* **32**(5) 260—6.
[70]   Lynam, D.R., Pierce, J.O., and Cholak, J. (1969). *Am. Ind. Hyg. Ass. J.* **30**(1) 83—8.
[71]   Tebbens, B. (1970). *Am. Ind. Hyg. Ass. J.* **31**(1) 44—51.
[72]   Clements, H.A. *et al.* (1972). *J. Air Pollut. Control Ass.* **22**(12) 955—8.
[73]   Heard, M.J. and Wiffen, R.D. (1972). *Atmos. Environ.* **6**(5) 343—51.
[74]   Gussman, R.A., Sacco, A.M. and McMahon, N.M. (1973). *J. Air Pollut. Control Ass.* **23**(9) 778—82.
[75]   Burton, R.M. *et al.* (1973). *J. Air Pollut. Control Ass.* **23**(4) 277—81.
[76]   Buchanan, L.M. *et al.* (1972). *Appl. Microbiol.* **23**(6) 1140—4.
[77]   Lippman M. (1970). *Am. Ind. Hyg. Ass. J.* **31**, 138—59.
[78]   Aerosol Technology Committee (1970). *Am. Ind. Hyg. Ass. J.* **31**, 133—7.
[79]   British Standard Method for the Measurement of Air Pollution, Part I — Deposit gauges, B.S. 1747: Part I, 1969.
[80]   British Standard Method for the Measurement of Air Pollution, Part 5 — Directional dust gauges, B.S. 1747: Part 5, 1972.
[81]   ASTM Annual Book of Standards (1973). Standard Method for Collection and Analysis of Dustfall (Settleable Particulates), D 1739-70, Part 23.
[82]   Recommended Standard Methods for Continuing Dustfall Survey (APM, Revision 1) (1966). *J. Air Pollut. Control Ass.* **16**(7) 372—7.
[83]   Boubel, R.W. (1971). *J. Air Pollut. Control Ass.* **21**(12) 783—7.
[84]   Morrow, N.L., Brief, R.S. and Bertrand, R.R. (1972). *Chem. Eng. (NY)* **79**(2) 84—98.
[85]   Wolfe, E.A. (1966). *Source Test Methods used by Bay Area Air Pollution Control District*, Bay Area Air Poll. Control District San Francisco, California.
[86]   Devorkin, H. *et al.* (1963). *Air Pollution Source Testing Manual*, Air Pollution Control District of the County of Los Angeles, L.A., California.
[87]   Hawksley, P.G.W., Badzioch, S. and Bleckett, J.H. (1961). *Measurement of Solids in flue gases*, British Coal Utilization Research Association, Leatherhead.
[88]   Hemeon, W.C.L. and Black, A.W. (1972). *J. Air Pollut. Control Ass.* **22**(7) 516—18.
[89]   Martin, R.M. (1971). *Environmental Protection Agency*, Pub. No. APTD-0581.
[90]   Hirschler, D.A. *et al.* (1957). *Ind. Eng. Chem.* **49**(7) 1131—42.
[91]   McKee, H.C. and McMahon, W.A. (1960). *J. Air Pollut. Control Ass.* **10**(6) 456—62.
[92]   Mueller, P.K. *et al.* (1964). Symp. on Air Poll. Measurement Methods, Special Tech. Publication no. 352, ASTM.
[93]   Habibi, K. (1971). *Environ. Sci. Technol.* **4**(8) 679.
[94]   Habiki, K. (1973). *Environ. Sci. Technol.* **7**(3) 223—4.

[95] Campbell, K. and Dartnell, P.L. (1971). *Symp. of Institution of Mechanical Engineers*, **14**, London.
[96] Ter Haar, G.L. *et al.* (1972). *J. Air Pollut. Control Ass.* **22**(1) 39–46.
[97] Sampson, R.F. and Springer, G.S. (1973). *Environ. Sci. Technol.* **7**(1) 55–60.
[98] Sexton, R.W. (1964). *Am. Ind. Hyg. Ass. J.* **25**, 346.
[99] Reid, F.H. and Halpin, W.R. (1968). *Am. Ind. Hyg. Ass. J.* **29**(4) 390–5.
[100] ASTM Annual Book of Standards (1973). Standard Method of Test for Concentration of Odorous Vapours (Adsorption Method), D 1354-60, Part 23.
[101] Fraust, C.L. and Hermann, E.R. (1969). *Am. Ind. Hyg. Ass. J.* **30**(5) 494–9.
[102] Jeltes R. (1969). *Atmos. Environ.* **3**(5) 587–8.
[103] Mann, J.B. *et al.* (1974). *Environ. Sci. Technol.* **8**(6) 584–5.
[104] Cantuti, V and Cartoni, G.P. (1968). *J. Chromat.* **32**, 641–7.
[105] Aue, W.A. and Teli, P.M. (1971). *J. Chromat.* **62**, 15–27.
[106] Perry, R. and Twibell, J.D. (1973). *Atmos. Environ.* **7**, 927.
[107] Cropper, F. and Kominsky, S. (1963). *Analyt. Chem.* **35**, 735.
[108] Williams, I. (1965). *Analyt. Chem.* **37**, 1723.
[109] Bellar, T.A. *et al.* (1963). *Analyt. Chem.* **35**, 1924.
[110] Stephens, E.R. and Burleson, F.R. (1967). *J. Air Pollut. Control Ass.* **17**, 147.
[111] Williams, F.W. and Umstead, M.E. (1968). *Analyt. Chem.* **40**, 2232.
[112] Novak, J., Vasak, V. and Janak, J. (1965). *Analyt. Chem.* **37**, 660.
[113] Dravnieks, A. *et al.* (1971). *Environ. Sci. Technol.* **5**, 1220.
[114] Shepherd, M. *et al.* (1951). *Analyt. Chem.* **23**, 1431.
[115] ASTM Annual Book of Standards (1973). Tentative Method of Test for C_1–C_5, Hydrocarbons in the atmosphere by Gas Chromatography, D 2820-69T, Part 23.
[116] Desbaumes, E. and Imhoff, C. (1971). *Staub.* **31**(6) 36–41.
[117] Schuette, F.J. (1967). *Atmos. Environ.* **1**, 515–19.
[118] Conner W.D. and Nader, J.S. (1964). *Am. Ind. Hyg. Ass. J.* **25**(3) 291–7.
[119] Oord, F. (1970). *Am. Ind. Hyg. Ass. J.* **31**, 532–3.
[120] Curtis, E.H. and Hendriks, R.H. (1969). *Am. Ind. Hyg. Ass. J.* **30**, 93–4.
[121] Lang, H.W. and Freedman, R.W. (1969). *Am. Ind. Hyg. Assoc. J.* **30**(5) 523.
[122] Meader, M.C. and Bethea, R.M. (1970). *Environ. Sci. Technol.* **4**(10) 853–5.
[123] Van Houten, R.V. and Lee, C. (1969). *Am. Ind. Hyg. Ass. J.* **30**(5) 465–9.
[124] British Standard Methods for the Sampling and Analysis of Flue Gases, Part I, Methods of Sampling, B.S. 1756-1: 1971.
[125] Berger, A.W., Driscoll, J.N. and Morgenstern, P. (1972). *Am. Ind. Hyg. Ass. J.* **33**(6) 397–404.

CHAPTER TWO

Analysis of particulate pollutants

2.1 Introduction

Particulate pollutants are emitted by a great many sources, both stationary and mobile. Additionally, particles are formed in the atmosphere by chemical and physical conversion from both natural and anthropogenic gaseous substances. In this chapter the methods of sampling atmospheric particulates are critically examined and techniques of physical examination are described. Analysis of the chemical composition is given comprehensive coverage in subsequent chapters.

Particulate pollutants may be sampled either from suspension in the air, by filtration for example, or by collection of deposited particles as they fall out from the atmosphere under gravitational influence, known as dustfall. Consequently, the study of particulate pollution is simplified by division of particles into two categories: (a) suspended matter; and (b) depositable matter. This division is not clear cut and is dependent upon such factors as meteorological conditions and size, density and terminal velocity of the particles. Most workers consider a diameter of $10\,\mu$m as the dividing line between suspended and depositable matter. This division should not be taken too literally however, and when, later in the chapter, estimates are made of deposition from chimneys, it will be seen that deposition rates in the open of particles with diameter in the region of $20\,\mu$m are negligible. This is not difficult to understand when it is considered that a $20\,\mu$m particle of unit density has a falling speed of approximately $1.2\,\mathrm{cm\,s}^{-1}$, so that a wind speed of $1\,\mathrm{mile\,h}^{-1}$ ($0.45\,\mathrm{m\,s}^{-1}$—almost calm conditions) is high enough to maintain $20\,\mu$m particles airborne for long distances. The picture is further complicated by the fact that turbulent fluctuations of the wind can maintain particles in suspension for far longer and for these reasons it is not proposed to go into greater detail on the subject of particle sizes. Another complication which will arise in practice is that at any site, pollution from different directions is likely to contain particles of radically differing densities, and consequently particles in any one size range, from one source, could well remain suspended whilst becoming deposited from another.

Because of these problems, when operating in the field, sampling techniques should always be well-defined.

2.1.1 Emission of particulate matter

There are few industrial processes which do not emit particulate matter and the amounts emitted vary considerably from process to process as well as in different parts of any particular process. Formulae exist for estimating the behaviour of pollution emitted from discrete sources such as chimneys; but when emissions are from diffuse sources such as coke heaps, or from materials handling processes such as grabs, excavators or conveyor belts, methods of estimating overall emissions are more difficult and have to be designed for each individual case. It is true that a conveyor belt can be regarded as a line source of pollution, and a mechanical grab, unloading cargo from a ship's hold, as a generator of discrete puffs of pollution. Formulae do exist for estimating ground level concentrations from such sources but in practice it is seldom possible to apply them because such industrial sources are generally installed in heavily congested areas where nearby buildings affect dispersion of the pollution and make any accurate estimation of likely concentrations from these sources almost impossible to carrry out.

2.1.2 Emission factors for particulate matter

Since different industrial processes emit differing quantities of particulate pollutants, data relating to such pollution are best presented as pounds of particulate matter per ton of finished product. Most industrial processes, however, employ methods, such as filtration, for reducing emissions and the efficiencies of these will, for obvious reasons, vary from factory to factory. In a recent paper Vandegrift et al. [1] looked at the particulate emission factor, i.e. the weight of particulate matter emitted per ton of raw material, and tried to assess the average efficiency of the control techniques in use. For easy reference his estimates are quoted in Table 2.1 together with the net emission of particulate material to the atmosphere for various industries after control. These give some idea of the relative importance of some industries as emitters of particulate pollution. Although the list is not complete and substantial differences exist between different installations performing the same process, for those sampling in the field it should serve as a useful guide not only as to which sources are potential offenders but also which particulate materials it will be necessary to sample. For example some

Table 2.1 Emission of particulate matter from industrial processes

Source	Emission factor without control (lb ton^{-1})	Net efficiency of applied control*	Net emission factor (lb ton^{-1})
(1) *Fuel combustion*			
Power stations—coal			
Pulverized	190	0.89	20.9 (as coal)
Stoker	146	0.70	43.8 (as coal)
Cyclone	35	0.64	12.6 (as coal)
Industrial boilers—coal			
Pulverized	170	0.81	32.3 (as coal)
Stoker	133	0.52	63.8 (as coal)
Cyclone	31	0.75	7.8 (as coal)
(2) *Agricultural operations*			
Grain elevators—grain	27	0.28	19.4 (as grain)
Cotton gins	12†	0.32	8.2 (lb bale^{-1})
Alfalfa dehydrators—dry meal	52	0.42	29.0 (as dry meal)
(3) *Iron and steel*			
Materials handling—steel	10	0.32	6.8 (as steel)
Sinter plant—sinter	42	0.90	4.2 (as sinter)
Blast furnace—iron	130	0.99	1.3 (as iron)
Steel furnaces—steel			
Open hearth	17	0.40	10.2 (as steel)
Basic oxygen	40	0.99	0.4 (as steel)
Electric arc	10	0.78	2.2 (as steel)
Scarfing	3	0.68	1.0 (as steel)
(4) *Cement*			
Kilns	167	0.88	20.0 (as cement)
Grinders, etc. (wet)	25	0.88	3.0 (as cement)
Grinders, etc. (dry)	67	0.88	8.0 (as cement)
(5) *Pulp mills*			
Recovery furnace	150	0.91	13.5 (as pulp)
Lime kilns	45	0.94	2.7 (as pulp)
Dissolving tanks	5	0.30	3.5 (as pulp)
(6) *Lime works*			
Crushing, screening—rock	24	0.20	19.2 (as rock)
Rotary kilns—lime	180	0.81	34.2 (as rock)
Vertical kilns—lime	7	0.39	4.3 (as rock)
Materials handling—lime	5	0.76	1.2 (as rock)
(7) *Clay*			
(a) *Ceramic*			
Grinding	76	0.60	30.4
Drying	70	0.60	28.0
(b) *Refractories*			
(1) *Kiln fired*			
Calcining	200	0.64	72.0
Drying	70	0.64	25.2
Grinding	76	0.64	27.4
(2) Castable	225	0.77	51.8
(3) Magnesite	250	0.56	111.0

Table 2.1 Continued

Source	Emission factor without control (lb ton^{-1})	Net efficiency of applied control*	Net emission factor (lb ton^{-1})
(4) *Mortars*			
Grinding	76	0.60	30.4
Drying	70	0.60	28.0
(5) *Mixes*	76	0.60	30.4
(8) *Primary non-ferrous*			
(a) *Aluminium*			
Grinding of bauxite	6	0.80	1.2 (as bauxite)
Calcining of			
hydroxide–alumina	200	0.90	20.0 (as alumina)
Reduction cells–aluminium			
H.S. Soderberg	144	0.40	86.4 (as aluminium)
V.S. Soderberg	84	0.64	30.2 (as aluminium)
Prebake	63	0.64	22.7 (as aluminium)
Materials handling	10	0.32	6.8 (as aluminium)
(b) *Copper*			
Ore crushing–ore	2	0	2.0 (as ore)
Roasting–copper	168	0.85	25.2 (as copper)
Reverb furnace–copper	206	0.81	39.1 (as copper)
Converters–copper	235	0.81	44.7 (as copper)
Materials handling–copper	10	0.32	6.8 (as copper)
(c) *Zinc*			
Roasting–zinc			
Fluid bed	2000	0.98	40.0 (as zinc)
Ropp, multihearth	333	0.85	50.0 (as zinc)
Sintering	180	0.95	9.0 (as zinc)
Materials handling	7	0.32	4.8 (as zinc)
(d) *Lead*			
Sintering	520	0.86	72.8 (as lead)
Blast furnace	250	0.83	42.5 (as lead)
Materials handling	5	0.32	3.4 (as lead)
(9) *Asphalt*			
Paving materials			
Dryers	32	0.96	1.3 (as material)
Secondary sources	8	0.96	0.3 (as material)
(10) *Ferroalloys*			
Blast furnaces	410	0.99	4.1 (as alloy)
Electric furnace	240	0.40	144.0 (as alloy)
Materials handling	10	0.32	6.8 (as alloy)
(11) *Iron foundries–metal*			
Furnaces	16	0.27	11.7 (as metal)
Materials handling	5	0.20	4.0 (as metal)
(coke, limestone etc.)			
(12) *Secondary non-ferrous*			
(a) *Copper*			
Materials preparation–scrap			
Sweating furnace	15	0.19	12.2 (as scrap)

Table 2.1   Continued

Source	Emission factor without control (lb ton^{-1})	Net efficiency of applied control*	Net emission factor (lb ton^{-1})
Blast furnace	50	0.68	16.0 (as scrap)
Smelting and Refining	70	0.57	30.1 (as scrap)
(b) *Aluminium*			
Sweating furnace–scrap	32	0.19	25.9 (as scrap)
Refining furnace–scrap	4	0.57	1.7 (as scrap)
Chlorine fluxing–chlorine	1000	0.25	750 (as chlorine)
(c) *Lead–scrap*			
Pot furnace	0.8	0.90	0.1 (as scrap)
Blast furnace	190	0.90	19.0 (as scrap)
Reverb furnace	100	0.90	10.0 (as scrap)
(d) *Zinc–scrap*			
Metallic scrap sweating	12	0.19	9.7 (as scrap)
Residual scrap sweating	30	0.19	24.3 (as scrap)
Distillation furnace	45	0.57	19.4 (as scrap)
(13) *Sulphuric acid*			
New acid–contact process–100% acid	2	0.85	0.3 (as 100% acid)
Spent acid concentrators–acid	30	0.80	6.0 (as spent acid)
(14) *Phosphoric acid*			
Thermal process–P_2O_5	134	0.97	4.0 (as P_2O_5)

* The overall level of control in the US; the product of the application of control and the efficiency of control.

of the particulate matter mentioned will have to be treated as suspended whilst other material can be treated as depositable. Factors such as these will determine the sampling technique used.

2.1.3 Dispersion of atmospheric pollutants from a point source

As a considerable amount of work has been carried out on the subject of atmospheric dispersion, it is not proposed in this book to go into full details of the various theories involved but instead to concentrate on the purely practical problems arising as a consequence of these theories, which have to be solved in order to perform effective sampling. The aim of this section is to give some idea of the extent to which pollution may be dispersed, such that when sampling is undertaken sites will be selected which will give meaningful results. It must be remembered that the formulae and methods presented refer to conditions in open, level country. They will be only a very approximate guide to the dispersion of pollutants in a built-up environment.

Most pollution is emitted into the atmosphere from hot stacks and under such conditions the behaviour of the plume is governed by four main factors:

(a) its velocity as it leaves the stack;

(b) the difference in the densities of the stack gases and the surrounding atmosphere (a factor which generally results from the temperature difference);

(c) the atmospheric temperature gradient; and

(d) the wind speed.

The combination of these factors causes the buoyant plume to rise above the top of the stack and it is thus necessary to know just how far the plume rises before any estimates can be made of how much pollution reaches ground level.

2.1.3.1 Plume rise

A number of formulae are available which may be used to determine the amount by which a plume will rise above the point of emission; one such is that given by Holland in the *Workbook of Atmospheric Dispersion Estimates* [2]

$$\Delta H = \frac{V_s d}{u}(1.5 + 2.68 \times 10^{-3} p\frac{T_s - T_a}{T_s} \times d)$$

where ΔH is the plume rise above the stack in m; V_s is the stack-gas velocity in $m\,s^{-1}$; d is the inside diameter of the stack in m; u is the wind speed in $m\,s^{-1}$; p is the atmospheric pressure in mb; T_s is the stack-gas temperature in K; T_a is the air temperature in K.

The effluent is also subject to atmospheric stability conditions and where this formula is applied corrections should be made: for unstable conditions a value of 1.1 to 1.2 ΔH and for stable conditions a value of 0.8 to 0.9 ΔH should be used.

2.1.3.2 Dispersion of the plume

An excellent paper by Pasquill [3] describes the problems involved in the determination of the effect of atmospheric stability upon the dispersion of pollution. Briefly, as it travels downwind, pollution from a given source disperses in both the horizontal and vertical planes and the amount of mixing which takes place depends on the stability of the atmosphere at that particular time. A suitable index of the dispersion may be obtained by the use of a bivane which consists of two meteorological vanes mounted on one shaft; one vane oscillates in a horizontal plane giving the horizontal component of the turbulence whilst the other oscillates in a vertical plane giving the vertical component (see section on meteorological instruments). If such instruments are available, the amount of vertical spread of a plume may be estimated from the formula

$$h = 2150 \, d\sigma_\phi$$

where h is the height in m to which the plume will rise above the height of

Table 2.2   Key to stability categories

Surface wind Speed at 10 m (m s⁻¹)	Sunlight			Night	
	Strong	Moderate	Slight	≥ 4/8 cloud	≤ 3/8 cloud
< 2	A	A–B	B	–	–
2–3	A–B	B	C	E	F
3–5	B	B–C	C	D	E
5–6	C	C–D	D	D	D
> 6	C	D	D	D	D

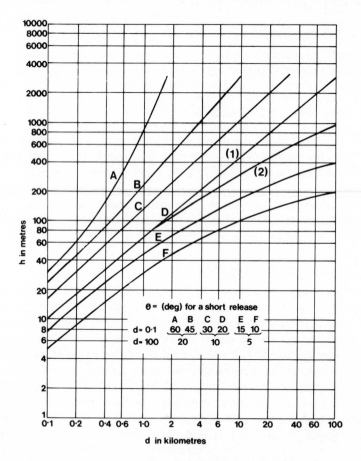

Fig. 2.1   Estimates of vertical (h) and lateral (θ) spread of a plume [3]. The extent of h and θ are defined by pollutant concentratons one-tenth of those on the axis of the plume.

emission; d is the distance from the source in km; and σ_ϕ is the S.D. of the vertical fluctuations of the wind in radians.

Although vertical mixing is also affected by the atmospheric mixing height which can restrict dispersion, this need not be considered at this point, as most field measurements will be concerned with dispersion comparatively close to the source.

If bivanes are not available it is possible to estimate the amount of mixing which will take place from the Stability Categories given in Pasquill's paper [3]. These are shown in Table 2.2.

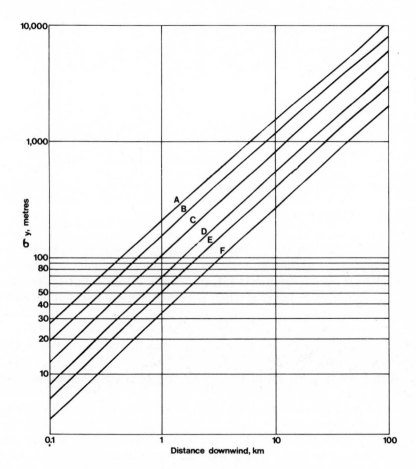

Fig. 2.2   The horizontal dispersion as a function of downwind distance from the source. The function σ_y is the standard deviation of the plume concentration distribution in the horizontal [2].

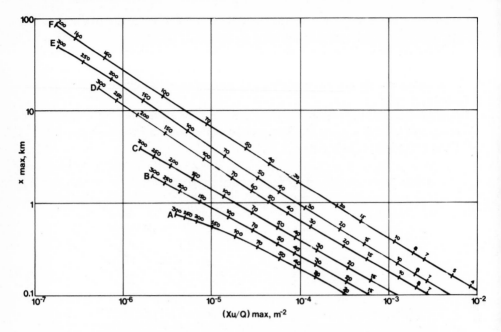

Fig. 2.3 Distance of maximum ground-level concentration and maximum $\chi u/Q$ as a function of stability (curves) and effective height (m) of emission (numbers) [2].

Once the stability category has been assessed it is possible, using Fig. 2.1 which gives curves showing probable heights of dispersion at different distances and different stability categories [3], to estimate the vertical extent of plume dispersion. This figure also gives estimates of the horizontal spread of the plume, in degrees, for the six stability categories at two distances, 0.1 and 100 km. A graph, extracted from the *Workbook on Dispersion Estimates* [2], which gives the horizontal dispersion in more detail is shown in Fig. 2.2. In this case the plume width is given in metres.

After having determined the physical dimensions of the plume the worker in the field must determine the point of maximum ground-level concentration of pollution and its distance from the source. This is shown in Fig. 2.3 [2]. When the effective chimney height (equal to the chimney height plus the plume rise) has been estimated and a stability category allocated, the distance of the peak ground-level concentration may be read off the y-axis of Fig. 2.3 by selecting the appropriate height on the correct stability curve and reading off the distance on the y-axis. The value shown on the x-axis for the height (the function $\chi u/Q\,\mathrm{m}^{-2}$) should also be determined. The mean ground-level concentration is then found by multiplying this value by Q/u (where Q is the rate of emission of the pollutant in $g\,s^{-1}$ and u is the wind speed in $m\,s^{-1}$). This formula gives concentrations maxima representative of approximately 10 min sampling periods and cannot be used to estimate mean concentrations over longer periods.

If sampling sites are selected in areas which are off-centre of the line of maximum pollution the values found will be lower since, generally speaking, there is a Gaussian distribution of pollution concentrations across a plume. When planning the siting of instruments, however, it is more important to know the likely maximum concentrations rather than off-axial concentrations.

Although this information should make possible the estimation of likely plume dimensions and concentrations, in actual practice it will be very difficult to locate points of maximum ground-level concentration because of continuously varying conditions.

2.1.3.3 Worked example

Plume rise

A source emits $72\,g\,s^{-1}$ of SO_2 from a stack 30 m high and of diameter 1.5 m. The effluent gases are emitted at a temperature of $250°\,F$ (394 K) with an efflux velocity of $13\,m\,s^{-1}$. Atmospheric pressure is 970 mb, ambient air temperature is $20°\,C$ (293 K) and the wind speed is $2.1\,m\,s^{-1}$. What is the effective chimney height?

$$\Delta H = \frac{V_s d}{u}\left(1.5 + 2.68 \times 10^{-3}\,p\,\frac{T_s - T_a}{T_s} \times d\right)$$

$$= \frac{13 \times 1.5}{2.1}\left(1.5 + 2.68 \times 10^{-3} \times 970 \times \frac{394 - 293}{394} \times 1.5\right)$$

$$= 9.29\,(1.5 + 1)$$

$$= 23.2\,m$$

i.e. the effective chimney height is 53.2 m.

Distance and value of the maximum ground-level concentration

For a stability category C give the maximum ground-level concentration and its distance from the source in example (a).

In Fig. 2.3, using stability curve C, at a height of 53 m the maximum ground-level concentration, read off the y-axis, is 0.56 km and on the x-axis the value given for $\chi u/Q$ is 5.2×10^{-5}.

The peak (30 min) concentration is therefore: $5.25 \times 10^{-5} \times Q/u$

$$= 5.25 \times 10^{-5} \times 72/2.1$$

$$= 1\,800\,\mu g\,m^{-3}.$$

2.1.3.4 Problems of short-term sampling

When short-term sampling is planned the requirements must be clearly understood before work starts otherwise the results may be of little value. An example might be to settle whether the pollution being monitored is from a particular source or whether it concerns area-wide pollution. In the latter case sites should be avoided which are under the influence of nearby sources. If on the other hand it concerns emissions from a particular source it may be desirable to locate the maximum ground-level pollution from such a source so that a suitable site may be found. In an earlier section a method was given for determining the location of peak ground-level concentrations from a specific source. On paper this is a simple problem but in practice, even in open country with no buildings to complicate dispersion, the actual location is not a simple task even using continuous recorders. The reasons are explained by the meteorological parameters which control dispersion.

If, in an attempt to locate high concentrations, sampling is carried out close to a source the pollution levels in the plume will be found to fluctuate wildly from peak values to zero in a few seconds as the plume shifts constantly in direction due to variations in wind direction, etc. The amount of these fluctuations is largely determined by the stability category and some useful figures given by Pasquill show that at a distance of 0.1 km from a source, under very stable conditions (stability category F) a plume of about 10^0 can be expected; under unstable conditions (category A) plumes can be about 60^0 in width at this distance. Further downwind concentration fluctuations will be found but they will not be quite so extreme as close to the source. Although these fluctuations will not be quite so severe it will still be difficult in practice to locate the centre and, since the crosswind concentrations of pollution in a plume are roughly Gaussian a slight error in the location of the site could give severely reduced concentrations.

When short-term sampling has to be started it is important that any measurements should be adequately supported by the relevant meteorological data, such as wind speed, direction, stability category, etc.

Another factor which has to be taken into consideration is the pattern of emissions. Many pollution sources emit different amounts of pollution, not only dependent on the time of day, the day of the week, but there may also be seasonal variations e.g. from summer to winter. Some cyclic changes in pollution emissions can be very short-term indeed, such as the effect on emissions from road traffic affected by traffic lights. When the effect of these variations is added to the variations due to meteorological conditions it will be seen that care will have to be taken with such short-term determinations.

2.1.4 Deposition of particulate matter from chimney stacks

The previous section has dealt with basic problems associated with the behaviour of gaseous and suspended particulate matter. When dealing with particulate material which has definite settling properties, however, the settling velocities must be taken into account when one is looking for areas into which such pollution will fall. Many papers have been written on this subject: amongst others Nonhebel [4] and Bosanquet et al. [5] have quoted suitable formulae. For the purpose of this book the formula by Bosanquet is used. This is:

$$\text{Average rate of deposition} = \frac{3.27 \times Wb \times 10^4}{H^2} \times F\left(\frac{f}{v}, \frac{x}{H}\right) \text{g m}^{-2} \text{month}^{-1}$$

where W is rate of dust emission in g s^{-1}; b is fraction of time during which wind is in the direction of the $45°$ sector being considered; H is the effective height of the plume in ft; f is the free falling speed of particles in ft s^{-1}; v is the wind speed in ft s^{-1}; x is the distance in ft from the stack.

If short-term sampling is being undertaken in order to determine deposition in a given direction then b is assumed to be 1. For long-term sampling, however, it will be necessary to determine the frequency of occurrence of winds in the area from synoptic data.

In order to simplify the calculation of dust deposition the function $F(f/v, x/H)$ has been plotted in Fig. 2.4 for various values of f/v and x/H.

No general advice can be given on the pattern of deposition likely to be encountered downwind of a chimney as this will be affected by not only the height of emission but also the size range and density of the particles emitted. The latter factor will, of course, be dependent on the efficiency of cyclones and filters installed to reduce particulate emissions.

Later in this section a worked example using this formula is given. Fig. 2.5 and 2.6 show idealized depositions from a source 372 ft high emitting 30 g s^{-1} of a dust of specific gravity 2. In this case, to illustrate the pattern, it has been assumed that all size fractions of the dust are of equal weight. Fig. 2.5 shows the deposition pattern in a wind of speed 4.6 m s^{-1} (15 ft s^{-1}) and Fig. 2.6 that in a wind of speed 1.5 m s^{-1} (5 ft s^{-1}).

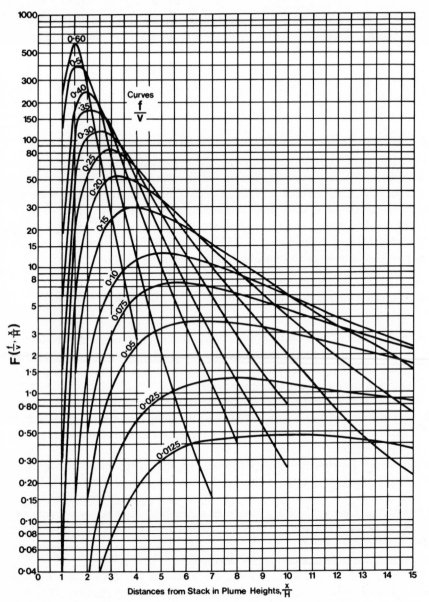

Fig. 2.4 Variation of rate of deposition with distance from stack for particles of various sizes [5].

$$\text{Rate of deposition} = F\left(\frac{f}{v}, \frac{x}{H}\right) \times \frac{3.27 \times W \cdot b \times 10^4}{H^2}$$

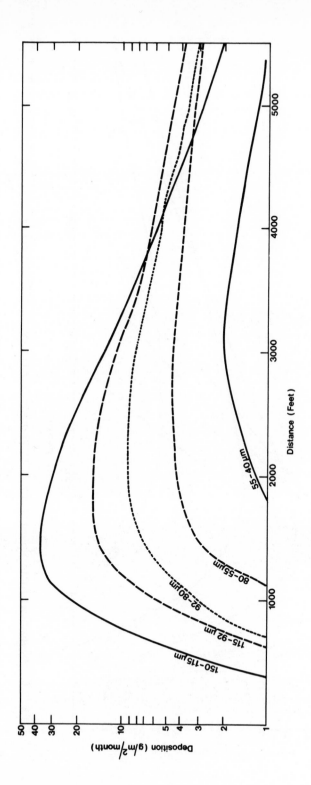

Fig. 2.5 Deposition pattern of dust from a 372 ft chimney. Wind speed is 15 ft s⁻¹ (4.6 m s⁻¹).

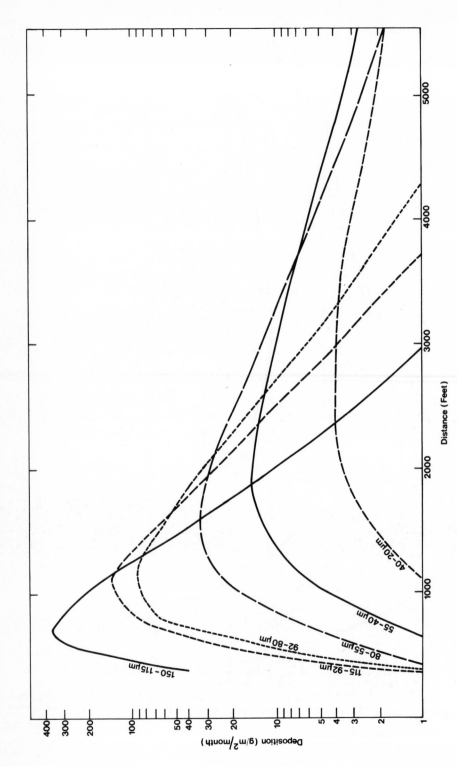

Fig. 2.6   Deposition pattern of dust from a 372 ft chimney. Wind speed is 5 ft s^{-1} (1.5 m s^{-1})

2.1.4.1 Worked example [5]

A chimney with an effective height of 372 ft emits $30\,\mathrm{g\,s^{-1}}$ of a dust, the specific gravity of which is 2, in a wind of $15\,\mathrm{ft\,s^{-1}}$. The particle size distribution of the dust is:

Grade sizes (μm)	Fraction by weight of grade
150–115	0.23
115– 92	0.15
92– 80	0.08
80– 55	0.17
55– 40	0.10
40– 22	0.12
22– 0	0.15

Calculate the rate of dust deposition at different distances downwind.

The method of calculation is simple and the stages are shown in Table 2.3. Each column, showing the successive stages, is lettered for easy reference. The size limits of each fraction and the fraction by weight in each grade are shown in columns (a) and (b) respectively. It is then necessary to estimate the mean falling speed for each size range and this is shown in column (c). Data may be obtained from the graph in Fig. 2.7. The values of f/v have then to be worked out for each size range at the given wind speed of $15\,\mathrm{ft\,s^{-1}}$ and are shown in column (d). In order to calculate the deposition at each chimney height downwind of the source select this chimney height (x/H) on the x-axis on Fig. 2.4 and read off the value of the function $F(f/v, x/H)$ on the y-axis for each f/v value shown in column (d). This value should be written in column (e). It is then necessary to multiply the values shown in column (e) by the fraction in grade values [column (b)] to obtain the product. All the values in column (f) should be added together and then multiplied by the first part of the equation, $(3.27\,Wb \times 10^4)/H^2$, which in this case is 7.09, to obtain deposition rates. A general formula, based on Stokes law, for estimating falling speeds of particles in general, is also given in Fig. 2.7.

2.1.5 Relevant meteorological instruments

When carrying out air pollution measurements it is desirable to have access to meteorological data to assist in the analysis of data collected. It is possible to make use of existing meteorological instruments in the area but if this is done, care should be taken to ensure that the measurements used are meaningful. Meteorological instruments, particularly wind vanes and anemometers, are very susceptible to interference and for general work should not be sited at a height lower than 10 m. The siting of instruments on roofs should be avoided

Table 2.3 Example of the estimation of dust deposition from a chimney stack

(a) Grade limits (µm)	(b) Fraction by weight	(c) Mean 'f'	(d) Mean f/v	(e) $\dfrac{x/H = 2}{F\left(\dfrac{f}{v},\dfrac{x}{H}\right)}$	(f) (b) × (e)	(e) $\dfrac{x/H = 4}{F\left(\dfrac{f}{v},\dfrac{x}{H}\right)}$	(f) (b) × (e)	(e) $\dfrac{x/H = 8}{F\left(\dfrac{f}{v},\dfrac{x}{H}\right)}$	(f) (b) × (e)	(e) $\dfrac{x/H = 10}{F\left(\dfrac{f}{v},\dfrac{x}{H}\right)}$	(f) (b) × (e)
150–115	0.23	2.30	0.15	7.0	1.61	30.0	6.90	12.0	2.76	6.0	1.38
115– 92	0.15	1.60	0.10	1.5	0.23	12.0	1.80	9.0	1.35	6.0	0.90
92– 80	0.08	1.20	0.08	0.5	0.04	6.5	0.52	8.0	0.64	5.0	0.40
80– 55	0.17	0.70	0.05	0.13	0.02	2.6	0.44	3.6	0.61	3.0	0.51
55– 40	0.10	0.45	0.03	0.05	0.01	0.7	0.07	1.5	0.15	1.3	0.13
40– 22	0.12	0.20	0.01	–	–	0.15	0.02	0.4	0.04	0.3	0.04
22– 0	0.15	–	–	–	–	–	–	–	–	–	–
Total $F\left(\dfrac{f}{v},\dfrac{x}{H}\right)$					1.91		9.75		5.56		3.36
Average rate of deposit $= 7.09 \times$ mean $F\left(\dfrac{f}{v},\dfrac{x}{H}\right)$ (g m^{-2} month^{-1})					13.5		69.1		39.4		23.8
At distances (ft) of					744		1488		2976		3720

Note: (1) It is worthwhile looking at the values given in column (f) for each size range at different distances. In the heavier size ranges the highest values occur at about four chimney heights, but, as would be expected, with smaller particles the highest values tend to occur further out at eight or ten chimney heights demonstrating the wider dispersion of the smaller particles.
(2) A value of $b = 1$ has been used. As explained above this will only be valid for very short periods of sampling and over longer periods smaller values will apply.

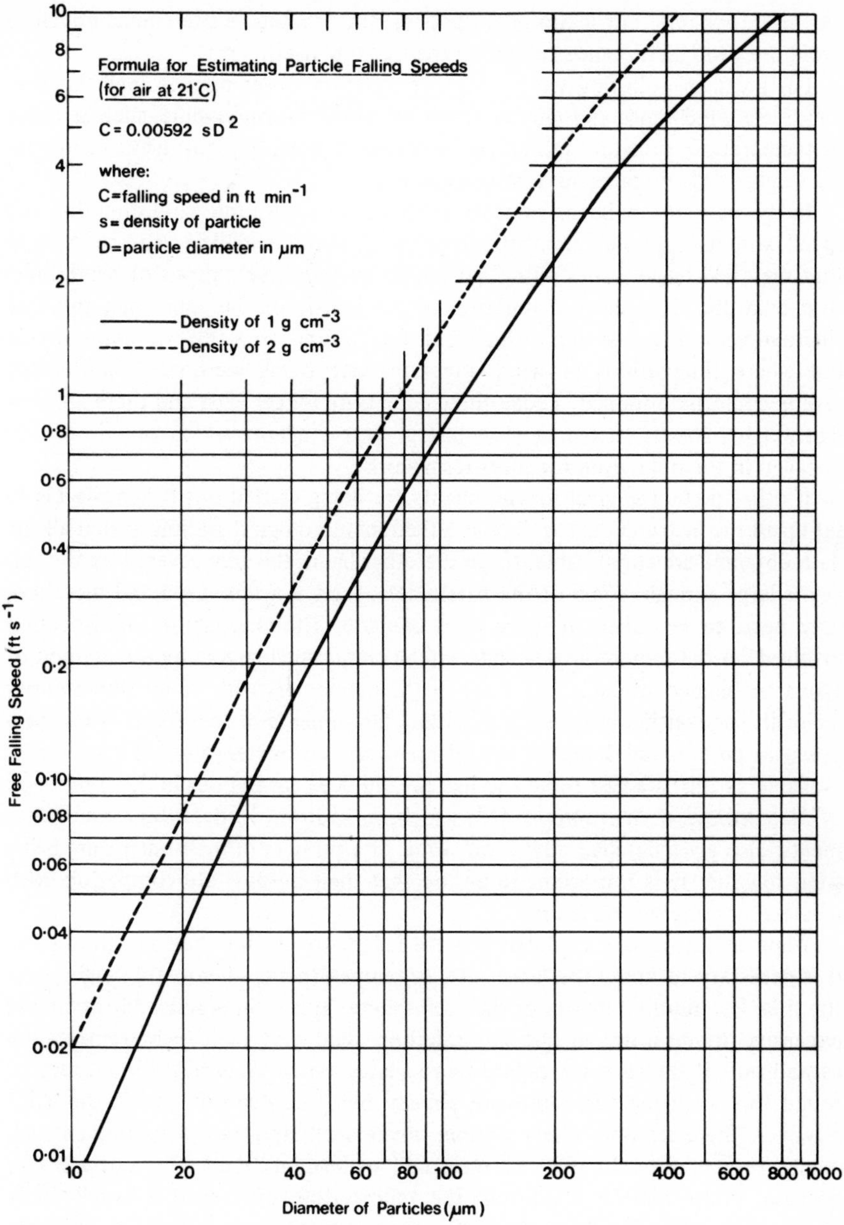
Formula for Estimating Particle Falling Speeds
(for air at 21°C)
C = 0.00592 sD 2
where:
C=falling speed in ft min-1
s= density of particle
D=particle diameter in μm
Density of 1 g cm-3
Density of 2 g cm-3
Free Falling Speed (ft s-1)
Diameter of Particles (μm)
10
8
6
4
2
1
0·8
0·6
0·4
0·2
0·10
0·08
0·06
0·04
0·02
0·01
10 20 40 60 80 100 200 400 600 800 1000

Fig. 2.7 Free falling speed in air of particles of two densities [5].

whenever possible, but if this is not possible they should be sited high enough to avoid the wind disturbance caused by the building itself.

For most air pollution work the most commonly used meteorological instruments are wind vanes and anemometers but other measurements, such as those of temperature, pressure, rainfall, solar radiation, humidity, turbulence and mixing heights or lapse rates, may be considered.

In the past, the only instruments available for measuring wind direction and speed were heavy, bulky and difficult to install in the field. They also suffered in that they had heavy vanes with high inertia and the fluctuations of wind direction recorded were more a function of the inertia of the vane than the true fluctuations in the direction of the wind. For routine work this was unimportant but where fluctuations of wind direction were being considered, misleading results could be obtained. Today these defects no longer exist and there are now lightweight, robust meteorological instruments available which can be readily installed in the field, even for short-term work.

If other meteorological measurements are being carried out it is advisable to set up all the instruments together as a field meteorological station so that all the data obtained are comparable. If, on the other hand, the area covered by the survey is large enough, some of the parameters, such as wind speed and direction, may need to be taken at more than one site. The amount of sophistication involved on the meteorological side will be determined largely by the amount of effort or finance available and most of the parameters may be measured either manually or continuously using recording instruments or loggers. It is not proposed to go into full details of the different types of meteorological instruments available as this is a vast topic and has already been carried out in the *Handbook of Meteorological Instruments* [6], which, in addition to describing the instruments, also gives detailed advice on siting. If a number of instruments are being sited together it is important to ensure that their outputs are compatible with whatever logging system is used.

When it is necessary to determine the height to which pollution will disperse it is necessary to know the lapse rate, or temperature gradient with height, over the first few hundred metres of the atmosphere. This is, in practice, difficult and expensive to measure; in the air pollution field few such measurements are carried out. If the survey area is close to a large meteorological station it may be found that such measurements are already being carried out; but in the U.K., however, there are only a few stations where such lapse rate measurements are made. Temperatures have to be measured with an accuracy of about $\pm 0.1°$ C because, under neutral condition, the temperature difference is usually only about $1°$ C $100 \, \text{m}^{-1}$. Because of the accuracy required, tall, solid buildings should be avoided not only because of the effect of the building on the wind structure, but also because the heat capacity of the building itself can affect the results. The effect of the warming of the framework of the building by solar radiation can completely ruin any measurements. This leaves the alternatives of using either a lattice structure mast or temperature measuring devices attached to balloons. Using the latter technique data from the instruments may be telemetered to ground level using radio transmitters.

The measurement of mixing height is now being made possible by the use of acoustic radar [7]. This device emits a signal in the form of a short duration pulse of sound at 1600 Hz vertically upwards and the back-scattered signal received gives some indication of the altitude of the mixing height. Although the use of this instrument is still in its infancy, the technique would appear to be ideally suited for air pollution work, particularly where mixing height measurements are required over extended periods.

When measurements of turbulence are required in the field, a bivane mounted on a 10 m mast with an anemometer is used. This instrument measures horizontal and vertical fluctuations of the wind. For obvious reasons the siting of a bivane is extremely critical and care should be taken to avoid areas which might be in the wake of any local obstruction.

For general turbulence analysis the standard deviation of the fluctuations is required and, therefore, some thought will have to be given to the method of recording the data. Any form of chart recording for this work can be discounted because, in order to record the individual fluctuations, chart recorders have to be run at a high chart speed and this results in amounts of charts too large to be analysed effectively, even if only short period operations in the field are being carried out. For short period work it is possible to record turbulence for subsequent computer analysis on continuously moving magnetic tape loggers. Because of the amount of magnetic tape used, however, this system can only be used when the field site is manned.

For unmanned operation it is possible to transmit turbulence data continuously by telephone using a multichannel high speed data transmission system. In this way data may be recorded by remote control at a central station where it can be conveniently logged. This technique was explained by Parker [8]. Another method is the use of a system of electrical filter circuits which filters out the individual fluctuations and puts out a slowly variable turbulence intensity index [9]. One drawback of this method is that once the original data have been filtered they are lost and cannot be retrieved if required for a different analysis at a later date.

2.2 Suspended material

2.2.1 Sampling techniques

2.2.1.1 Filter paper techniques

Probably the most common method of sampling particulate pollution (this includes 'smoke') is by collecting it on filter paper by drawing a sample of contaminated air through a filter which is held in a clamp as illustrated in Fig. 2.8 [10]. This figure illustrates the equipment used in the National Survey of Air Pollution in the U.K. The suction pump draws approximately $2\,m^3$ of air per day through the equipment; particulate matter is filtered out by the filter, and the Drechsel bottle which contains dilute H_2O_2 removes SO_2 contamination [10].

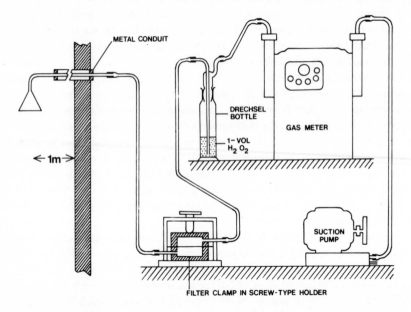

Fig. 2.8 Schematic arrangement of standard daily smoke and SO_2 sampling apparatus [10].

It is not possible to give firm advice on the type of filter material which should be used for any particular work since this will depend on the purpose for which the sampling is being conducted. For example, cellulose and glass fibre filters of suitable porosity will be suitable for routine determination of the total mass of particulate pollution. If subsequent chemical analysis is required they may be suitable if the filter background is negligible for the particular material being sought. It is always necessary to check the background contamination of the filters to be used in any analytical work. For microscopic work cellulose and glass fibre filters are unsuitable, because the atmospheric particulate matter tends to become trapped amongst the fibres of the filters, and thus membrane and Nuclepore filters are preferable. At this stage it will be useful to look at some of the more common types of filters in general use and describe the advantages of each.

Cellulose filters

Cellulose filters are probably the most frequently used for routine air pollution work. In most cases, where the density of the collected matter is determined by light reflectance methods, Whatman No. 1 is the most common but with Whatman No. 4 in use in the U.S.A. Whatman No. 4 offers less resistance to air flow but against this the penetration of particles into the filter is greater than with No. 1 grade. This is best illustrated by the graph shown in Fig. 2.9 which was taken from a technical bulletin published by the manufacturers [11]. This figure shows the particulate penetration of a number of the more common filters against increasing dust loading. It is interesting to note the difference in penetration between the cellulose filters and the glass fibre filters. The penetration figures shown were obtained by drawing air containing NaCl particles (0.6 to 1.7 μm diameter) through the filter and measuring the concentration of sodium downstream by means of flame photometry.

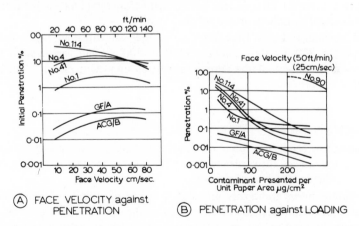

(A) FACE VELOCITY against PENETRATION

(B) PENETRATION against LOADING

Fig. 2.9 The efficiency of Whatman cellulose and glass-fibre filters.
(a) The effect of face velocity upon penetration (= 100% − efficiency).
(b) The effect of filter loading upon penetration at a face velocity of 25 cm s^{-1} [11].

The purpose of the sampling will largely determine the type of filter used. For example, for analytical work where the filter has to be ashed before analysis, one must take account of the fact that both Whatman No. 1 and No. 4 have an ash content of 0.06%, whereas the ash content of No. 41 grade is 0.01%. Cellulose filters also contain trace impurities and this may, depending of course on the materials being examined, preclude their suitability for work where chemical analysis of the sample is to be carried out.

Glass fibre filters

These filters offer less resistance to air flow than cellulose filters and, as may be seen in Fig. 2.9, the penetration of particles through the filter is much less than with cellulose filters, being less than 0.1% for the size of particles tested. Thus more of the particulate matter is collected on the surface of the filter than is the case with cellulose. Since the particles are more heavily concentrated on the surface this filter will appear blacker with a given concentration of material than would appear with a cellulose filter. In fact, if the light reflectance method is used to determine suspended matter collected, the sample on glass fibre would appear to be about 2.5 times more dense than an equal amount on cellulose.

One serious disadvantage of glass fibre filters is that they are rather fragile and extreme care should be taken when using them for gravimetric work to ensure that fibres do not break away in handling, particularly after being clamped into a sampling apparatus. In all cases, as with cellulose filters, the background contamination from the filter should be checked before sampling if subsequent sensitive analytical work is to be carried out.

One advantage of glass fibre filters is that they can withstand higher temperatures (up to 800° C) than cellulose (250 to 300° C) and vaporization of the carbonaceous content of the sample may be an important means of improving analytical resolution.

Membrane and Nuclepore filters

Membrane and Nuclepore filters collect the particles predominantly on the surface and this makes them ideally suited for microscopic work, for which glass fibre and cellulose filters are unsuited.

Membrane filters are mostly prepared from one of several cellulose esters and the pore size of the filter can be controlled by the manufacturing process. These filters may be used to trap particles down to far less than 0.1 μm in diameter. One of the main drawbacks of membrane filters is that they have a high resistance to air flow and tend to be brittle if flexed. In use it is therefore advisable to support them by means of a porous carbon plate or very fine stainless steel mesh.

Table 2.4 Comparative efficiency of filters at 5 cm s⁻¹ face velocity [13]

Analytical filter	Nominal pore size (μm)	Pressure drop (mm H$_2$O)	Efficiency for particles of 0.03 μm	Efficiency for particles of 0.3 μm
Nuclepore	0.50	701.0	0.987	0.993
	0.80	698.0	0.946	0.619
	1.00	127.0	0.868	0.522
	2.00	56.0	0.433	0.283
	5.0	35.0	0.184	0.144
	8.0	65.0	0.057	0.101
Membrane				
VUFS – Synthesia	0.30	903.0	0.998	0.991
VUFS – Synthesia	0.8	553.0	0.999	0.987
HA – Millipore	0.45	418.0	0.999	0.988
AA – Millipore	0.8	260.0	0.999	0.973
Acropore–Gelman	0.8	51.0	0.993	0.970
OH – Millipore	1.5	607.0	0.872	0.511
SS – Millipore	3.0	97.0	0.999	0.912
AUFS – Synthesia	1.4	28.0	0.992	0.932
RUFS – Synthesia	2.4	27.0	0.945	0.792
PUFS – Synthesia	7.0	15.0	0.949	0.688
OS – Millipore	10.0	28.0	0.634	0.312
Fibre				
AGF – Gelman		688.0	0.999	0.990
PF – 41 – Whatman		48.0	0.589	0.224

Nuclepore filters are polycarbonate sheet which has been irradiated by neutrons and etched to give pores of constant dimensions. These bear certain simi-
larities to membrane filters and the properties of both have been extensively
investigated by Spurny and co-workers [12–14]. The mechanism of filtration of
suspended particles is a product of several processes including impaction and
diffusion precipitation and electrostatic attraction. Hence, efficiency is a complex function of many variables, including pore size, particle size, face velocity
and filter loading. Some comparative efficiency data appear in Table 2.4.
Although some workers have reported low filtration efficiencies at high face velocities [15], this appears to be by no means a general finding and no clear
recommendation on this point can be made. In general, the smaller the pore size
of the filter, the greater the efficiency but the lower the rate of flow readily
achieved [14, 16]. Efficiency increases with filter loading and hence for long
sampling periods a more effective collection of particulates will be achieved [13,
17]. Since particles passing the filter represent a loss in analytical accuracy, care
should be exercised in selection of a filter. In general, membrane and Nuclepore
filters are of higher efficiency than cellulose and glass fibre, and of the former
the membrane filter is generally the better. For near-quantitative collection of
particles under almost all conditions, the use of membrane filters of pore size

$\leqslant 0.6\,\mu m$ is recommended, whilst those requiring more comprehensive data upon collection efficiencies are referred to the experimental studies of Spurny *et al.* [14] and Lui and Lee [16].

Membrane filters can be made transparent for microscopic work by the addition of a few drops of immersion oil or other oils with a refractive index of 1.56. In addition, samples may be prepared for electron microscopy by taking a small piece of the exposed filter and floating it, sample uppermost, on a small quantity of acetone. The cellulose ester material will dissolve leaving a thin film holding the dust. An uncoated electron microscope specimen grid may then be inserted carefully under the film by the use of forceps. When the film is lifted away from the container the acetone on the grid will evaporate leaving the specimen ready for examination.

Membrane and Nuclepore filters have a very low ash content and low levels of background impurities and are thus ideally suited for some of the more sophisticated analytical work.

2.2.2 Determination of total particulate pollutant concentrations

There are two main methods used in Europe for determining the concentrations of total particulate matter collected on filter papers. These are direct gravimetric determination of collected particles, and measurement of the soiling of a filter paper by the dimunition of light reflectance from it. In the U.S.A. smoke density is also measured by transmitted light and is referred to in terms of coefficient of haze (COH) units. Useful comparisons of techniques have been reported by Dalager [18] and Kretzschmar [19]. In the latter work, a good correlation was found between the results obtained by a gravimetric method, an integrating nephelometer (*vide infra*) and light reflectance, although the three methods do not give identical measurements of particulate concentrations.

2.2.2.1 Light reflectance method

In this method the darkness of the stained filters is assessed using a reflectometer. This instrument consists of a light source and a photosensitive element mounted together in a measuring head. In the model used in the U.K. the light beam passes through a circular hole in the centre of the photo-sensitive element. The measuring head fits snugly into a masking disc which restricts the illumination to an 0.5 inch diameter hole through which the stained filter is examined. Light reflected back from the filter paper falls on the sensitive surface of the photosensitive element and the current generated is measured by means of a milliammeter. The darker the stain the lower the intensity of the reflected light.

This method is usually calibrated by running two sampling lines in parallel;

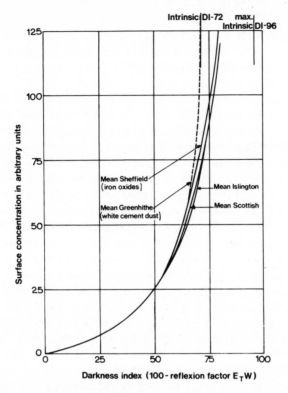

Fig. 2.10 British smoke calibration curves. E.E.L. reflectometer; Whatman No. 1 filter paper, 1 inch diameter.

one operating at the standard air flow of approximately $70 \, ft^3$ $(2 \, m^3)$ per day and a second line operating at an air flow sufficiently high to enable the amount of collected particulate matter to be weighed. Calibrations have been carried out in different regions where there was a predominance of particulate emissions from sources, such as cement works and steel works, which could impart a different colour to the particulate matter. The calibration curves are shown in Fig. 2.10. The light reflectance method is suitable for most routine air pollution work but too much reliance should not be placed on its accuracy. For example, in areas where there is a high percentage of crystalline matter this could affect the amount of light reflected back to the photo-sensitive element. Some doubt has also been cast on the reliability of the calibration because of suspicions that owing to the bad design of the inlet orifice and the use of higher sampling rates, the calibration instrument could sample a particle size distribution different from that of the standard instrument.

2.2.2.2 Gravimetric techniques

Because of doubts on the reliability of the light reflectance method for determining the amount of particulate matter sampled, the International Standards Organization is considering the possibilities of using a gravimetric technique as the standard method for the determination of particulate concentrations.

This technique is self-explanatory in that a sample of particulate material is collected at a sampling rate sufficiently high for a reliable weighing of the particulates to be carried out. The higher sampling rate, however, brings in added complications in that a totally different size range of particles may be collected. When considering sampling systems, importance should be given to the design of the inlet orifice to ensure that the size fraction of the sampled particles is known. In some routine instruments air is drawn into the system through an inverted funnel. This may effectively prevent rain being drawn into the instrument but it is of too poor a design to enable an accurate particle size distribution to be reliably sampled.

When considering inlet orifice design, attention should be paid to work by Davies [20] on the correct design of dust sampling instruments. One instrument which would appear to fulfil the requirements is that designed by the Landesanstalt für Immissions und Bodennutzsschutz in Essen [21]. In this unit, air is drawn into the instrument through a sharp edged, parallel-sided orifice. One advantage of this instrument is that the filter paper is situated inside the orifice at the top, thereby eliminating the possible deposition of particles on the tubing. Despite a poor inlet design the Hi—Vol sampler (see Chapter 1) is widely used for the collection of large particulate samples for gravimetric analysis.

When examining instruments for long period sampling of particulates, care should be taken to ensure that there is an accurate method of measuring the air flow through the instrument. Normally, after a few hours operation, the dust loading on filters increases their resistance to flow causing a subsequent drop in the air flow. In all cases, gas meters or other accurate flow monitoring devices, should be incorporated. Pressure drop methods of estimating air flows can cause serious errors and should be avoided.

Before weighing the particulate matter collected on a filter, it should be equilibrated to constant weight in a controlled temperature and humidity environment. It should be noted that some filter materials, especially membrane filters, are themselves substantially sensitive to humidity variations.

2.2.2.3 Other filter paper devices

8 Port sampling instrument

Strictly speaking this is not merely a filter-using device but is rather a change-over device for converting the simple smoke and SO_2 instrument into a form suitable for semi-automatic use.

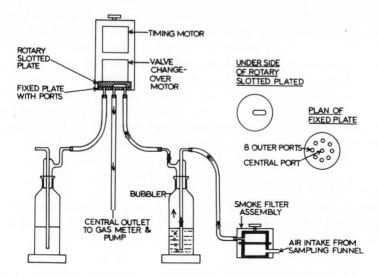

Fig. 2.11   Schematic diagram of a semi-automatic sampler [10].

In practice, when using the simple instrument, difficulty may be found especially at week-ends in arranging to change filter papers and Drechsel bottles daily and this instrument has been designed to eliminate the difficulty. A diagrammatic sketch of the unit is shown in Fig. 2.11. The timing motor shown selects a number of predetermined periods (e.g. 8 × 24 h periods, or 8 × 3 h periods) when it will automatically change over sampling lines by switching on the valve change over motor. This motor moves the upper slotted disc through 45° (1/8 of a revolution) which connects the air intake in sequence to a number of standard daily instruments for predetermined periods. At the end of the sampling sequence, say once per week, the Drechsel bottles and filter papers are removed and analysed. This is probably one of the simplest automatic devices in use in air pollution analysis.

Automatic smoke instruments

Quite frequently in air pollution work there is a necessity for short period samples which cannot easily be obtained with simple instruments operated manually; the problem may be the study either of the diurnal variation of suspended particulate matter or how the variations in particulate concentrations near busy streets can be related to traffic flows. Such sampling can be achieved by the use of an automatic tape sampler. This instrument (see also Chapter 1) uses filter paper in the form of a reel. The filter strip is held in a 1 inch diameter clamp which can be released automatically by means of a solenoid operated by a timing motor. The timing motor period may be varied as desired but normal times of operation are for periods of 1, 3, or 24 h. Air is drawn through the filter clamp for a predetermined period after which the solenoid automatically raises the clamp. At the same time a second motor drives the take-up spool through one complete revolution thus winding up the filter reel and exposing fresh filter under the clamp. The solenoid is then de-energized, clamping the filter paper again. After exposure the strip of filter paper is removed for evaluation by light reflectance methods.

β-Radiation monitors

Another method which has been used to monitor the concentration of particulate material in the atmosphere involves the use of the attenuation of intensity of β-radiation passing through solid material.

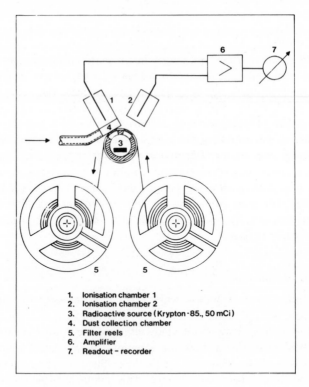

1. Ionisation chamber 1
2. Ionisation chamber 2
3. Radioactive source (Krypton-85., 50 mCi)
4. Dust collection chamber
5. Filter reels
6. Amplifier
7. Readout – recorder

Fig. 2.12   β-Radiation monitor.

The instrument is basically an automatic smoke recorder in which ambient air is drawn through a high efficiency glass-fibre filter. This filter is in the form of a reel which can be automatically moved forwards at intervals. A diagram of the basic system is shown in Fig. 2.12 showing the radioactive krypton-85 source mounted opposite two ionization chambers. The radiation is measured as ionization current in chamber 2 where it is partly absorbed by the chamber and the clean filter. In chamber 1 there is further attenuation caused by the dust on the filter. The two chambers have opposite polarity and a voltage drop proportional to the amount of dust is produced across the resistance connected to both chambers. The output, which is proportional to the amount of smoke on the filter, is amplified and recorded continuously on a chart or other type of recorder.

As already mentioned, the instrument samples continuously and the filter tape is moved forward automatically at predetermined intervals. As new filter tape appears before ionization chamber 1 this causes a voltage drop which is compensated by an automatic zero balance. The current from ionization chamber 2 is used as a reference current for the current from chamber 1 which continues to decrease as the dust concentration builds up on the tape. This eliminates variations which could be caused by variations in the thickness of the filter. Some deviations can occur where there is a high proportion of heavier elements but allowance can be made for them.

Short-term portable sampling

A major constraint on air pollution work in the field occurs when sampling needs to be carried out in locations where mains power is either not available or not convenient for short periods. Lightweight portable, battery operated equipment [22] may be used for such work. Today there is a plentiful supply of small battery driven pumps but before they can be used in air pollution work they require meters for monitoring gas flows and in the field this is not always convenient. One unit which may be used is a small positive displacement suction pump, for example, a converted toy steam engine running in an oil bath and driven by a 6 V electric motor. The number of revolutions of the piston is monitored using a digital mechanical counter driven by the piston or the pump. Before use the units may be calibrated against wet gas meters.

2.2.2.4 Mass monitors

Another technique which has also recently appeared is the continuous microbalance. In this instrument air is drawn through a chamber containing a small electrostatic precipitator which precipitates particles from the air onto the surface of an oscillating crystal. This crystal is one of a matched pair and the precipitated particles cause its osillatory frequency to change. This change is proportional to the weight of precipitated material. One drawback to this method, and also to the preceding β-radiation method, is that both are affected by moisture. In addition the mass monitor has shown that weights of collected material can change if some of the material collected is volatile. This is not strictly a disadvantage of the method but rather gives some indication of the sensitivity of the method, because such slight changes would not be noticeable with simpler instruments. A detection limit of $1\,\mu g$ is claimed. An example of the loss of volatile matter was the collection of a sample of cigarette smoke on the crystal followed by continued operation of the monitor on clean filtered air. After about $4\,h$ of operation on filtered air the effective weight recorded by the crystal dropped by approximately 17%, probably caused jointly by the evaporation of moisture and some of the tarry matter. A similar test carried out after sampling normal atmospheric pollution produced no such drop in weight.

One disadvantage of the mass monitor is that periodically the crystal has to be removed for cleaning but this difficulty can be reduced by sampling intermittently over long periods. When using the more sophisticated techniques described in the last section provision may have to be made to record other parameters such as humidity if high accuracy is required.

2.2.3 Cascade impactors

The size distributions of particles collected on filters may be determined microscopically, but this is very laborious, and larger particles tend to obscure the smaller, making counting difficult. The cascade impactor shown schematically in Fig. 2.13, separates particles into fractions on the basis of their aerodynamic impaction properties (see also Chapter 1).

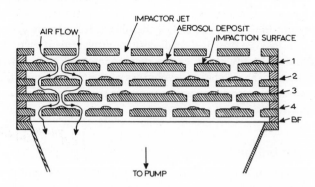

Fig. 2.13   Schematic representation of a cascade impactor [27].

There are several types of cascade impactors in use, probably the best known being the Andersen sampler [23, 24]. This instrument consists of as many as 7 stages, backed by a membrane filter. Each stage contains accurately drilled holes (360 in each of the top 3 stages and 400 in each of the remaining 4), and the stages are arranged so that beneath each hole there is a solid section of plate which is part of the next stage. The holes in each successive stage are smaller than those in the preceding, and as air is drawn through the instrument at a constant rate of $1 \, \text{ft}^3 \, \text{min}^{-1}$, the effective velocity through each stage increases. Large particles of high inertia are aerodynamically impacted on the backing plate at the first stage, and progressively smaller particles are impacted at each stage as velocities increase. The theory of impaction processes is given by Butcher and Charlson [25], and although other cascade impactors are different in form and appearance (see Chapter 1), the basic principle is the same.

Because the aerodynamic impaction properties of particles are dependent upon several factors including density and shape and are also affected by the possibility of a particle being hollow, microscopic examination of particles collected on impactor plates will not show a high uniformity of physical size. If such examination is to be performed, sampling should not be extended over too long a period, or too large a particulate load will be collected, as particles appear only as small colonies under each hole in the preceding stage. If the sampled particles will later be examined by electron microscope techniques, it is possible to grind a circular groove in the impaction plates to fit either an electron microscope sample grid or a small palladium-covered copper disc to enable the collected sample to be transferred to such instruments without the tedious operation normally required for sample preparation.

Another impactor in use is the high volume fractionating cascade impactor. This is a four stage instrument employing 12 inch diameter plates, each with 300 jets [26]. Impaction of particles is onto glass-fibre filters or aluminium sheets which are themselves perforated, and this is a useful instrument for collection of larger samples. Details of the particle size fractions collected are given in Chapter 1. It suffers from several defects, however. It is based on the Hi–Vol sampler and no positive measurement of air flow is made; the design of air intake leaves room for improvement, and if glass-fibre filters are used, extreme caution must be exercised to prevent fragments of filter breaking away.

Measurements of particle size distribution are extremely valuable, as they give an indication of the source of given particulate materials, and they allow determination of the proportion of particles in the respirable size range, these being the greatest hazard to health. Caution should, however, be exercised in the use of impactors and the interpretation of results, as distorted size distributions, attributed to particle bounce from dry impaction surfaces, have been reported [27]. Such errors may be avoided by the use of a virtual impactor [28] which

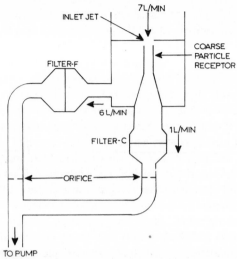

Fig. 2.14 Schematic view of a dichotomous sampler containing a single stage virtual impactor. Coarse particles are collected on filter C and fine particles on F [28].

divides particles into 0 to $2\,\mu$m (respirable) and 2 to $10\,\mu$m size ranges. In this instrument (Fig. 2.14), the large particles impact into a void, and both size ranges are collected, each on a separate filter.

2.2.4 Light scattering techniques

2.2.4.1 The integrating nephelometer

Another method for analysing airborne particles is the light-scattering method. This method is becoming increasingly used because of the employment of automatic monitoring networks throughout the world and it is now the standard method for sampling suspended particulate matter in Japan. One of the drawbacks of this system is that it suffers from interference by high humidity. A number of designs are available [29] but the basic principles are fairly similar. One, called an integrating nephelometer is shown diagrammatically in Fig. 2.15 [30]. The theory of such instruments is described by Butcher and Charlson [25].

In this instrument (Fig. 2.15) the photomultiplier 'looks' down the centre of the main body of the instrument through holes in the discs forming the collimator and light trap and the cone of observation is defined by the first and fourth discs. The light source illuminates the sample air in the centre section of the pipe and any material causing light scatter will be seen by the photomultiplier.

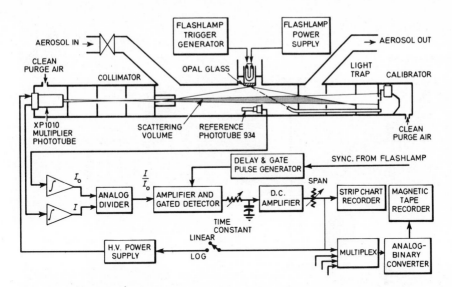

Fig. 2.15   Diagrammatic sketch of an integrating nephelometer [30].

The unit illustrated has provision for a rapid field calibration. The normal air supply is sealed off and clean purging air is introduced through orifices and fills the instrument with particle-free air with a sea-level Rayleigh scattering coefficient of $2.8 \times 10^{-5}\,\text{m}^{-1}$ at approximately 460 nm. The upper scale calibration is provided by a white surface which is illuminated by light from the flash lamp through a flexible light-pipe. A solenoid-actuated shutter uncovers a 1 mm hole in the end of the light trap through which the photomultiplier views this white surface. The signal thus introduced into the photomultiplier is about equal to that produced by the scattering of Freon-12; this is usually set at about half scale on the lowest range of the instrument which is 0 to $10^{-4}\,\text{m}^{-1}$. These two points – i.e. clean air and a half scale signal – provide a check of calibration in the field in less than 1 min.

The papers by both Garland [29] and Ahlquist and Charlson [30] give details of methods used to eliminate the drift in sensitivity in the circuitry, caused mainly by the flashtube. The unit described by the latter workers has been designed for mobile work to measure, among other things, horizontal profiles of visibility in a city.

2.2.4.2 Aerosol particle counters

Other optical instruments of importance in the air pollution field are aerosol particle counters such as that manufactured by Royco Instruments. These

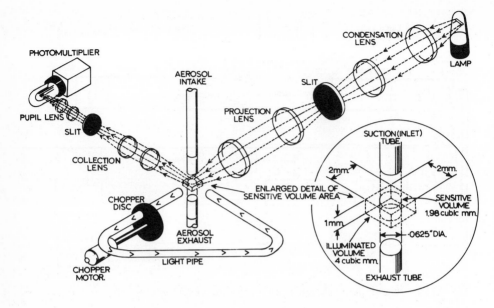

Fig. 2.16   The optical sensor of an aerosol particle counter.

instruments are of particular use in clean rooms etc. where limitations are laid down not only on total dust loading but also on size ranges. A diagrammatic sketch of the optical sensor of such an instrument is shown in Fig. 2.16. The particle sensor is a right-angle scatter, dark field optical system as can be seen. A sampling pump draws air through the optical viewing cell where it is intensely illuminated by a projecting lens system, a slit and a lamp. The sample air stream is viewed through a collection lens, a slit and a photomultiplier which are positioned at 90^0 to the main projection axis. Thus a small volume in the centre of the sensor head is optically defined by the lamp. When there is no particle passing through this volume, the photomultiplier sees a dark field. If a particle is present, light is scattered into the photomultiplier during the time that the particle is passing through the illuminated volume.

The amount of scattered light is a function of the particle diameter, and the photomultiplier produces an output pulse train where the height of each pulse is dependent upon the particle diameter. An electronic pulse height analyser is used to determine the number of particles that should be classified into each size range [25].

Because the projection lamp can vary in intensity and the photomultiplier gain can vary with temperature the instrument is provided with a built-in calibrator. When this is used a portion of the light from the lamp passes through the projection lens system, through the mechanical chopper, through a light pipe to the cell and photomultiplier. The calibrator simulates pulses of light from actual particles. The energy in the pulse train formed by the chopper is integrated and measured with a front panel calibration meter. The amplification of the photo-multiplier is adjusted with a front panel control until the calibration voltmeter indicates the proper values.

2.2.5 The directional sampler

When field investigations of pollution levels are being carried out, one of the most common complaints to be investigated is of excessive pollution from a given direction affecting a particular area. With the simpler instruments described in this chapter it is very difficult to make such an assessment because of the extreme variability of meteorological conditions. The problem of comparing pollution levels can be solved by the use of an instrument in which sampling is controlled by wind direction; this makes it possible to give more meaningful pollution comparisons. The directional sampler performs such a function. A wind vane and anemometer, apart from giving a record of wind speed and direction, also control the operation of sampling equipment, and this is of course applicable to the sampling of gaseous pollutants as well as to suspended particulates.

The meteorological vane consists of a very low torque potentiometer of conducting plastic with a vane fixed to an extension to the potentiometer shaft and a stabilized voltage fed across the potentiometer. The anemometer is basically a small d.c. tachogenerator, calibrated in a wind tunnel, with anemometer cups fitted to the shaft. The outputs from both meteorological instruments are fed to a control box which contains three pairs of gate circuits. These circuits can be adjusted in such a way that the unit may be preset to switch on external instruments when the voltage outputs and consequently the angles of the vane are between certain preset values. In this way pollution may be measured automatically when the wind is blowing from given directions. Up to three different sectors may be selected.

When the wind speeds are low, however, directions tend to be extremely variable and directional sampling tends to be unsatisfactory. Under these conditions a limit switch is fitted to the anemometer output which eliminates directional sampling when the wind speed drops below a preset but variable value. If required, a socket may be provided for sampling to be carried out separately under these calm conditions.

In one survey, such an instrument was used to examine pollution entering an airport. Two such instruments were set up, one at the western end of the airport and one at the eastern end. The comparative values of smoke and SO_2 obtained at these two sites are shown in Table 2.5.

Table 2.5 Pollution levels at airport (μg m^{-3})

Western site		Eastern site	
Smoke	Sulphur dioxide	Smoke	Sulphur dioxide
From the west		From the airport	
16	152	15	59
From the airport		From the east	
26	91	29	208

This gives some idea of the complication of air pollution work. In easterly winds, for example, the mean concentration of SO$_2$ entering the airport was 208 μg m^{-3} and the level of SO$_2$ leaving the airport at the western site was 91 μg m^{-3}. Since the airport itself was emitting little SO$_2$ a large percentage of this 91 μg m^{-3} must have come into the airport from the east. At the western site the mean level of SO$_2$ entering the airport was 152 μg m^{-3}, less than that which came in from the east. This would be expected as the airport is less heavily built up on the western side. Since little SO$_2$ is emitted at the airport, the results serve to illustrate the dilution of SO$_2$ in passing over the airport. In westerly winds concentrations decreased from 152 to 59 μg m^{-3}, whilst in easterly winds they dropped from 208 to 91 μg m^{-3}, in both cases a decrease of approximately 100 μg m^{-3}. The drop in smoke concentrations is less marked, suggesting some emission of particles at the airport.

2.3 Dustfall sampling

2.3.1 Introduction

A method for estimating dust deposition from stacks was given in an earlier section and it is important to know the pattern of deposition when attempting to measure it. The problems of such measurements are two-fold: firstly, if it is required to monitor general deposition in an area it is important to know how to avoid siting gauges at positions where the results could be seriously affected by one particular source; and secondly if it is required to monitor deposition from a particular source the general deposition pattern must be known.

Deposit gauges to monitor the deposition of particulate matter are probably the earliest instruments used for air pollution measurement. Very little basic research has ever gone into the design of such gauges however, and most of the designs leave much to be desired. For routine work they all fall into the same broad category, namely a bowl or cylinder with a horizontal, upward facing, collecting surface. Their major drawback is that the very presence of the gauge introduces disturbance into the falling pattern of particles, and it is conceivable that the results obtained may bear no relationship to true deposition figures in the area. In addition, it must be pointed out that the roughness of a particular area will affect the amount of deposition. Even in an ideal, open, grass-covered site the rate of deposition could change from day to day, all other things being equal, merely because of differences in the length of the grass. A concrete surface will also present a different collection efficiency to dust. Despite these drawbacks, deposit gauges still have their uses, but in recent years greater attention is being paid to collecting dust samples with the aim of identifying and isolating nuisance sources, rather than for the routine collection of data for statistical purposes.

When looking at dust deposition it must be remembered that when dust particles fall, only the very largest and densest fall at an angle even approaching that of the vertical. To illustrate this point further, some examples are given of angles of fall of particles of two different densities under wind speeds of 5 miles h^{-1} (2.2 m s^{-1}) (taken as average) and also 10 miles h^{-1} (4.5 m s^{-1}) (Table 2.6).

Table 2.6   Angle of fall of particles

Particle size (μm)	Density	Angle of fall 5 mile h^{-1}	(Degrees to horizontal) 10 mile h^{-1}
1000	1	61	42
800	1	54	34
600	1	45	27
400	1	37	20
200	1	20	10
100	1	8	4
50	1	2	1
1000	2	75	61
800	2	70	54
600	2	64	45
400	2	56	37
200	2	35	20
100	2	15	8
50	2	4	2

It can be seen that the heavier particles will fall close to the stacks, or sources, and if a gauge is sited to monitor general dustfall in an area well away from local sources, only the smaller particles will be collected in anything but higher wind speeds. Most of the particles which are being collected will be arriving at a very shallow angle to the horizontal, and under these conditions it is doubtful whether a horizontal collecting surface is really ideal for the purpose. However, provided it is realized that deposits collected in horizontal deposit gauges represent only the particulate matter collected in the configuration offered by the gauge, and that such results may only represent a very approximate index of the dust loading in the neighbourhood, then it is acceptable to use them with reservation.

2.3.2 Designs of national deposit gauges

There are, in various parts of the world, standard deposit gauges of different designs. As has already been mentioned, the very presence of the gauge affects the pattern of dust fall, an effect well illustrated by wind tunnel work by Pestel [31] who tested various types of gauges in a wind tunnel and using an artificial smoke tracer discovered the flow patterns generated round the gauges shown in Fig. 2.17.

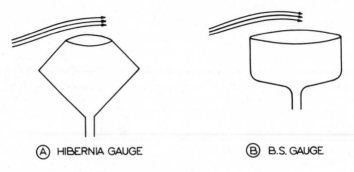

A HIBERNIA GAUGE   B B.S. GAUGE

Fig. 2.17 Flow visualization over gauges [31].

It is not proposed to go into the detailed differences of all the different designs but to select three major types of design: (a) the British Standard gauge; (b) the French Standard gauge; and (c) the Norwegian N.I.L.U. Standard gauge, since they broadly represent the main design differences.

2.3.2.1 The British Standard deposit gauge

Details of the specifications of this gauge are given in B.S. 1747: 1951 [32] and a sketch is shown in Chapter 1. The gauge has been in use for many years but, because of doubts as to the value of the results obtained from it, its use is now becoming less common. It is now mainly used in areas where there are already in existence data collected over a number of years. The instrument consists of a glass or plastic collecting bowl into which dust falls. Dust washed down by rain falls into a large collecting bottle under the bowl. In periods of dry weather, however, the collected dust could remain in the top collecting bowl for long periods and there is a suspicion that in periods of high winds this dust could be blown out of the gauge because of the shallow nature of the collecting bowl. (The dimensions are approximately 315 mm diameter; with a depth of 105 mm near the circumference, increasing to 180 mm near the centre. These dimensions give a ratio of 1.75 : 1 for diameter versus depth). As far as is known no checks have ever been carried out on this possible defect.

At the end of one month the collecting bowl is washed down carefully with a measured volume of water to wash any dust in the bowl into the collecting bottle below. The analysis of the samples is very simple indeed and is usually restricted to the determination of the total amount of water collected and its pH. This latter measurement is, however, of importance because recently there has been increased world-wide interest in the acidity of rainfall in projects dealing with the long-range drift of pollution. Routine analysis is restricted to the determination of the total weights of undissolved solids and the total weights of dissolved substances, such as chlorides, etc. These are carried out after extraneous matter such as twigs, leaves and dead insects have been removed from the samples. If the gauges are situated in regions, such as near steelworks, where there may be specific emissions, then the total weights of, say, iron collected should also be determined.

One main disadvantage of the standard deposit gauge is that, because of the extended exposure times, their ability to provide any directional resolution is limited. This is a serious disadvantage today because there is increased interest in using deposit gauges as 'troubleshooters' in identifying sources of a nuisance rather than for the routine collection of data over long periods.

2.3.2.2 French Standard deposit gauge (Ref. NF, X43 − 006(1972))

This gauge, shown in Fig. 2.18(a) is a development from the earlier British Standard deposit gauge and incorporates a modification added to prevent dust being blown out. Firstly the angles of the lower half of the collecting bowl are much steeper than the British gauge and this should result in less dust resting in the collecting bowl and being in danger of being blown out again in high winds. In order to reduce the danger even further, an inverted, truncated conical section has been added to the top and this should reduce this possible fault. It is not known whether the French authorities have any data comparing their improved gauge with the original British unit but it should be less liable to loss of sample.

2.3.2.3 Norwegian N.I.L.U. deposit gauge

This is the design which has been adopted in Norway and it is shown in Fig. 2.18(b). This gauge has been shown as typical of the simpler, cylindrical types of

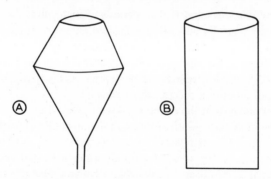

Fig. 2.18   Deposit gauges. (a) French Standard gauge; (b) N.I.L.U. gauge.

deposit gauges, such as the Bergerhoff (Germany) or the U.S.A., A.S.T.M. Dust Collector (Ref D.1739-70). In all these designs the ratio of the diameter of the orifice to the depth is of the order 1 : 2 and this has been done in an attempt to reduce loss of sample by being blown out again. The top of the cylinder of the N.I.L.U. gauge is shaped and bevelled at an angle of 45° downwards and out-wards. In order to make doubly certain of reducing the loss of sample in high winds, it has also been suggested that enough water should be placed in the gauge at the start of the exposure period to ensure that there is liquid in the gauge throughout the exposure period thereby preventing dust from blowing out. The amount will be determined by local climatic conditions. The other problems which arise with dust samples in liquids for long periods is the development of algae in the container and also the danger that the water could freeze in cold weather. Copper sulphate has been the material used to inhibit the growth of algae, but some workers favour a 5% initial concentration of 2-methoxyethanol which is not only an effective bactericide and algicide, but will also prevent frost damage. The 2-methoxyethanol may also be easily removed by evaporation (b.p. 124° C) prior to analysis.

Care should be taken in siting deposit gauges: there should be no object within 3 m of the gauge because such obstructions can affect the wind flow in the vicin-ity of the gauge. When gauges are sited in the vicinity of tall buildings, trees, etc., the angular height of such objects from the collecting surface should be less than 30°. Siting in the proximity of sources of dust should be avoided unless this is the primary reason for the tests. In order to avoid ground effects and to reduce interference from dust re-entrained into the air from the ground, collecting sur-faces should be at a height of at least 1.5 m.

2.3.3 Short-term surveys

In order to obtain useful data on dust deposition it is necessary to carry out measurements for long periods using horizontal deposit gauges. Quite frequently, however, complaints are received of a dust nuisance which either comes from a particular direction or which occurs intermittently. If such determinations were made over periods of one month, the true effect of the nuisance would not be found because of variations in wind direction.

In the past, many methods have been suggested for carrying out short-term surveys using jam jars or Petri dishes because of the high cost of the standard gauge. Both of these containers are too small to be really effective and the Petri dish suffers from the added disadvantage that it is too shallow, so that the dust collected can easily be blown out again. Others have suggested the use of greased plates. In practice the greased plate is too effective a collector to give results comparable with actual values and in addition the grease can complicate subsequent examination of the sample. If organic solvents are used to dissolve away the grease, attack on the atmospheric deposits may occur, especially if they contain a high percentage of carbonaceous matter.

Cheap polythene containers, such as washing-up bowls, provide an extremely simple compromise. There may, however, be objections to these because of the wide collecting surface in relation to the depth, (diameter 0.4 m; depth 0.15 m). There is, however, a large variety of polythene containers available and one of suitable dimensions can easily be found. If exposure is limited to periods of light winds the objection to the possibility of dust being blown out may not apply. Such polythene containers have been successfully tried in a number of field surveys for locating the source of a dust nuisance. The bowls are sited downwind of a suspected source for periods of one working day. Restricting the length of exposure to a period of steady wind direction increases the directional resolution of the sampling. If sufficient deposit is not collected in the space of one working day with bowls such as these, which have a diameter of approximately 0.4 m, it may be assumed that the dust problem is minimal. For such tests it is advisable to choose times when the meteorological conditions can be expected to be constant during exposure. If possible, sampling should not take place in wind speeds in excess of 15 miles h^{-1} (6.7 m s^{-1}), otherwise the dust may be blown out of the gauges. It is also advisable to avoid exposure during periods of rain because rain affects the deposition of solid material and can cause agglomeration, which makes subsequent identification difficult. After collection the dust should be carefully brushed out of the bowl and weighed prior to subsequent identification or analysis.

From such short period tests it is possible, knowing the time of exposure and the area of the collecting surface, to estimate the rate of deposition in mg m^{-2} day^{-1}, which is the standard unit used for quoting deposition results.

Because of the short period of exposure, however, such values should be treated with caution. Although it may not be required in detail, it is useful to determine the approximate size range of the particles collected because, if the wind speed is known, approximate estimates can then be made of the likely distance over which the particles have travelled and this can assist the location of a source. This distance of travel will also depend on the height of emission, but once the material has been identified, all possible local sources can be checked to determine which one is likely to be the potential one after taking into account the different heights of emission and the likely distances over which the respective emissions will travel under the existing meteorological conditions.

Short-term deposition problems may be tackled in two ways. If the problem is simply to identify what appears to be a single source causing a nuisance, this can usually be solved by the use of one or two deposit bowls. If, on the other hand, the problem is to try to estimate the dust loading and major emission sources in a particular area, then a more extensive survey will be required. These will be discussed in turn.

2.3.3.1 Single bowl surveys

When complaints of dust nuisance are received in an area the first important point to be determined is the wind direction in which the nuisance occurs. When this is known one or two bowls can be sited in the affected area during these wind directions. The bowls should be sited at a height of about 1.5 m above the ground, and, where possible, there should not be an obstruction within 3 m. If there are tall buildings or trees in the neighbourhood, the angle between the line from the collecting surface of the bowl to such obstructions and the ground should be less than 30°.

The bowls should be placed in position early in the morning, when conditions are suitable, and exposed for a period of about 8 h. They should then be collected and the sample carefully brushed out and weighed. Subsequently, the sample may be analysed as required. If it is not possible to isolate the main offending source, it may be necessary to repeat the tests when the wind is blowing from other directions, changing the sites of the bowls such that they remain downwind of the suspected source.

2.3.3.2 Larger surveys

Where a major source of nuisance exists in an area, it will sometimes be useful to determine the effective dust loading in the area. For such work to be effective, it is important that dust from the source should be readily distinguishable microscopically or analytically, from general background dust in the area.

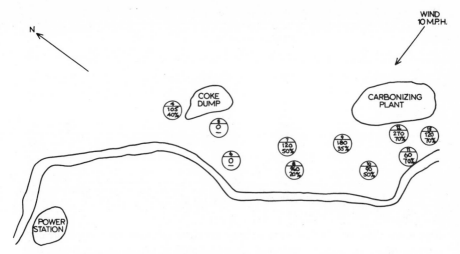

THE CIRCLES ARE CENTRED ON THE DEPOSIT BOWL SITES. THE NUMBERS ON THE CIRCLES REPRESENT—

TOP   – IDENTIFICATION NUMBER OF THE SITE
MIDDLE – RATE OF DEPOSIT IN mg m^{-2} day^{-1}
BOTTOM – % COKE IN THE DEPOSIT

Fig. 2.19   Illustration of a short-term survey using a number of deposit bowls.

Probably the best way of describing this type of survey is to consider a typical project carried out in an industrial area where there was one major souce of dust emission, namely a carbonizing plant. Ten sampling bowls were located in the area as shown in Fig. 2.19 which also gives the site numbers and the percentage amount of carbonized coal collected in each bowl. The method of operation was similar to that explained in the preceding section: when the meteorological conditions were suitable the bowls were set out in the morning and collected in the evening. After collection, the dust was weighed and, the surface area of the collecting bowls being known, the rate of deposition was estimated. The results of one typical test are shown in Table 2.7.

As would be expected the proportion of carbonized coal from the carbonizing plant decreased with distance from the plant, but increased again at sites 7 and 4, which were affected by a nearby coke storage area. Another interesting result which emerged from this one day survey was the one from site 8, which showed a very high rate of deposition. Microscopic determination of the deposit indicated that the sample did not come from the carbonizing plant but from the nearby chimney of an oil fired installation, which was not only very low but was emitting large quantities of smoke: both these factors affected the deposit at this site. This one test illustrates the technique and shows what may be obtained from even one day's work. It must be realized, however, that to obtain reliable

Table 2.7   Results of short-term survey

Site no.	Rate of deposition (mg m^{-2} day^{-1})	% of carbonized coal
15	120	70
12	270	70
11	60	70
10	90	50
9	180	35
8	960	20*
7	120	50
6	0	–
4	105	40

* At this site 80% of the deposit consisted of cenospheres from an oil fired chimney.

data more than one day's results would have to be obtained for them to be conclusive. In a case such as this it is also advisable to obtain results from other directions, for example when the wind was blowing from the direction of the power station.

This illustration also serves to demonstrate that it is not advisable to make assumptions of expected pollution levels in an area, without being supported by actual measurements. This was demonstrated by the exceptionally heavy deposits found at site 8 in the last example.

2.3.4 British Standard directional deposit gauge

As explained in an earlier section the early British standard deposit gauge has limitations, particularly when it is required to be used to locate a potential source of nuisance in an area. In order to improve the directional resolution for the determination of deposited matter for work in the vicinity of power stations, the Central Electricity Research Laboratory designed a directional deposit gauge [33]. A diagram of this gauge is shown in Chapter 1. The unit consists of four plastic tubes which are mounted as shown in the illustration with the vertical slots facing outwards at right angles to its neighbour. Dust trapped in each of the tubes is collected in a removable container at the base. After collection, a suspension of the dust is placed in a water-filled glass cell, and a measure of the dust loading (the 10 day percentage obscuration) estimated by the amount of obscuration of a beam of light passing through the cell. The unit is fully described in B.S. 1747: Part 5 [34]. The method advised in the British Standard which was designed to measure dinginess has limitations when applied to the identification of the source of a nuisance in a complex industrial area. Basically the B.S. method of analysing the deposit could only be applied where there is one source in an area, for example, a power station in open country. In such a case, a very large percentage of the total dust will come from the one source and that dust will have approximately reproducible composition and density. If however, such a directional deposit gauge were installed in an industrial area, the density of the dusts collected can vary considerably (this will be illustrated in a later section on density separation) and this could affect the results obtained by the method of light obscuration mentioned earlier, because some of the dusts would tend to settle out much more rapidly than the less dense dusts. Secondly, if there is more than one source in the area, a simple obscuration test will not give useful data on the contribution from the different sources and some form of more sophisticated analysis will have to be carried out, such as microscopic examination.

When discussing the deposit gauges with horizontal collecting surfaces in an earlier section, it was mentioned that only very large particles fall out at angles approaching the vertical. Most airborne dust in the size range which will be found away from the immediate vicinity of emission sources will be falling at a fairly shallow angle to the horizontal and it could be argued that in such cases the vertical collecting surface of the directional deposit gauge should prove a more efficient collector than the horizontal collecting surfaces of the earlier gauges. However, the reverse should be the case with heavier particles.

When the directional deposit gauge was first tried in the field some comparisons were carried out between the two instruments by siting them close together to monitor dust deposition in certain types of areas;
(a) In an area containing a cement works;
(b) In a typical industrial area;
(c) Near a granite quarry where frequent blasting operations were carried out. An approximate particle sizing of the deposits collected showed the following ranges:

    (a) Cement works area   50– 100 μm;
    (b) Industrial area   200– 500 μm;
    (c) Granite quarry   100–1000 μm.

In these tests, in the three areas, the ratios of the weights collected in the directional deposit gauge to the weights of deposits collected in the older type of horizontal gauge were as in Table 2.8.

From the results in Table 2.8 there would appear to be some positive confirmation of the earlier statement that the smaller dust size fractions are collected preferentially by the directional deposit gauge.

When the directional deposit gauge is installed to monitor emissions from one major source the orifice should be sited in such a way that it is pointing directly to the source. However, in more congested areas with multiple sources, it is not possible to install the gauge ideally and the position arises where dust from

Table 2.8

Area	Ratio directional gauge to B.S. horizontal gauge	No. of monthly readings averaged
(a) Cement works	2.49 : 1	25
(b) Industrial area	0.65 : 1	12
(c) Granite quarry	0.70 : 1	5

particular sources is arriving at the site from a direction between the openings of two adjoining collectors in the gauge. In such cases the dust will collect in the two collectors and a large part of the directional resolution is lost. In practice the position appears to be more complex and dust tends to appear in collectors sited at angles pointing in the opposite direction. This is probably caused by turbulence generated by the configuration of the gauges. No evidence is available to date on this aspect but some preliminary tracer work in a wind tunnel appears to confirm this theory. Methods of analysing dusts collected in directional deposit gauges are discussed in a later section.

The points raised in this section have attempted to show that the pattern of dust deposition, particularly in a complex industrial area, can be very involved and it is worthwhile taking great care, not only in the choice of the correct instrument, but also in siting, and the analytical techniques which are used, to obtain the best results in this field of air pollution research. It is surprising just how little effective work has been done recently to improve our ability to understand dust deposition problems.

2.4 Physical techniques for classification of particulates

2.4.1 Density gradient separation

One of the simplest techniques in the analysis of dust samples is to make use of differences in density which occur throughout the whole range of atmospheric particles. The method is applicable to dusts above $10\,\mu m$ in diameter and with densities up to $5\,\mathrm{g\,cm^{-3}}$. In this technique dust samples are inserted into a tube containing two density gradient liquids mixed in such a way that the density of the liquid in the tube decreases almost linearly with height. When the tubes are centrifuged the dust separates out into its respective density bands and these bands may, if desired, be physically removed one by one and analysed separately.

Table 2.9   Properties of density gradient liquids

Liquid	Density ($g\,ml^{-1}$ at 20° C)	Vapour pressure (mm Hg at 20° C)
Bromoform	2.89	5
Tetrabromoethane	2.96	< 1
Di-iodomethane	3.32	1.25
Bromobenzene	1.52	5.5
Di-n-butyl phthalate	1.05	–
Acetone	0.79	285

2.4.1.1 Density gradient liquids

There are a number of suitable liquids which can be used for this work and some of them are shown in Table 2.9. In addition, Clerici solutions (50 : 50 thallium malonate and thallium formate in water) extend the range to the majority of materials likely to be encountered.

Dibutyl phthalate has been found very suitable as a diluent as it has a low vapour pressure similar to that of tetrabromoethane and lower than that of di-iodomethane so that, when it is used as a diluent the resulting composite liquid is stable with respect to density over relatively long periods. When accurate control of the density of a composite liquid is not essential, or when an inherently stable liquid is not required, acetone may be conveniently substituted as a diluent.

The three heavy liquids are sensitive to light and heat and the breakdown products produce a red to dark red discoloration. The liquids should therefore be stored in the dark, in darkened containers.

2.4.1.2 Recovery and cleaning of liquids

When dibutyl phthalate is used as a diluent there is no simple method for the recovery of the heavy liquids, most of which are expensive. However, di-iodomethane may be recovered from dibutyl phthalate by placing the liquids in a suitable separating funnel and rapidly freezing the liquid using dry ice, or preferably liquid nitrogen (− 193° C) if available, or alternatively a mixture of ice, with brine, acetone or alcohol. After freezing, the solid is allowed to melt at room temperature; on melting the resulting liquid begins to separate out into its two component liquids. Repeated freezing and thawing (say 3 to 5 times) eventually leads to the separation of about 90% of the original di-iodomethane which forms a clear-cut layer in the bottom of the funnel. This process has the advantage of removing any free iodine in the di-iodomethane into the phthalate so that the pure di-iodomethane is recovered as a light straw coloured liquid. If acetone is used, this may be completely removed by streaming water through the mixture. The purified heavy liquid may then be separated off from the excess water and shaken with calcium chloride to remove the final traces of water.

Another way to clean the liquids is to use small quantities of Fuller's earth. The earth should be vigorously shaken or stirred with the liquid for a few minutes followed by filtration. This technique is simple to apply and in addition to decolorizing the liquid also removes traces of water and therefore may be used as an alternative method of drying a washed purified liquid.

2.4.1.3 Preparation of the gradient

One of the earlier methods of preparing a density gradient, described by Peters [35], was to take a number of glass tubes (12 by 1/4 inch) with one end sealed and, using a mixture of bromoform and bromobenzene, to make up the columns as follows: the bottom inch of the tube was filled with 100% bromoform, the next inch with 90% bromoform/10% bromobenzene, the next inch with an 80/20 mixture and so on until the last inch contained 100% bromobenzene. The columns were allowed to stand for at least 6 h to allow diffusion to occur between the density boundaries. The dust samples were then inserted and allowed to stand for about 24 h to allow the dust to settle out to its correct density level.

For air pollution work the long delay in preparation was considered too long for practical purposes, and there was some indication, that, while there might be some mixing near the density boundaries, the gradient tended to be more stepwise than linear. In such a case where the dusts contained flat, or plate-like particles, these could settle out on a density boundary which was not representative of their true densities. The third, and most serious criticism of this technique was that, where dust contained carbonaceous matter, most of the liquids suitable for density gradient work would attack the deposits dissolving out some of the tarry matter and ruining the identification.

One of the most simple methods of preparing a density gradient, although control of the actual gradient is less certain, is to take a test tube half-filled with the heavier liquid and then fill the upper half carefully with the lighter liquid. If the tube is held between the thumb and forefinger and is rhythmically shaken in a pendulum fashion the formation of the gradient can be followed by the behaviour of the striae in the liquid. This is a useful technique for rapid preliminary examination.

A technique used for routine work is that described by Muller and Burton [36] in which the entire operation of making up the gradient and centrifuging the sample could be done in 30 min, thus allowing much less time for the attack of the organic materials in the sample.

The unit used for the preparation of the columns consists of a small glass filter funnel with the stem inserted into a small separating funnel through an airtight bung. This is the mixer unit and it is completed by the addition of an intermittently operated electromagnetic stirrer. This unit is energized twice per s to actuate a number of glass coated pins contained in the separating funnel. When a 50 ml centrifuge tube is being used the mixer is loaded by placing 25 ml of the denser liquid in the separating funnel and 25 ml of the lighter liquid in the filter funnel. The lower separating funnel stopcock is adjusted to give a rate of discharge of liquid such that the centrifuge tube is filled in about 5 to 10 min. At the same time the intermittent magnet is switched on. As the denser liquid begins to drip from the separating funnel it is replaced by an equal volume of the less dense liquid from the upper funnel. This small amount of less dense liquid will be mixed with the dense liquid by the agitation of the pins, thereby slightly reducing the density. In this way the density of the emerging liquid continuously decreases and, when the centrifuge tube is filled, the resultant density will be found to decrease almost linearly. It is not necessary to start with pure liquids. With pure di-iodomethane and acetone the density difference will be from 3.32 to 0.79 g ml^{-1}. However both liquids can be diluted before preparing the gradient and thus expand any intermediate section of the gradient. For example a gradient tube varying, say from a density of 2.50 to 2.00 g ml^{-1} may be prepared by diluting the two liquids suitably before starting.

After the columns have been prepared, weighed samples of dust are added to the liquid. Prior to centrifuging, the columns are made up in pairs with the weights carefully adjusted and balanced beforehand to avoid damage to the centrifuge. For most work centrifuging at 3 000 rev min^{-1} for 10–15 min is adequate to separate the dust into its respective density bands.

In the case of naturally occurring mineral ores it will be found that the size of the particles can affect the density separation. Fig. 2.20, illustrative of work at Warren Spring Laboratory, demonstrates this point. The tube on the left contains mineral dust in the size fraction 10 to 14 mesh, the second 14 to 25; the third 25 to 32; the fourth 52 to 72 and the last 72 to 100. The mineral ore is separated out into its component fractions only when it is ground to around 100 mesh. When such liberation sizes are obtained the resolution of the density technique is very satisfactory. However, in air pollution work, as has already been mentioned, a great many atmospheric particles are hollow and this can complicate density resolution, as in the case of silica. Naturally occuring silica will settle out at a particular density level, but silica from a particular combustion process may be in the form of hollow spheres and these will settle out at other levels. Far from being a disadvantage this can be very useful. A similar separation occurs with carbonaceous matter; if there is a coal-fired source in the vicinity, samples of dust will be found containing particles varying from almost completely

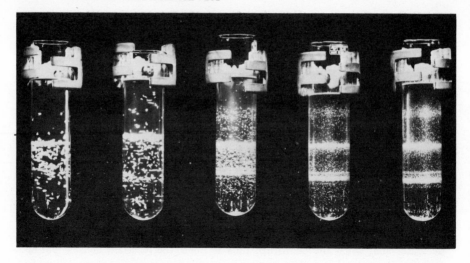

Fig. 2.20   The effect of particle mesh size upon the resolution of density gradient fractionation.

unburnt coal through all the stages of carbonization to coke, resulting in a very broad band of carbonaceous matter.

When further detailed analysis is required, it is possible to centrifuge the sample in a co-axial centrifuge tube, shown in Fig. 2.21. When the sample has been centrifuged the splitter unit is placed over the ground glass joint in the top of the co-axial centrifuge tube. If pure di-iodomethane or mercury is carefully introduced into the side arm of the co-axial tube it will gradually force the entire density column up into the splitter unit. As each dust band passes through the two flat glass discs the splitter unit should be rotated through 180°, at which point the dust band suspended in the upper part of the splitter unit will drip through the stopcock into a collecting jar. The density liquid may then be removed by washing in acetone, and the dust simply evaporated to dryness for subsequent analysis. Each of the density bands may be collected separately if the process is repeated.

Some examples of this technique applied to air pollution work are now shown. Fig. 2.22 illustrative of work at Warren Spring Laboratory, shows dust from two types of power stations; on the left, dust from an old-fashioned coal fired power station, and on the right, the deposit from a more modern pulverized coal power station. The broad band of carbonaceous matter may be seen in the left-hand sample.

Fig. 2.23, also from work at Warren Spring Laboratory, shows three of the four samples taken from a directional deposit gauge exposed in an industrial area. In the left-hand tube which faced open country, the only visible band

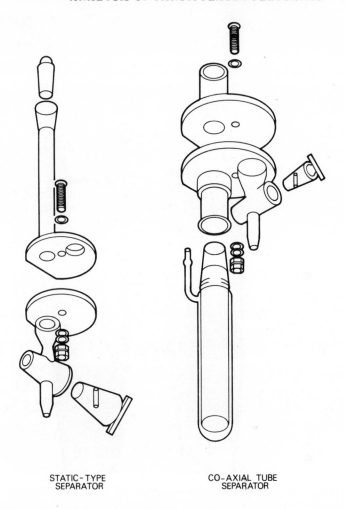

Fig. 2.21 Co-axial centrifuge tube and splitter unit.

appears at a density of about $2.6\,\mathrm{g\,cm^{-3}}$ and this is composed of wind-blown silica. The centre tube shows a much stronger band at this density, together with traces of carbonaceous deposits just about half-way up the tube. The main deposit near the base of the tube is silicon carbide (density $3.2\,\mathrm{g\,cm^{-3}}$) and the deposit at the bottom of the tube, alumina (density $4.0\,\mathrm{g\,cm^{-3}}$). Both of these deposits come from a nearby factory making grinding wheels. This sector of the directional deposit gauge was on the boundary of the factory. The right-hand tube, containing the sample taken directly in line with the factory, shows much heavier deposits of the three main types of dust shown in the centre tube. In

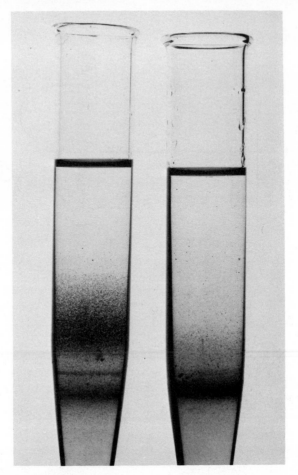

Fig. 2.22 Density gradient fractionation of dusts from two power stations.

addition there are traces of cenospheres from an oil fired chimney near the top of the tube. These spheres, although of carbonaceous matter, are hollow and therefore tend to float if undamaged.

2.4.2 Dispersion staining

A large percentage of atmospheric particulates are carbonaceous and, therefore, opaque; however there is a sufficient abundance of transparent particulates in the atmosphere to justify the use of this technique [37, 38] which is an optical method of imparting a colour to transparent substances to simplify

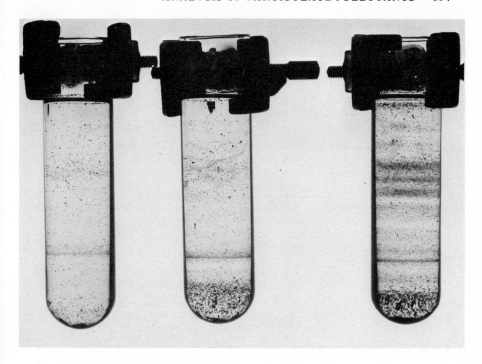

Fig. 2.23   Density gradient fractionation of dusts from a directional deposit gauge.

identification. It is of considerable value, for example, in determining the percentage of quartz, or other dangerous dusts, in air samples. The technique is based on the difference in the refractive index of transparent solid-particles and the liquid medium in which they are immersed. This difference in refractive index causes refraction to take place at the crystal boundaries; the greater the difference, the greater the refraction. A diagrammatic sketch, Fig. 2.24, shows the principle of the technique. In this sketch the refracted blue and red radiation are shown on the left and right respectively. After being focussed by the objective lens the rays pass to the back focal plane where either an annular or central stop is placed as shown in the illustration. Each of the stops gives coloured particle boundaries. The annular stop shows a colour containing those wavelengths near that at which both particle and liquid show a match in refractive index. The central stop shows colours complementary to those shown by the annular stop, i.e. light which has been highly refracted by the particle in that particular liquid medium.

With this technique, which uses white light, good centralized axial illumination is essential. The condenser lens should be accurately centred with respect to the objective and stopped down to give optimum colour development.

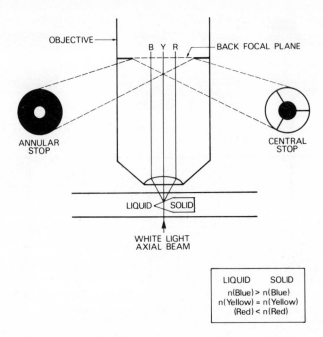

Fig. 2.24 Dispersion staining.

In the work reported in the literature [37, 38] a large number of atmospheric particles were examined and dispersion curves drawn for them. A typical series of such curves, extracted from these papers are shown in Fig. 2.25. If this figure is used to explain the technique, let it be assumed that the unknown material is lead nitrate. If crystals of this compound are immersed in a dispersion liquid of refractive index 1.777, i.e. where the two materials have a matching refractive index, the material will appear to be greenish in colour, and this will be found to have a wavelength of 486 nm. If, on the other hand, a liquid with a refractive index of 1.783 is used, the colour of the crystal will appear red, at 620 nm.

There are a number of uses for this technique in the air pollution field, for example, the determination of the percentage of quartz in a sample. In such an application a dust sample containing quartz dust is sized microscopically using standard illumination. If the preparation is then examined using a dispersion staining objective, the quartz particles may be made to appear coloured and it is then a matter of carrying out a particle size distribution analysis of the coloured particles to obtain the necessary details of the quartz content.

The technique has also been used in conjunction with samples from the directional deposit gauge. For example, in the illustration shown earlier in the vicinity

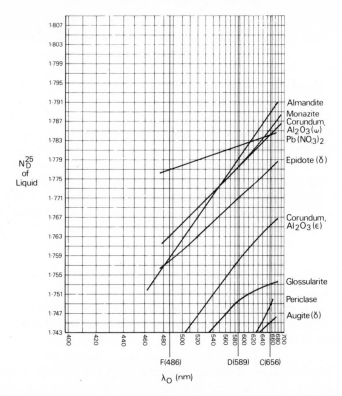

Fig. 2.25 Typical dispersion curves (from Brown and McCrone [37, 38]).

of a factory making grinding wheels the contribution from the factory consisted of mainly alumina and silicon carbide, the amounts of which can be determined by dispersion staining.

2.4.3 Microscopic techniques

The normal technique for the identification of atmospheric dusts is by their characteristic shape, structure and colour. This is a technique which can be learned only by experience, and some necessary background information is given in this section. Although the composition of dusts will change from area to area, it will usually be found that in urban areas a large percentage of dusts come from combustion sources. There are, of course, innumerable industrial processes emitting specific dusts, but it will initially be found preferable to concentrate on the identification of combustion products.

The composition of atmospheric dusts will be found to vary considerably at any one site under different wind directions. Because of such changes it is, therefore, important that satisfactory sampling be carried out and not to base dust identifications on dust swept up from flat surfaces such as window sills.

If samples are collected from chimney stacks then the proper sampling technique detailed in the relevant British Standard should be followed [39]. However, if it is desired to perform a preliminary examination of dusts from a particular source and provided care is used, this may be carried out using grits collected in chimney arresters. Such samples will tend to contain larger particles than will be found in the atmosphere. It is possible to build a small but effective dust collector, but it must be emphasized that this simple unit is for the purpose of collecting a preliminary sample and it should not be assumed that the dust collected is representative. Briefly the method is to draw a sample of dust from a chimney into a small deposition chamber by means of a suction pump. A filter paper should be inserted in the line to the pump to protect it from contamination by the dust. If the tube from the chimney is approximately 1/4 inch diameter the deposition chamber should have a diameter of about 6 inch. The chamber should have a removable base so that a Petri dish may be inserted. As air is drawn from the chimney into the chamber the reduction in gas velocity will cause particles to be deposited out and they will be collected in the Petri dish. The length of sampling will depend on the dust loading in the chimney and the size range of the particles.

After collection the dust samples may be removed from the Petri dish and mounted on microscope slides for examination or to be used as reference samples. Initially it may be found difficult to pick up the individual particles to transfer them to slides but this operation can be simplified by the use of a very sharply pointed needle which can be prepared from a short length of tungsten wire heated to red heat and then plunged into a dish containing sodium nitrite. During the subsequent reaction the tungsten will be oxidized and it is possible to burn the metal away to a very fine point. If this point is drawn across the fingers after it is cooled, a fine film of skin grease will be deposited on the needle which will be adequate to cause adhesion of individual particles.

2.4.3.1 Mounting samples

In the past it was common practice to mount samples in liquids such as Canada Balsam. Unfortunately, with atmospheric particles containing carbonaceous matter, it will be found that the mounting liquid will attack the particles and if the mounted sample is retained for any length of time the preparation will turn dark brown. However, a simple but effective method can be used to mount samples for reference purposes, but the method is not suitable for more sophisticated optical work such as dispersion staining. The method involves placing a drop of Durafix diluted in amyl acetate (30% Durafix to 70% amyl acetate by volume) on a microscope slide in the centre of the cover-slip ring. The droplet will spread out inside the ring and when the liquid becomes 'tacky', the dust sample should be carefully sprinkled over the slide. With experience it is possible to gauge the correct time to add the dust. If the operation is successful, the particulates will be held on top of a very thin film of Durafix after the liquid has evaporated. If the particles are placed on the slide too early they will become embedded in the liquid and the dried film of Durafix over the particles will make subsequent microscopic examination almost impossible.

If more sophisticated optical work is planned, some of the more modern thermoplastic mounting media may be used. A number of different grades are available, but for air pollution work it is suggested that a grade be used which is solid at room temperatures. For mounting, the medium is heated until it melts. After mounting, the preparation is allowed to cool and solidify. These thermoplastic media have a light yellowish tint but this does not affect identification.

2.4.3.2 Identification of dusts and reference library

There are a number of aids for identifying atmospheric particles such as two atlases of atmospheric dusts; one by the C.E.G.B. [40] and one by McCrone [41]. However, many beginners will find it difficult to identify these photomicrographs with actual dust samples under the microscope. Difficulties arise from the limited depth of focus of the microscope which tends to complicate identification for beginners, and from the difficulty of accurate colour reproduction. These problems may be circumvented by making up a reference library of dusts from known sources for comparison with samples for identification. The work involved in building up this library gives the beginner experience in recognizing the different types of dust and also the component parts of dust samples from particular sources. Initially the library should concentrate on the more common dusts in the area, and continue with less common samples as experience is gained.

2.4.3.3 Description of dusts from different combustion and industrial sources

Mineral matter in British coals can be divided into three main classes, i.e. shales and clays, sulphide minerals and carbonate minerals. In most coals the mineral matter will fuse at temperatures between 1200 and 1600° C under oxidizing conditions, but may fuse at lower temperatures under reducing conditions. Generally speaking, the flame or fuel-bed temperatures of small appliances will be below the fusion temperature. In larger installations, where the ash is heated beyond its fusion temperature while 'airborne', as in a pulverized fuel plant, it will tend to become spherical, whereas if it reaches the same temperature when resting on the fuel bed, it is unlikely to become spherical but will tend rather to be rounded. It is, therefore, possible to determine approximate fuel bed conditions from the physical characteristics of the ash particles.

Domestic fires

Typical fly ash from domestic sources consists of small unfused particles, mostly colourless or white but with some pale yellow material. In addition, there may be small black and red iron oxide particles. Generally the dust emitted from domestic sources will not be distinguishable from the constituents in typical dust samples because of the small quantity involved compared with other sources. Owing to the low flue gas flow in domestic chimneys, these particles emitted rarely exceed $10\,\mu m$.

Shell type boilers

This type of small, hand fired boiler is still extensively used in the U.K. in industry for steam raising. Although the fuel bed temperatures are in the range at which many coal ashes fuse, only a small percentage of the ash becomes completely spherical; the bulk of the particles are merely rounded. Coal or coked coal particles are also to be seen in deposits from such boilers. Deposits from this type of boiler are usually in the range $10–100\,\mu m$.

Stoker fired boilers

The carbonaceous matter from such boilers consists of irregularly shaped coke or partially coked coal particles. Most of the ash consists of spherical particles with a dull olive green or brown colouration; these are usually referred to as 'smoky' spheres. These 'smoky' spheres may be regarded as clearly indicative of stoker fired boilers. The colouration of the spheres is not strong, and the percentage of spheres in dust samples from this source tends to vary. Particulates from small economic boilers contain mainly semi-fused mineral matter, coke particles, and only a few spheres. The dust is similar in appearance to that from the large hand fired boilers. On the other hand, dusts from the larger boilers contain a large percentage of spheres which are mainly colourless, pale yellow or orange, together with the 'smoky' spheres. Puffy particles of partially carbonized coal are also found and these are generally larger than the ash particles with which they are associated. When ash from this type of boiler is compared with ash from a pulverized fuel boiler, it will be found to contain more coloured material.

Pulverized fuel boilers

Fuel burned in these boilers reaches maximum temperatures of $1500-1600°$ C. The fuel is burned in suspension, and, because of this, most of the mineral matter emitted emerges as fully spherical, colourless spheres. In addition, black magnetite spheres may be found. The magnetite is formed by the decomposition of the iron sulphide minerals in the coal. Upon heating, these minerals are oxidized, releasing SO_2 and leaving a black oxide of iron (Fe_3O_4) residue. As this occurs in suspension, the magnetite fuses into spheres. In pulverized fuel boilers, coloured spheres are rare and there is also usually a small amount of combustible matter, usually angular coke fragments or puffy coke coal.

Oil fired boilers

Since fuel oils have a low mineral content, the fly ash content from industrial boilers should contain little ash. However, the small amount of mineral matter there is, is usually more highly coloured than that of coal and consists of green, yellow, orange, red and black particles about $1-5\,\mu m$ in diameter. The green coloration of the ash, which is specific to oil firing, is due to nickel and vanadium compounds. This is not the best way of identifying a particular dust as coming from an oil fired source. The main deposit from such boilers consists of cenospheres with little ash and it is this combination which will point to the oil fired boiler as the culprit. The cenospheres vary in appearance, between a fine lace-like structure, similar to those from a strongly swelling coal, to a black matt surface without any apparent structure. In the case of oil fired boilers which are badly run, samples of shiny, dark green spheres around $50\,\mu m$ in diameter may also be found. These are almost completely unburnt oil.

The carbonization industry

Owing to conversion to natural gas, dust from this industry is not very common today. Coke particles produced by carbonization of a swelling coal are harder and denser than those found in normal fly ash and, on occasions, it is only by careful examination that the porous nature of the coke can be seen.

Coke ovens

Dust emissions from this process comprise two different types: particles of unburnt coal from the oven charging process, and swollen, rounded particles of partially carbonized coal. During discharging of the ovens and the subsequent quenching of the coke, angular fragments of coke and small ash particles appear. Dust emission during quenching is small.

Horizontal retorts

This method of coke production is now obsolete. Dust emission during the charging process consists of particles of puffy, swollen coal. A typical dust from a horizontal retort would contain small particles of coal at various stages of carbonization. Angular fragments of coke and occasional coke cenospheres can also be found.

Intermittent vertical retorts

This operation is similar to that for coke ovens and the dust produced is similar in appearance to that from coke ovens and horizontal retorts.

Continuous vertical retorts

This is a less dusty operation than other methods of carbonizing coal, because the coal passes through a seal before reaching the hot zone of the retort. The only dust produced by this technique is that arising from the coal and coke handling operations.

The cement industry

Cement is manufactured by heating limestone or chalk with a small percentage of clay. In the wet process the mixture is fed as a slurry into the top of a long inclined rotary kiln. During its passage down the kiln the mixture is heated to 1450° C. The sintered product is cooled and ground with a small amount of gypsum. Most of the dust produced in a cement works is removed by electrostatic precipitators. The dust consists of fine, colourless or pale-yellow particles with a small proportion of typical pulverized fuel fly ash. The dust tends to be hygroscopic and it helps to heat it to about 110° C for 1 h prior to examination. If this is inadequate and further segregation is required, this may be achieved by preparation of a slurry in acetone. Water should not be used for this purpose, because some of the dust is water soluble.

The ceramic industry

Raw materials are heated in kilns which can be fired by a number of methods. The general features of the emissions depend on the kiln temperature. In small bottle kilns, where the temperature of the fuel bed is around 1150° C, the major part of the dust consists of amorphous soot with small, rounded or semi-fused ash particles. Kilns fitted with mechanical stokers have higher fuel-bed temperatures and the ash emission contains semi-fused particles, but no spherical particles. However, the amount of dust emitted from large kilns is small, unless smoke is being emitted. Even so, the dust will not be emitted in quantities large enough for identification.

The other type of emission from this industry is caused by grinding and handling of raw materials. This emission, however, tends to be very localized and should not cause any nuisance outside the works area, except in high winds.

Iron and steel industry

Blast furnaces

These are used to produce basic iron, and the charge, consisting of iron ore, metallurgical coke and limestone is fed into the top of the furnace where the temperature reaches about 1500° C in the hottest zone. Most of the particulate emissions from this process are caused by 'hanging', i.e. where there is a sudden drop of part of the furnace charge, which causes the rapid ejection of large quantities of gas heavily laden with dust. The dust emitted has a size range up to 30 μm and consists of black, sharp-edged matt-surfaced iron particles, with similar amounts of coke, as well as pale-yellow and colourless limestone and slag. The iron particles, being denser than either coke or slag particles, settle out more quickly than the other constituents.

Sintering plant

This plant is usually associated with blast furnaces, and makes some use of dust collected by dedusting the blast furnace, which can contain up to 25% iron oxide. The dust is mixed with fine iron ore and coke breeze and burned to produce a sintered mass. The dust emission tends to be similar to that produced by blast furnaces.

Open hearth furnaces

This process converts pig iron or scrap iron into steel. The furnaces may be fired either by producer gas or heavy fuel oil, and the temperatures can reach 1800° C. The dust emitted consists mainly of iron oxides and minerals such as limestone which is used as a flux to form slag. The dust is yellowish brown in colour and is composed of black magnetite spheres and yellow, orange and colourless material from the flux and slag. The size range of the particles from this process is below 5 μm, and it is unlikely that dust from this process will settle readily.

Electric arc furnaces

These furnaces operate at high temperatures to produce alloy steels, and they create a fume of iron oxide with particles up to 5 μm. The bulk of the emission however, is in the form of ferric oxide fume with a size range below 1 μm, which does not have any distinguishable features.

Deseaming mills

In the deseaming mills the surface layer of scale is removed from the ingots by burning off with oxyacetylene torches. The emissions consist of black, shiny spheres of magnetite and also yellow fume. The magnetite spheres are usually between 10 and 20 μm but they can be as large as 100 μm in diameter. Owing to their high density, the larger spheres will not be expected to travel far, and only the smaller spheres are likely to be found at any distance from the mills.

Cupolae

These installations are used for melting cast iron or convertor iron, and the emission varies from a coarse grit of 100 μm to a fine fume of 0.5 μm. The coarse grit consists of angular, matt-surfaced, black iron fragments with slightly reddened, oxidized surfaces in places. Smaller quantities of yellow and orange translucent 'slag' material and coke are also found. Despite the high temperatures, the particles show no sign of rounding or sphere formation.

The sections dealt with here represent some of the major sources of dust which are likely to be encountered. However, in some areas, these dusts may be in the minority. In such a case it will be necessary to compile a list to cover such materials.

2.4.3.4 Dust identification table

An attempt has been made to summarize the different types of dust by arranging them as shown in Fig. 2.26. This figure has been arranged so that the dust may be identified by a process of successive elimination. The presence or absence of particular physical characteristics in dusts lead by stages to the final conclusion. Although dusts will be found which are not dealt with here, the illustration will assist in the identification of a large percentage of the dusts encountered.

2.4.4 Determination of asbestos

Recently, great concern has arisen over the hazards of exposure to asbestos dust. Three conditions, all of which may take many years to develop, have been

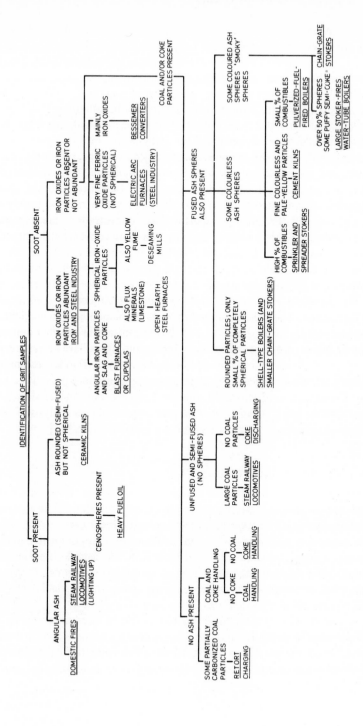

Fig. 2.26 Table for identification of dust samples.

identified in workers occupationally exposed. These are fibrosis of the lung, known as asbestosis; cancer of the lung consequent upon fibrosis; and mesothelioma, a rare malignant tumour of the pleura or peritoneum [42]. Two forms of asbestos have been used predominantly: crocidolite (blue asbestos) and chrysotile (white asbestos), but the former is not now in use due to the greater hazard associated with it, but will still exist in older installations.

Sampling asbestos dust is normally performed using membrane filters as these have a high collection efficiency and are suitable for the subsequent analytical procedures. Because of the chemical complexity of asbestos, no simple analysis is available, and optical techniques have been recommended for determination of asbestos in the occupational environment [43]. Detection is based upon the characteristic fibrous nature of asbestos particles which may be readily recognized and counted. Frequently, asbestos dust concentrations are expressed in terms of fibres per unit volume of air, although more conventional units of mass per unit volume are also in use. After collection on membrane filters, the fibres may be counted by light [43], or electron microscopy [44, 45], following dissolution or low temperature ashing of the filter material for the latter technique.

In the industrial atmosphere, such counting techniques are applicable, as is X-ray diffraction. The concentrations of asbestos encountered in ambient urban atmospheres are far lower, however, and the concentrations of other minerals are frequently higher. Samples collected on membrane filters have been successfully analysed by electron microscopy [44, 45]. An X-ray diffraction technique for urban samples, after collection by electrostatic precipitation, proved too insensitive for the detection of chrysotile asbestos however, even in the vicinity of an asbestos factory [46]. Another method used successfully in analysis of ambient air utilizes the i.r. absorption of chrysotile for quantification [47].

2.4.4.1 Membrane filter method [43]

This method is fully discussed in a leaflet issued by the Asbestosis Research Council and is designed specifically for measuring concentrations encountered occupationally in factories, workshops and constructional sites.

The method is restricted to a count of fibres of a length greater than 5μm and having a length/breadth ratio of at least $3:1$, The sample is collected on a membrane filter $(0.8 \mu$m pore size) fixed in an open filter holder. The length of sampling will be determined by the expected concentrations and some recommendations are given in Table 2.10.

When the sample has been taken the dust should be fixed immediately. There are several ways of doing this and one recommended method is to drip a solution of poly(methyl methacrylate) (perspex) in chloroform onto the filter, while clean air is being drawn through it. On return from field sampling sufficient

Table 2.10

Anticipated asbestos dust concentration (fibres ml^{-1})	Sample volume (ml)
< 2	10 000–20 000
2– 4	5 000–10 000
4–12	2 000– 5 000
> 12	Pro rata

triacetin (glycerol triacetate) should be placed on a microscope slide to give a circle the same diameter as the filter and to serve as a base for it. The filter should be carefully removed from the sampling head using forceps and placed dust-side uppermost on top of the triacetin. After about 3 min, place a clean cover-slip of the same diameter over the filter and press gently to remove air bubbles which may be trapped underneath.

The dust should be examined under transmitted light with phase contrast at a magnification of × 500. With an even distribution of fibres, counting can proceed by counting the fibres in random fields; the number of fields needed for a total count of 200 fibres should be noted. If the deposit is sparse or unevenly distributed then 100 fields are observed. It is suggested that a multi-stage method of choosing fields be used, i.e. 10 filter grids are selected scattered over the sample area and within each grid 10 random fields are counted, giving 100 fields in all.

The formula for estimating the fibre counts (i.e. fibres longer than $5 \mu m$ is:

$$\text{Concentration} = \frac{D^2}{d^2} \times \frac{N}{n} \times \frac{1}{V} \text{ fibres ml}^{-1}$$

where: D = diameter of entire dust deposit; d = diameter of field of view; N = number of fibres counted; n = number of fields examined; V = volume of sample (ml).

2.4.4.2 Infrared technique for ambient atmospheres

Concentrations of asbestos likely to be found in ambient air are very much lower than may be encountered in industry as already discussed. This rules out a counting technique because of the large number of fields which have to be examined, even to find one fibre. In addition, sampling carried out in the past has shown most asbestos fibres in ambient air to be in the region of $0.3 \mu m$ in length, which eliminates optical microscopy.

A method has been reported which makes use of the strong infrared absorption band at $2.72 \mu m$ possessed by chrysotile, the most commonly used of the asbestos minerals [47]. Atmospheric samples are collected on membrane filters and ashed at $450°$ C for about 1 h, during which process the material of the filter and any carbonaceous matter present in the sample is oxidized or volatized. Under these conditions decomposition of the chrysotile is barely detectable, and no correction to the calibration is necessary. The residue is incorporated in a potassium bromide disc and the amount of chrysotile present determined from the i.r. spectrum.

One possible defect of the method is that serpentine minerals also exhibit strong absorption bands at the same wavelength and it may be necessary to detect whether these minerals are present in the samples to avoid interference.

2.4.5 Determination of particle size distribution

In any estimation of the particle size distribution of a sample of atmospheric particulates it is advisable to take into account some measure of density as well as of size, particularly if the sample is taken from a general urban or industrial area where a wide range of densities can be expected to be found. Depending on the overall distribution of sizes it may be necessary to carry out an initial separation of the particles in an instrument such as the cascade impactor to avoid having large and small particles on the one slide, where the larger particles could obscure the small particles.

It is not possible to lay down hard and fast rules on sampling for a particle size distribution because the length of sampling will depend not only on the concentration of particulates in the area. If a cascade impactor is used it may be necessary to take the sample in two parts: (a) a shorter sampling period for the smaller size fraction and (b) a much longer period for the upper fraction of the dust, where the number of particles is likely to be less in comparison and thus a particle count would be less accurate. Whatever sampling system is used it should be ensured that the sample taken is meaningful and time should not be wasted, for example, analysing the size distribution of particles which have collected on surfaces like window sills under unspecified conditions, because samples taken from such surfaces can yield misleading results.

2.4.5.1 Sieve techniques

With coarser materials a size distribution can be carried out using a number of standard sieves. The range of sizes of sieves varies from country to country, but the sieves available in the U.K. to British Standard specifications have the mesh sizes shown in Table 2.11.

Size distribution by sieves is normally confined to the size determinations of larger dust and grit from industrial chimneys to check the efficiency of control techniques for dust abatement.

Table 2.11   British Standard sieve sizes

B.S. Sieve No.	Mesh size (μm)
16	1000
18	840
22	710
25	590
30	500
36	420
44	350
52	297
60	250
72	210
85	177
100	149
120	125
150	105
170	88
200	76
240	62
300	53

For smaller sizes U.S. standard sieves are available which extend the range down to:

270	53
325	44
400	37

2.4.5.2 Microscope techniques

For a full description of particle size distribution work by the use of the microscope the reader is referred to the work of Chamot and Mason [48]. This section will be confined to description of the usual method used in air pollution work.

Because of the irregular shaped particles normally found in air pollution work, one of the first problems which arises when carrying out determinations of size distribution is how the dimensions or diameter of such particles are to be expressed. The most frequently used method is known as 'Martin's' diameter. This is the horizontal, or west-to-east dimension of each particle which divides the projected area of the particle into two equal halves and this is illustrated diagrammatically in Fig. 2.27. This dimension is obtained by moving a sample on a microscope stage past the lens system which contains a calibrated scale in such a way that as each particle is bisected by the scale the dimensions are estimated. This may appear to be a very approximate method of measuring the size of irregular shaped particles, but provided a large number of particles is counted, it will be found to be sufficiently accurate. Another method is to use the circles on one of the special air pollution graticules discussed later and to estimate the

Fig. 2.27   Illustration of Martin's diameter.

size most closely approximating to the irregularly shaped particle being viewed. Depending on the irregularities in particle shapes the total count can vary from 200 to 2000.

Micrometers and graticules

Before sizing can be carried out it is necessary to calibrate any micrometers or graticules in use at each range of magnifications. This is done using a stage micrometer, which in its simplest form is an etched scale, 1 mm in length and divided into 100 equal divisions, i.e. each division is $10 \mu m$ in length. The eyepiece graticule may also be a simple etched scale which is calibrated against the stage micrometer. However, for air pollution work a number of special graticules have been designed to simplify these size determinations. These are shown in Fig. 2.28 (a–c). The Patterson graticule consists of nine rectangles and a series of numbered globes and circles of progressive diameters. In use, the graticule should be moved over a given particle until the circle which most closely represents the irregular dimension is found.

The Porton graticule has a wide range of applications and is of later design than the Patterson. The right-hand side of the rectangle is divided vertically as shown. The diameters of the globes and circles increase in a geometric progression of two.

The third most commonly used graticule is the B.S. graticule [49] which is also illustrated in Fig. 2.28 (c). The main rectangle of the grid is subdivided in such a way that the rectangles formed give two each of 1/4, 1/8, 1/16 and 1/32 and four of 1/64 of the total area of the grid. In addition, the diameters of the seven different sized circles are in the ratio:

Circle no.	Numerical value of diameter
1	1.00
2	1.41
3	2.00
4	2.83
5	4.00
6	5.66
7	8.00

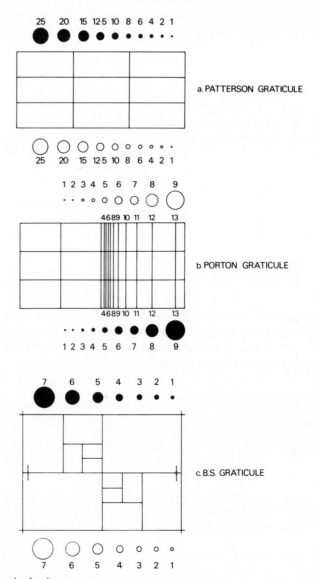

Fig. 2.28   Graticules in common use.

Counting is carried out by moving the slide preparation containing the particles past the eyepiece graticule to enable the respective particle diameters to be measured. The movement of the microscope stage may be done either mechanically or by setting up the preparation off-centre on a rotating stage and then rotating it. Where possible a combination of eyepiece and objective magnification should be chosen to give about 10 or 12 size classes. During the count, tallies should be kept of the numbers of particles in the different size classes and Table 2.12 gives a typical table of results of a count.

Table 2.12   Particle size count

Eyepiece graticule size class	Total no. of particles per class
1	8
2	70
3	130
4	105
5	90
6	72
7	42
8	21
9	12
10	2

The next stage in the preliminary calculations is to take the product of the number of particles in each size class (n), and (a) the diameter class, (b) the square of the diameter class, (c) the cube of the diameter class and (d) the fourth power of the diameter class as shown in Table 2.13.

Table 2.13   Preliminary calculations of size distribution

Eyepiece size Class (d)	n	nd	nd^2	nd^3	nd^4
1	8	8	8	8	8
2	70	140	280	560	1 120
3	130	390	1 170	3 510	10 530
4	105	420	1 680	6 720	26 880
5	90	450	2 250	11 250	56 250
6	72	432	2 592	15 552	93 312
7	42	294	2 058	14 406	100 842
8	21	168	1 344	10 752	86 016
9	12	108	972	8 748	78 732
10	2	20	200	2 000	20 000
Total	552	2 432	12 554	73 506	473 690

If the calibration of the eyepiece graticule is taken as $1.8\,\mu$m per division, then the following average values may be determined from the above calculations.

Mean particle diameter $= \Sigma nd/\Sigma n = 2430/552 \times 1.8\,\mu m = 7.9\,\mu m$

Area mean diameter $= \Sigma nd^3/\Sigma nd^2 = 73\,506/12\,554 \times 1.8\,\mu m = 10.5\,\mu m_1$

Mass mean diameter $= \Sigma nd^4/\Sigma nd^3 = 473\,690/73\,506 \times 1.8\,\mu m = 11.6\,\mu m$

Other calculations, such as cumulative percents by number, surface or mass, etc. may be made from the figures shown in Table 2.13. A discussion of the meaning of different expressions of particle diameter is given by Ledbetter [50].

As was mentioned earlier in this section, some correction should be made where there are obvious differences in density of the particles. In such cases the data shown in Table 2.12 should be duplicated for each density class. If this is not done the figure obtained for the mass mean diameter in particular will be seriously in error.

References

[1] Vangegrift A.E., Shannon L.J., Sallee E.E., Gorman P.G. and Park W.R. (1971). *J. Air Pollut. Control Assoc.* **21**, 321–8.
[2] Turner D.B. (1971). *Workbook of Atmospheric Dispersion Estimates*, U.S. Environmental Protection Agency, Publication AP-26, Research Triangle Park, N.C.
[3] Pasquill F. (1961). *Met. Magazine* **90**, 33–49.
[4] Nonhebel G. (1960). *J. Inst. Fuel* **33**, 479–511.
[5] Bosanquet C.H., Carey W.F. and Halton E.M. (1950). *Proc. Inst. Mech. Eng.* **162**, 355–67.
[6] H.M.S.O. (1956). *Handbook of Meteorological Instruments*.
[7] Beran D.W. and Hall F.F. Jr. (1973). Paper at *Second Joint Conference on Sensing of Environmental Pollutants*, I.S.A: Washington D.C.
[8] Parker J. (1969). Paper No. 149 at Clean Air Congress, Düsseldorf.
[9] Jones J.I.P. (1963). *Br. J. Appl. Phys.* **14**, 95–101.
[10] Warren Spring Laboratory (1966). *National Survey of Smoke and Sulphur Dioxide Instruction Manual*, Ministry of Technology: London.
[11] Whatman Leaflet AP-69, W. & R. Balston Ltd: Maidstone, Kent, England.
[12] Spurny K. and Lodge J.P. (1968). *Staub-Reinhalt. Luft*, **28**, 1–10.
[13] Spurny K.R., Lodge J.P., Frank E.R. and Sheesley D.C. (1969). *Environ. Sci. Technol.* **3**, 453–64.
[14] Spurny K.R., Lodge J.P., Frank E.R. and Sheesley D.C. (1969). *Environ. Sci. Technol.* **3**, 464–8.
[15] Seeley J.L. and Skogerboe R.K. (1974). *Analyt. Chem.* **46**, 415–21.
[16] Lui B.Y.H. and Lee K.W. (1976). *Environ. Sci. Technol.* **10**, 345–50.
[17] Biles B. and Ellison J.Mc.K. (1975). *Atmos. Environ.* **9**, 1030–2.
[18] Dalager S. (1975). *Atmos. Environ.* **9**, 687–91.
[19] Kretzschmar J.G. (1975). *Atmos. Environ.* **9**, 931–4.
[20] Davies C.N. (1954). *Dust is Dangerous*. Faber and Faber: London.
[21] Herpetz E. (1969). *Staub-Reinhalt. Luft*, **29**, 12–18.
[22] Barrett C.F. and Parker J. (1963). *Int. J. Air Water Pollution*, **7**, 995–8.

[23] Anderson A.A. (1958). *J. Bact.* **76**, 471–84.
[24] May K.R. (1964). *Appl. Bact.* **12**, 37–43.
[25] Butcher S.S. and Charlson R.J. (1972). *An Introduction to Air Chemistry.* Academic Press: London.
[26] Burton R.M., Howard J.N., Penley R.L., Ramsey P. and Clark T.A. (1973). *J. Air Pollut. Control Ass.* **23**, 277–81.
[27] Dzubay T.G., Hines L.E. and Stevens R.K. (1976). *Atmos. Environ.* **10**, 229–34.
[28] Dzubay T.G. and Stevens R.K. (1975). *Environ. Sci. Technol.* **9**, 663–8.
[29] Garland J.A. (1972). Proceedings, *Conference at Inst. Mech. Eng.*
[30] Ahlquist N.C. and Charlson R.J. (1968). *Environ. Sci. Technol.* **2**, 363–6.
[31] Pestel E. (1963). *Strömungstechnische Untersuchungen von Staubnieder-schlagmessgeräten*, Westdeutscher Verlag, Köln, Nr 1183.
[32] British Standards Institution, Deposit Gauges, B.S. 1747: Part 1, 1969.
[33] Lucas D.H. and Moore D.J. (1964). *Int. J. Air Water Pollut.* **8**, 441–53.
[34] British Standards Institution, Directional Dust Gauges, B.S. 1747: Part 5, 1972.
[35] Peters D.W.A. (1962). *The Examination of Soil with Particular Reference to the Density Gradient Test*, British Academy of Forensic Sciences Teaching Symposium No. **1**, 27–36, Sweet and Maxwell: London.
[36] Muller L.D. and Burton C.J. (1965). *The Heavy Liquid Density Gradient and its Application on Ore Dressing Mineralogy*, 8th Commonwealth Mining and Metallurgical Congress, Melbourne.
[37] Brown K.M. and McCrone W.C. (1963). *Microscope and Crystal Front*, **13**, 311–22.
[38] Brown K.M. and McCrone W.C. (1963). *Microscope and Crystal Front*, **14**, 39–54.
[39] British Standards Institution, Simplified Methods for the Measurement of Grit and Dust Emissions from Chimneys, B.S. 3405: 1961.
[40] Hamilton E.M. and Jarvis W.D. (1962). *Central Electricity Board Monograph No. RD/P/21*, London.
[41] McCrone W.C. (1967). *The Particle Atlas*, Ann Arbor Science Publ. Inc: Ann Arbor.
[42] Bruckman L. and Rubino R.A. (1975) *J. Air Pollut. Control Ass.* **25**, 1207–15.
[43] Asbestosis Research Council (1971). Tech. Note No. 1: *The Measurement of Airborne Asbestos by the Membrane Filter Method.*
[44] Holt P.F. and Young D.K. (1973). *Atmos. Environ.* **7**, 481–3.
[45] Nicholson W.J., Rohl A.N. and Ferrard E.F. (1970). *Proc. Second International Clean Air Congress*, 136–9, Washington D.C.
[46] Rickards A.L. and Badami D.C. (1971). *Nature* **234**, 93–4.
[47] Gadsden J.A., Parker J. and Smith W.L. (1970). *Atmos. Environ.* **4**, 667–70.
[48] Chamot E.M. and Mason C.W. (1959). *Handbook of Chemical Microscopy, Vol I*, 3rd Edn. Wiley, New York.
[49] British Standards Institution, Eyepiece and Screen Graticules for the Sizing of Particles, B.S. 3625: 1963.
[50] Ledbetter J.O. (1972). *Air Pollution, Part A: Analysis*, Dekker, New York.

CHAPTER THREE

Metal analysis

3.1 Introduction

Metal pollutants are emitted into the atmosphere from numerous sources including combustion of fossil fuels (including leaded petrol) metal smelters and alloy refineries, cement manufacturing plants and municipal incinerators [1]. Metals and metallic compounds exist in the atmosphere in three distinct physical forms: solid particulate matter, liquid droplets (mists) and vapours. The size range of airborne particulate matter is broad [1], and particles represent by far the most common form of metallic air pollution. Table 3.1 indicates the levels of metals in suspended particulate matter as measured by the U.S. National Air Sampling Network [2]. Figures such as these give an indication of levels likely to be encountered during ambient air monitoring and may thus be used to estimate the volume of air which needs to be sampled to allow collection of a quantity of a metal which is compatible with a proposed analytical technique.

3.2 Analysis of particulate matter

3.2.1 General sampling considerations

Airborne particles may be collected by impingers, electrostatic precipitators or filters. Filters are currently much favoured because of their ease of use and high efficiency of collection for small particles. Organic membrane filters are suitable for the collection of airborne particles as small as $0.03\,\mu$m in diameter, although efficiency is reduced with increasing pore size of the filter and with increasing particle velocity at the face of the filter and thus sampling rate [3–5]. Glass fibre filters, although not having such a high collection efficiency [6–8], allow

Table 3.1 Average and maximum concentrations of airborne pollutants measured at urban stations by the U.S. National air sampling network, 1964–1965 [2]

| Pollutant | Concentration (μg m^{-3})* | |
	Arithmetic average	Maximum
Total suspended particulate matter	105	1254
Antimony (Sb)	0.001	0.160
Arsenic (As)	0.02	
Beryllium (Be)	< 0.0005	0.010
Bismuth (Bi)	< 0.0005	0.064
Cadmium (Cd)	0.002	0.420
Chromium (Cr)	0.015	0.330
Cobalt (Co)	0.0005	0.060
Copper (Cu)	0.09	10.00
Iron (Fe)	1.58	22.00
Lead (Pb)	0.79	8.60
Manganese (Mn)	0.10	9.98
Molybdenum (Mo)	< 0.005	0.78
Nickel (Ni)	0.034	0.460
Tin (Sn)	0.02	0.50
Titanium (Ti)	0.04	1.10
Vanadium (V)	0.050	2.200
Zinc (Zn)	0.67	58.00

* Bi-weekly 24 h samples.

very fast passage of air and are used with high volume (Hi–Vol) samplers. Several other filter materials are also available.

An important consideration in selecting a filter medium, whatever the subsequent method of analysis, is the background metal content of the filter material. Table 3.2 shows levels of metallic impurities in various filter materials [9–11]. Dams et al. [10] evaluated a number of filter materials and impaction surfaces for their impurity content, flow properties, retention properties and tensile strength (where appropriate) and their general suitability for neutron activation analysis of aerosols. It was concluded that Whatman No. 41 cellulose was the preferred filter medium for neutron activation work, and Vogg and Haertel [12] have similarly recommended the use of cellulose paper filters because of the low levels and homogenous distribution of the metallic impurities in them. Zoller and Gordon [11], also using neutron activation analysis, report the levels of impurities in Millipore (cellulose ester membrane) and Delbag (polystyrene) filters. Continuous washing of some cellulose paper filters may be used to remove water-soluble impurities, without adverse effect upon the filtration properties [13].

Sampling of particulate matter from gas streams such as stack gases should be isokinetic, but for sampling relatively stagnant ambient air this is not essential, and generally not feasible as wind velocity and direction are seldom constant.

Table 3.2 Metallic impurities in filter materials (ng cm^{-2})

	Filter material					
Metal	Polystyrene (Delbag) [10]	Cellulose Ester (Millipore 0.45 µm) [10]	Cellulose Paper (Whatman No. 41) [10]	Glass Fibre [9]	Organic membrane [9]	Silver membrane [9]
Ag	< 2		2			
Al	20	10	12			
As				80		
Ba	< 500	< 100	< 100			
Be				40	0.3	200
Bi					< 1	
Ca	300	250	140			
Cd					5	
Co	0.2	< 1	0.1		0.02	
Cr	2	14	3	80	2	60
Cu	320	40	< 4	20	6	20
Fe	85	< 300	40	4 000	30	300
Hg	1	< 1	0.5			
Mg	< 1 500	< 200	< 80			
Mn	2	2	0.5	400	10	30
Mo					0.1	
Ni	< 25	< 50	< 10	< 80	1	100
Pb				800	8	200
Sb	1	3	0.15	30	0.1	
Sn				50	1	
Ti	70	5	10	800	2 000	200
V	< 0.6	0.09	< 0.03	30	0.1	
Zn	515	20	< 25	160 000	2	10

Passage of an air sample through a bent tube prior to collection of particles may lead to deposition and loss of particles by impaction at the bend. Also, in a bent tube different flow patterns for smaller and larger particles result from differing particle momenta and this can lead to uneven deposition onto a filter which will introduce error if only a segment of the filter is analysed. Since airborne particles are subject to gravitational settling, a probe pointing upwards will collect more particles than one which points downwards, but the effect is only significant when sampling relatively large particles (greater than 30 µm diameter). Particles may be lost in a sampler by electrostatic attraction to surfaces within the sampler, particularly glass and plastics. Ideally, therefore, particles should reach the collector as soon as possible after entering the sampling apparatus. Additionally, all sampling equipment should be rigorously tested for leaks before use if quantitative sampling is required.

Although some methods of analysis of particulate matter for metals require

Table 3.3   Effect of ignition method on metal recovery [16, 17]

Metal	Recovery (%)	
	Low temp. ashing	Muffle furnace at 550° C
Ba	97	99
Cd	92	53
Co	96	97
Cr	112	100
Cu	98	92
Mo	98	116
Mn	99	107
Ni	97	99
Pb	101	46
Sb	99	46
Sn	95	87
Ti	95	92
Zn	96	39

no pretreatment of the collected sample prior to analysis, for most analytical methods it is necessary to destroy organic matter associated with the sample and, as a dissolved sample is normally required, to render the sample soluble. Wet ashing with acids is frequently a suitable procedure [14, 15], as is dry ashing in a muffle furnace followed by dissolution in strong acids. Losses, assumed to be a result of volatilization, may occur with some metals under normal dry ashing conditions (550° C for 1 h) and Thompson and co-workers [16, 17] report a low temperature ashing procedure, for glass fibre and membrane filters, which uses an oxygen plasma and leads to much reduced losses (Table 3.3). Instruments for low temperature ashing are now commercially available. The presence of other interfering substances in a sample of atmospheric particulate matter was found to have little effect upon metal recovery when using the low temperature ashing procedure [16, 17].

Kometani et al. [14] advance the theory that apparent losses of metals during dry ashing result from the formation of insoluble metal silicates. They show that ashing of paper filters at 500° C, with or without prior wetting with sulphuric acid (H_2SO_4), gives a recovery of metals from atmospheric particulates comparable to that achieved by wet ashing with nitric-perchloric acids or low temperature ashing. The use of glass fibre filters, or of silica, glass or porcelain-glazed crucibles leads to a low recovery of metals.

Once prepared, both sample solutions and standard solutions of metals may not keep satisfactorily. Acid solutions may leach lead from some types of glass, soda glass in particular, and from some plastics (e.g. PVC). Additionally, both glass and plastics may adsorb a number of metal ions. Struempler [18] showed absorption of Ag, Pb, Cd, Zn and Ni onto borosilicate glass surfaces, which was prevented, for all but Ni, by acidification to pH 2. Polyethylene containers absorbed Ag, Pb and Ni, but not Cd nor Zn, and Ag adsorption could be prevented by acidification to pH 2. Cadmium is lost from aqueous solution during storage in glass containers, but only at alkaline pH values [19]. Dilute aqueous solutions of mercury have poor keeping properties, even in acidic solution [20, 21], but may be kept satisfactorily by addition of 5% v/v HNO_3 + 0.01% dichromate. Silver is also adsorbed by borosilicate glass, with slightly reduced adsorption at acid pH values [22], and storage is assisted by the addition of sodium thiosulphate [23]. Although acidification may assist the preservation of dilute solutions of metals, lengthy storage cannot be guaranteed. It is, therefore, advisable to prepare fresh dilute standards daily, and to investigate the keeping properties of sample solutions if they are not to be analysed immediately upon preparation.

Polypropylene containers have been shown to be serious sources of Cd and Zn contamination [18]. Lubricants in ground-glass joints and stopcocks may also lead to contamination and for trace analysis, PTFE stopcocks should be used wherever possible.

The sampling of mists from the atmosphere is a problem occasionally encountered. Silverman and Ege [24] report that liquid impingers, electrostatic precipitators, filter papers and cotton plugs moistened with glycerol have all been used for collection of chromic acid mists from air. A comparative evaluation of these techniques showed a thick (0.026 inch), absorbent filter paper to be the best collection medium with an efficiency under the conditions used of 99.9% [24].

3.2.2 Analytical methods involving no pretreatment of the sample

Analytical methods for metals and applications of these to the analysis of air pollutants will be described. For a full explanation of the theory and practice of a given technique, the reader is referred to more comprehensive texts on the subject. Although an exhaustive review of the literature regarding each technique is not attempted, examples of the applications of techniques are given, and full experimental details of at least one procedure are described.

3.2.2.1 X-ray fluorescence spectroscopy

A metal target bombarded with accelerated electrons emits X-rays. The emitted spectrum is in two parts: a continuous spectrum with a definite short wavelength limit (white radiation), and a number of sharp high intensity lines characteristic of the target element. Bombarding electrons knock atomic electrons from their orbitals, and when other electrons 'jump' from outer orbitals to fill the vacant orbital, they lose energy by emission of radiation in the X-ray wavelength region. These effects arise from inner orbitals, and valency-level electrons have little effect on the energetics of the process. Hence, the chemical state of the emitting atom has virtually no effect upon the emitted wavelength. If, instead of using electrons, X-rays are used to bombard the metal, absorption occurs with a secondary emission of X-rays (fluorescence), characteristic of the irradiated metal. In this case, the continuous emission spectrum is absent, but the secondary emission is less intense than the primary emission caused by electron bombardment. In addition to the secondary X-rays emitted, scatter of the incident primary radiation also occurs, as well as scattering involving a change in wavelength (Compton scattering) and diffraction by crystalline substances, which obeys the Bragg equation.

Using appropriate instrumentation, the intensity of the secondary emissions may, by comparison with a suitable standard, be used to give a quantitative measure of a metal. In a mixture of elements the secondary emission of one element may be absorbed by another and re-emitted at a different wavelength. Hence complex inter-element effects arise in which the emissions of some elements are enhanced whilst others are reduced, and calculation or comparison with appropriate standards is necessary for estimation of the elemental composition of the sample. Macdonald has given an excellent account of both practical and theoretical considerations in X-ray analysis [25].

Gilfrich *et al.* [26] have compared several X-ray fluorescence techniques for the analysis of airborne particulate matter collected on filters. They indicate that X-ray analysis is rapid, needing no sample pretreatment, non-destructive and applicable to elements irrespective of their position in the periodic table, including all elements beyond atomic number 11 (sodium). Numerous elements may be analysed simultaneously and results are comparable to those obtained using neutron activation analysis [27] and atomic absorption analysis [15, 26, 28, 29]. In a comparison of wavelength dispersive and energy dispersive techniques, multichannel wavelength dispersive instruments were found to be the most satisfactory [26]. Typical detection limits for X-ray fluorescence analysis of particulate material collected on filter papers, using a 100 s or 10 min analytical period, appear in Table 3.4.

Because of the small values for the total quantities of samples collected by air filtration (typically up to $1 \, mg \, cm^{-2}$) the metals are in a 'dilute' form and the inter-element effects which are encountered in bulk samples are not normally significant. Size effects may, however, be significant for larger particles, in general those greater than $5 \, \mu m$ in diameter [26]. For known particle size distributions the effect is calculable, but for normal air pollution samples this is not possible. For optimum precision, samples collected by fractionating impactors are necessary.

Two calibration procedures are most commonly used. Dilute standard solutions of the metals being analysed may be spotted onto filter discs and allowed

Table 3.4   Detection limits for X-ray fluorescence analysis [26, 28]

Metal	*Detection limit* (ng cm^{-2})		
	Energy dispersive; radioisotope sources [28]	*Wavelength dispersive; X-ray tube excitation (Rh Tube)* [26]	*Energy dispersive; X-ray tube excitation (W Tube-Ni foil)* [26]
Al		85	
As	100		
Ca	60	29	140
Co	80		
Cr	240		
Cu	50	49	
Fe	80	30	120
K		18	220
Mn	120		
Mo	30		
Ni	60		
Pb	110	260	110
Se		150	81
Sr	40		
Ti	50		
V	30	33	90
Zr	30		
Zn	30	51	110

Gilfrich *et al.* [26] used a 100 s count, and Rhodes *et al.* [28] used a 10 min count.

to dry leaving a thin film. Calibration curves produced from such standards are linear up to a few hundred μg per cm^2 [26, 28, 30]. Alternatively, insoluble metal salts chemically produced in fine particle form by precipitation from aqueous solution, and deposited on filter discs by filtration of the aqueous suspension, may be used [15, 29]. Besides weighed lead metal samples vaporized onto aluminium foil, comparison with wet chemical procedures has also been used for calibration [31]. Data from X-ray fluorescence analyses may be speedily processed by the use of on-line computing facilities [28, 32].

X-ray techniques for the examination of individual particles are available. The electron probe microanalyser allows the bombardment of particles with an electron beam, typically about 1 μm in diameter. Primary X-ray emission is then analysed for characteristic emission wavelengths, allowing a quantitative analysis of the composition of the particle under the electron beam. Pure elements are used as standards. For samples of greater diameter than the electron beam, the technique may be used to give information on the distribution of an element over the sample, the electron beam being made to scan over an area of sample; monitoring scattered electrons or emitted X-rays allows a 'picture' of the sample to be built up. Combination with an optical microscope is also possible.

In order to avoid electric charge build-up or overheating under the electron beam, a thin coating of a conducting substance on the sample is frequently required. Ter Haar and Bayard collected airborne particulates on Millipore filters which were subsequently dissolved in acetone [33]. The centrifuged particles were mounted in a film of nitrocellulose on Be coated with a 300 Å layer of carbon. Use of the electron probe microanalyser allowed determination of the elemental content of particles of different sizes. Particulate matter emitted from vehicle exhausts was analysed in a similar fashion. Particulate matter deposited on the bark of trees growing near busy roads has also been examined [34]. Sections of dried bark were mounted on a carbon block and covered with a thin carbon coating prior to electron probe analysis.

Electron microscopy and microanalysis (EMMA), which combines high resolution electron microscopy with X-ray microanalysis allows size determination and semi-quantitative elemental analysis of particles as small as $0.05 \, \mu m$. Its application to the analysis of urban airborne particulate material has been described [35].

X-ray diffraction measurements allow the identification of homogeneous crystalline particles, and even allow discrimination between different crystalline forms of a compound. Modification of X-ray cameras allows the application of the technique to particles as small as 2 to $5 \, \mu m$ in diameter [36].

Experimental procedure for X-ray fluorescence analysis of Pb, Zn, Co, Cu, Ni, Fe, Mn *and* Cr [15]

Air sampling

Samples (24 h) of approximately $1500 \, m^3$ are drawn through 8 × 10 inch sheets of Schleicher and Schuell (S & S) No. 589 Green Ribbon filter paper in high-volume samplers. Discs (25 mm) are cut using a special die, and these are held firmly in the spectrometer by a clamping ring to an aluminium mask having an 11/16 inch diameter opening. The area exposed to the X-ray beam is $0.371 \, inch^2$.

Preparation of standards

Standard solutions ($1 \, mg \, ml^{-1}$) of the metals are prepared using chloride salts, if available, and sufficient acid to prevent hydrolysis. Transfer hydrochloric acid (HCl) (5 ml) and de-ionized water (50 ml) to a 100 ml volumetric flask and add appropriate amounts of the standard metal solutions. Solutions containing nitrates are first combined in a small conical flask, treated with perchloric acid ($HClO_4$) (1 ml) (Care needed), evaporated to white fumes to decompose all nitrates and, after cooling, quantitatively transferred to the volumetric flask. Fill the flask to the mark with de-ionized water. Samples collected from urban air using the above sampling procedure dictated that 1 ml of the multimetal solution should contain $50\,\mu g$ of Pb and Fe, $10\,\mu g$ of Zn and Cu, and $2\,\mu g$ of Ni, Co, Mn and Cr. Of necessity, these amounts are adjusted to the analytical problem at hand. Transfer an aliquot (1.0 ml) of the multimetal standard solution to a 50 ml beaker, add de-ionized water (10 ml), HCl (2 to 3 drops) and metacresol purple indicator (0.1%; 1 drop). Neutralize the solution by dropwise addition of NH_4OH $(1 + 1)$ and add one drop in excess. Add a clear solution of 2% sodium diethyl-dithiocarbamate (5 ml), swirl the mixture and allow to stand for 5 min. (The carbamate solution should be stored in a plastic bottle and refrigerated when not in use.) Filter the liquid slowly through a 25 mm $0.8\,\mu m$ Millipore disc and wash with de-ionized water (about 2 ml). Dry the disc thoroughly at room temperature, or more rapidly by placing it on a glass surface at 50 to 60° C using a lightweight plastic ring on top to prevent curling of the paper.

Instrumental

Luke *et al.* [15] used a General Electric XRD-6S air or helium path instrument equipped with a four-crystal changer, bulk sample holder, dual (Tungsten/Chromium) target X-ray tube, and dual (scintillation and flow proportional) counter tube detector system with solid-state electronics. The analytical line employed for X-ray counting is the K_α doublet for all elements except lead, for which L_α, line is used. The tungsten target of the EA75 X-ray tube is operated at 50 kVP and 75 mA. Other parameters include a lithium fluoride (200) analysing crystal, 10-mil Soller slit and helium path. Take a 10 s count at the goniometer setting for the X-ray line of each element to be determined on a clean Millipore disc, a blank S & S paper disc, the multimetal standard disc, and all S & S paper samples to be analysed. For each analytical line, subtract the total intensity measured on the Millipore blank from that obtained on the multi-element standard. Similarly, subtract the intensity on the S & S blank from those found on the pollution samples. The number of μg of each metal per inch^2 of sample may then be calculated from the formula

$$\mu g\,in^{-2} = 1/0.371 \times \frac{\text{Net counts sample}}{\text{Net counts standard}} \times \mu g \text{ of metal on standard disc.}$$

Experimental procedure for electron probe microanalysis [33]

Air sampling

Air is filtered through $0.45\,\mu m$ Millipore filters.

Sample preparation and analysis

Dissolve the filter in acetone and centrifuge the particles from the resultant suspension. Mount the particles in a film of nitrocellulose on an ultrapure polished Be substrate coated with about 300 Å of carbon. (Be was chosen because it has very low intensity white radiation and no line interferences with the elements of interest.) Ratios of elements, rather than absolute intensities are then measured.

3.2.2.2 Radioactivation methods

In radioactivation analysis, the sample is irradiated with neutrons or charged particles without pretreatment. Interaction of the sample with the bombarding particles causes the formation of different isotopes of the elements in the sample, or conversion to isotopes of neighbouring elements. Many of these isotopes produced are radioactive, and measurement of their activity and comparison with standards allows quantitative analysis of the elements in the sample. Gibbons and Lambie have reviewed the theory and practice of radioactivation analysis [37]. Advantages of the technique are speed, simplicity, specificity and non-destruction of the sample [38]; the major disadvantage is the need for a high intensity source of activating particles.

The process most commonly used in activation analysis is irradiation with a high thermal neutron flux, generally produced by a nuclear reactor. Measurement of the characteristic γ-ray emissions of the isotopes produced is then used for analysis of the sample. Several neutron activation techniques for the analysis of airborne particulate matter collected on filters have been described [10–12, 39–41]. Sample filters are each sealed in a polyethylene vial and irradiated for 5 min with neutrons. γ-ray counts after 3 and 15 min allow the determination of 13 elements. A further irradiation at a higher neutron flux for 2 to 5 h, followed by γ-ray counts after 20 to 30 h and 20 to 30 days of cooling allow the determination of a further 20 elements [39]. Standards are prepared by deposition of solutions containing a mixture of the elements to be analysed upon blank filters. After drying, the standards are sealed in polyethylene and irradiated in a manner identical to that used for the samples [27, 39].

Neutron activation analysis of airborne particulate matter has been criticized because, although numerous elements can be analysed, many cannot because of inadequate sensitivity, interferences or high blanks. Few low molecular weight elements can be analysed, the analysis of important elements such as Pb and Cd is not reported, and several other important metals may only be analysed after a 20 to 30 day decay period [36]. Hence, the application of neutron activation analysis is subject to severe limitations. The sensitivity of the technique for a given element is dependent upon the overall composition of the sample, owing to interference effects, but Dams *et al.* [39] quote typical detection limits,

Table 3.5 Sensitivity of neutron activation analysis for determination of trace elements in aerosols [39]

Element	Decay time before counting	Detection limit (μg)	Minimum concn. in urban air – 24 h sample (μg m^{-3})
Al	3 (min)	0.04	0.008
Ca	3	1.0	0.2
Ti	3	0.2	0.04
V	3	0.001	0.002
Cu	3	0.1	0.02
Na	15	0.2	0.04
Mg	15	3.0	0.6
Mn	15	0.003	0.000 6
In	15	0.000 2	0.000 04
K	20–30 (h)	0.075	0.007 5
Cu	20–30	0.05	0.005
Zn	20–30	0.2	0.02
As	20–30	0.04	0.004
Ga	20–30	0.01	0.001
Sb	20–30	0.03	0.003
La	20–30	0.002	0.000 2
Sm	20–30	0.000 05	0.000 005
Eu	20–30	0.000 1	0.000 01
W	20–30	0.005	0.000 5
Au	20–30	0.001	0.000 1
Sc	20–30 (day)	0.003	0.000 004
Cr	20–30	0.02	0.000 25
Fe	20–30	1.5	0.02
Co	20–30	0.002	0.000 025
Ni	20–30	1.5	0.02
Zn	20–30	0.1	0.001
Se	20–30	0.01	0.000 1
Ag	20–30	0.1	0.001
Sb	20–30	0.08	0.001
Ce	20–30	0.02	0.000 25
Hg	20–30	0.01	0.000 1
Th	20–30	0.003	0.000 04

Areas of filter of 0.8, 0.8, 1.6 and 13 cm^2 used for counts at 3, 15 min, 20–30 h and 20–30 day respectively.

shown in Table 3.5. For those metals determined by neutron activation analysis, limits of detection compare quite favourably with those for emission spectrographic and conventional atomic absorption methods [11, 39].

The scope and applications of charged particle activation analysis appear limited [42], although fairly sensitive methods for lead analysis have recently been developed [43, 44].

Experimental procedure for neutron activation analysis [39]

Air sampling

Air is drawn through 25 mm or 47 mm diameter polystyrene filters at up to 121 min^{-1} cm^{-2} for 24 h using a high vacuum pump.

Activation and analysis

(a) *Short-lived isotopes*. For the analysis of elements giving rise to short-lived isotopes, each sample is packaged in a polyethylene vial, and placed in a rabbit which carries it through a pneumatic tube to a position near the core of the reactor, where it is irradiated for 5 min at a flux of 2×10^{12} neutrons cm^{-2} s^{-1}. At the end of this period, the sample is returned to the laboratory where it is manually transferred to a counting vial and carried to the counting room. 3 min after irradiation, commence a count of 400 s live-time duration, and follow by a count of 1000 s live-time starting 15 min after irradiation. These and subsequent counts are performed on a 30 cm^3 Ge(Li) detector coupled to a 4096 chan-nel analyser. The detector is housed in an iron shield and operated, in an air-conditioned room, at a gain of 1 keV per channel. The observed resolution (reported by Dams *et al.* [39]) was 2.5 keV full-width at half maximum (FWHM) for the ^{60}Co 1332 keV photopeak and a peak to Compton ratio of 18/1. Table 3.6 includes the isotopes determined by the first 2 counts.

All spectra are recorded on 7-track magnetic tape for further analysis. Conver-sion of counting rates under the various peaks to concentrations is accomplished by subjecting a few standard solutions containing well-known mixtures of the same elements to the same irradiation and counting sequence. In order to avoid possible errors due to coincidence summing or to broadening of peaks at high counting rates, sample sizes are generally adjusted to make counting rates of sample and standard of comparable magnitude.

Small corrections for variations of both neutron flux and rabbit placement from irradiation to irradiation are accomplished by co-irradiation of a titanium foil flux monitor with each sample. It is counted for 20 s at 13 min after the end of the irradiation, between the 2 sample counts. If the analysis rate does not exceed 1 sample in 40 min, the same flux monitor may be used repeatedly with less than 1% of the original 5.8 min ^{51}Ti remaining in the next count. Net count-ing rates of the sample spectrum are normalized to an arbitrary titanium activity, equivalent to a reference neutron flux.

(b) *Long-lived isotopes*. The same sample or another portion of the same air filter is then irradiated in the reactor core at a higher flux (1.5×10^{13} neutrons cm^{-2} s^{-1}) for 2–5 h. Each is individually heat-sealed in a polyethylene tube and irradiated with 8 others, plus a standard mixture of elements in a

Table 3.6   Nuclear properties and measurement of isotopes [39]

Element	Isotope	Half-life	t (irradiate)	t (cool)	t (count) (s)	Gamma-rays used, (keV)
Al	^{28}Al	2.31 (min)	5 (min)	3 (min)	400	1778.9
Ca	^{49}Ca	8.8	5	3	400	3083.0
Ti	^{51}Ti	5.79	5	3	400	320.0
V	^{52}V	3.76	5	3	400	1434.4
Cu	^{66}Cu	5.1	5	3	400	1039.0
Na	^{24}Na	15 (h)	5	15	1000	1368.4; 2753.6
Mg	^{27}Mg	9.45 (min)	5	15	1000	1014.1
Mn	^{56}Mn	2.58 (h)	5	15	1000	846.9; 1810.7
In	^{116m}In	54 (min)	5	15	1000	417.0; 1097.1
K	^{42}K	12.52 (h)	2–5 (h)	20–30 (h)	2000	1524.7
Cu	^{64}Cu	12.5	2–5	20–30	2000	511.0
Zn	^{69m}Zn	13.8	2–5	20–30	2000	438.7
As	^{76}As	26.3	2–5	20–30	2000	657.0; 1215.8
Ga	^{72}Ga	14.3	2–5	20–30	2000	630.1; 834.1; 1860.4
Sb	^{122}Sb	2.75 (day)	2–5	20–30	2000	564.0; 692.5
La	^{140}La	40.3 (h)	2–5	20–30	2000	486.8; 1595.4
Sm	^{153}Sm	47.1	2–5	20–30	2000	103.2
Eu	^{152m}Eu	9.35	2–5	20–30	2000	121.8; 963.5
W	^{187}W	24.0	2–5	20–30	2000	479.3; 685.7
Au	^{198}Au	2.70 (day)	2–5	20–30	2000	411.8
Sc	^{46}Sc	83.9	2–5	20–30 (day)	4000	889.4; 1120.3
Cr	^{51}Cr	27.8	2–5	20–30	4000	320.0
Fe	^{59}Fe	45.1	2–5	20–30	4000	1098.6; 1291.5
Co	^{60}Co	5.2 (year)	2–5	20–30	4000	1173.1; 1332.4
Ni	^{58}Co	71.3 (day)	2–5	20–30	4000	810.3
Zn	^{65}Zn	245	2–5	20–30	4000	1115.4
Se	^{75}Se	121	2–5	20–30	4000	136.0; 264.6
Ag	^{110m}Ag	253	2–5	20–30	4000	937.2; 1384.0
Sb	^{124}Sb	60.9	2–5	20–30	4000	602.6; 1690.7
Ce	^{141}Ce	32.5	2–5	20–30	4000	145.4
Hg	^{203}Hg	46.9	2–5	20–30	4000	279.1
Th	^{233}Pa	27.0	2–5	20–30	4000	311.8

polyethylene bottle, 4 cm in diameter, lowered into the reactor pool. Cooling of the samples during irradiation is achieved by allowing the pool water to circulate through several holes cut in the container bottle. Standards are prepared by depositing 100 μl each of 2 well-balanced mixtures of the appropriate elements on to a highly pure substrate (ashless filter paper) and allowing to dry, then sealing inside polyethylene tubes.

After irradiation, the samples and standards are transferred to clean containers and counted once for 2000 s live-time after 20 to 30 h of cooling and then for 4000 s live-time after 20 to 30 days of cooling. Table 3.6 includes the elements determined from these counts. Errors resulting from thermal neutron flux gradients over the bottle dimension were less than 5%, provided the samples were confined to a single horizontal layer of vertically oriented tubes at the bottom of the bottle and the bottle was rotated through 180° half-way through the irradiation time. Fast neutron flux gradients were about twice as large as thermal gradients, but the only fast neutron reaction used is in the determination of nickel.

Interferences

Some prominent photopeaks are not used because of interferences by neighbouring peaks of other isotopes. In other cases corrections are necessary. The ^{75}Se (279.6 keV) interferes with the monoenergetic ^{203}Hg (279.1 keV), but a correction based on the spectrum of pure ^{75}Se can be applied since the interference in air pollution samples is usually less than 20% of the ^{203}Hg activity. The measurement of ^{64}Cu (511.0 keV) is complicated by interference caused by external pair production of high energy γ-rays. In typical samples 15 h ^{24}Na is the most important source of γ-rays after a decay period of 20 h, and a correction, usually less than 10%, may be applied to the apparent ^{64}Cu activity. Interferences by threshold reactions have been calculated and checked experimentally, and in typical aerosol samples the only reaction affecting a calculated concentration by greater than 2% is ^{27}Al(n, p) ^{27}Mg. Once the aluminium concentration is known, the appropriate correction can be applied to the magnesium concentration.

3.2.3 Methods involving pretreatment of the samples

3.2.3.1 Emission spectrography

At ambient temperatures, in the absence of external stimuli, almost all atoms exist in the electronic ground state. External stimuli, however, may be used to excite atoms or molecules electronically, such that the resultant excited species return to the ground state by emission of energy in the form of light. In emission spectrography, a sample is deposited upon a graphite electrode which is made of one of two electrodes between which a d.c. arc is passed. The energy liberated in the arc causes volatilization of the sample, which is excited and emits light at discrete wavelengths (lines) characteristic of the elements present. Dispersion of the light according to wavelength by a prism or grating allows separation of the emission lines which are recorded on a photographic plate, the intensity of light recorded by the plate from the various lines being a measure of the quantity of the different elements present in the sample.

Although the d.c. arc is still widely used for sample volatilization, numerous refinements are available which may improve sensitivity [45]. Standards of

Table 3.7   Limits of detection for emission spectrographic analysis [49]

Metal	Analytical line (Å)	Minimum concn. detectable in urban air (μg m^{-3})*	Minimum concn. detectable in non-urban air (μg m^{-3})
Be	2348.6	0.0008	0.000 16
Bi	3067.7	0.0011	0.000 2
Cd	2288.0	0.011	0.004
Co	3453.5	0.0064	0.002
Cr	2677.2	0.0064	0.002
Cu	2824.4; 3274.0	0.01	0.001 5
Fe	2457.6	0.084	0.006
Mn	2933.1	0.011	0.006 0
Mo	3170.3	0.0028	0.000 5
Ni	3003.6	0.0064	0.001 6
Pb	2663.2	0.04	0.01
Sb	2877.9	0.040	0.006
Sn	2840.0	0.006	0.001 8
Ti	3199.9	0.0024	0.000 48
V	3183.4	0.0032	0.000 48
Zn	3345.0	0.24	0.08

* Using 26% of a 24 h high-volume sample (2000 m^3 total).

composition similar to that of the sample are used to calibrate the instrument. Because of matrix effects, such standards should resemble the sample, both chemically and physically, as closely as possible. All possible variables which may affect the analysis should be controlled as closely as possible. Such factors include the uniformity of the graphite electrodes and photographic plates and chemicals, and the temperature and humidity of the room where the analysis is performed.

Emission spectrography has not been widely applied to the analysis of metallic air pollutants. Most notably, however, it has been used for many years by the U.S. National Air Sampling Network [2, 6, 46–49]. Samples must be ashed and taken up in acid prior to analysis [6]. Although the method allows simultaneous determination of many metals it is reported to be only semi-quantitative despite the application of considerable analytical skill, to lack sensitivity for a number of metals and to be limited by blanks in the filter substrate [36]. The sensitivity of the method in the determination of a number of metals is shown in Table 3.7. Extrapolation to other elements may be made by using the data of Zoller and Gordon [11] and Mitteldorf [45] for relative sensitivities.

The very high sensitivity of emission spectrography for analysis of Be has led to recommendation of the method for the determination of this metal in air [50]. Normal particle collection techniques are used, and the Be may be

Table 3.8   Detection limits for combined sampling–analysis method [55]

Element	Wavelength (Å)	Absolute detection limit (ng)	Detection limit for air (30 min sample) (μg m^{-3})
Al	3082.15	1	0.03*
Be	3130.41	0.1	0.003
Co	3453.50	1	0.03
Cr	2843.25	0.5	0.015
Hg	2536.51	0.5	0.015
Mg	2852.12	0.5	0.015*
Mn	2576.10	0.5	0.015
Mo	3798.25	1	0.03
Ni	3414.76	0.5	0.015
Pb	2833.06	3	0.09
Ti	3349.03	1	0.03
V	3183.98	2	0.06
W	4008.75	5	0.15
Zn	3345.02	3	0.10

* The calculated detection limits for aluminium and magnesium in air are not normally attainable as a result of high levels of these elements as impurities in the graphite cup electrodes.

determined , after pretreatment of the sample, using graphite rod electrodes and a d.c. arc [51], or a rotating graphite electrode and an a.c. current arc [52]. Using the latter technique, as little as $0.002\,\mu$g Be ml^{-1} may be analysed, corresponding to $0.001\,\mu$g m^{-3} in a 20 to 30 min air sample. Alternatively, if a Whatman No. 40 filter paper is used, it may be rolled and pressed into the crater of a graphite electrode. After charring thermally and addition of barium chloride as a carrier, a d.c. arc is used to determine Be with a high sensitivity [53].

A technique for continuous monitoring of Be in air involves drawing the polluted air through a spark between copper electrodes [54]. Photo-electric monitoring of the 3130.4 Å Be line and display on a chart recorder allows continuous recording of Be concentrations. The instrument has a range up to $20\,\mu$g m^{-3} of Be with a detection limit of $0.5\,\mu$g m^{-3}.

An ingenious procedure which uses a graphite cup electrode as a filter for atmospheric particulate matter has recently been reported [55]. Particulate material collected by the graphite cup is volatilized by a d.c. arc in a specially modified source of an emission spectrograph. Very high filtration efficiency was found, and the use of an indium internal standard allowed quantitative analysis of collected metals with a standard deviation in the 10 to 20% standard deviation range for the elements studied. Table 3.8 shows the reported analytical wavelengths and detection limits for a sampling period of 30 min. The workers did, however, fail to acknowledge or investigate the possible adsorption of organometallic compounds, such as lead alkyls, or of elemental mercury by the graphite cup. Hence, in the case of certain metals, there is uncertainty regarding the analysis when the metal is present in the air in other than particulate forms.

Experimental procedure for emission spectrographic analysis [6]

Air sampling

Samples of about 2000 m^3 of air are drawn through 8 × 10 inch flash fired glass fibre filters over 24 h using a high-volume sampler.

Standardization

Prepare a stock solution of the following metals, with the concentrations indicated: Be, Bi, Cd, Co, 1 μg/0.05 ml; Sb, Cr, 2 μg/0.05 ml; Mo, Ti, 4 μg/0.05 ml; Ni, V, Sn, 8 μg/0.05 ml; Mn, 20 μg/0.05 ml; Cu, Fe, Pb, 100 μg/0.05 ml; Zn, 200 μg/0.05 ml. Salts should be better than or at least analytical grade. Make working standard solutions by diluting the stock solution to give concentrations 1/2, 1/4, 1/8 ... 1/1024 of the original. One stronger standard is made by doubling the quantity on the electrode.

Add a HNO$_3$ extract of the glass filter (63 inch^2 of filter = 40 ml acid extract) directly to the electrodes in order to duplicate the conditions obtained in the particulate sample solution. This compensates for the acid soluble material present in the glass filter, some of which could interfere in the metal analysis, and eliminates the need for making separate corrections for the metals present in the filter.

Three replicate standard plates are made as follows. Warm electrodes of high purity graphite (V crater with centre post) for 10 min in an oven at about 80° C. Briefly immerse the crater portion into a hot 20% solution of paraffin in redistilled benzene, then oven dry for a few minutes. Then fill the crater 1/3 full with a lithium chloride–graphite mixture (1 part lithium chloride and 2.5 parts graphite), transfer 0.05 ml of the glass filter extract (equal to 1/800 of the filter) to the electrode, and add one drop of methanol to distribute the solution evenly. Partially dry the electrodes in an oven, then pipette 0.05 ml of the working standard (0.10 ml for the stronger standard) and add one drop of methanol. Dry the electrode at 80° C, then heat at 105° C for 1 h.

Arc the loaded electrodes as the anode for 30 s at about 7 A d.c. (N.A.S.N. use a Baird 3-meter-grating spectrograph [6]). Prior to arcing set the current at 8.25 A closed circuit and do not adjust further during the arcing period. Maintain gap at 3 mm, slit 6 mm × 25 μm. Sharpen pointed counter electrodes with a pencil sharpener.

A single-step sector, passing 1/8 of the uninterrupted beam, is interposed between the slit and the lens. Eastman 103-0 plates are used, and the first order of the region 2170 Å to 3590 Å recorded. Process the plates by standard procedures (3 min D-19, 20° C water rinse; 10 min in Kodak acid fix, and 25 min wash). Measure line intensity with and N.S.L. Spec. Reader.

Analysis lines for Cd, Be, Fe, Pb, Cr, Cu, Sn, Sb, Mn, Ni, Bi, Mo, V, Ti, Zn and Co shown in Table 3.7 are used. Plot calibration curves for each metal as a ratio of % transmittance of line plus background to % transmittance of adjacent background versus μg of metal on the electrode on log–log graph paper. Data from the curves can be consolidated in chart form.

Analysis

In order to permit more complete removal of inorganic materials than is possible by simple acid extraction, first muffle samples (26% of the original) at 500° C for 1 h to burn off organic matter. Then make two extractions with 40 ml portions of 1 : 1 redistilled HNO_3 at slightly below the boiling point for 1 h. After filtering the resulting solutions through Whatman No. 42 filter paper, evaporate them to 3 to 4 ml and make up to 10.4 ml (0.05 ml of this solution is equivalent to 1/800 of the sample).

Load electrodes (prepared as described above, except the addition of glass filter extract is omitted) with 0.05 ml of the sample solution. Arcing conditions, plate processing, and densitometry are the same as standard plate conditions. By use of the numerical value of the line plus background to background ratio, amounts of each metal on the electrode are obtained by reference to the chart.

$$\mu g \text{ of metal on electrode} \times 800/m^3 \text{ of air sampled } = \mu g \text{ of metal m}^{-3} \text{ of air}$$

3.2.3.2 Ring oven methods

The ring oven is a relatively cheap and simple piece of apparatus [56]. A clean piece of filter paper is fixed horizontally on the oven surface, and a sample is introduced at its centre. Addition of solvent at the centre of the paper causes the sample to be washed outwards in a circle by capillary movements (Fig. 3.1). At a certain distance from the centre, the sample solution reaches a ring, heated to about 15° C above the boiling point of the solvent used for washings. The solution is consequently dried, leaving a ring of sample. The filter paper may then be cut into segments, each of which is analysed for a given metal or radical. Analysis is by the addition of highly sensitive specific or selective organic colorimetric or fluorimetric reagents. Selectivity may be enhanced by the use of masking agents or other processes. Once the test colour has been developed, visual comparison with rings produced using standard samples

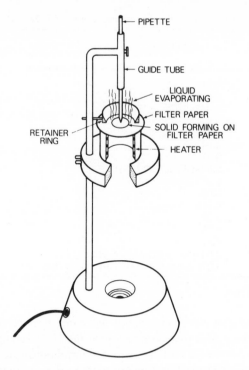

PIPETTE

GUIDE TUBE

LIQUID
EVAPORATING

FILTER PAPER

SOLID FORMING ON
FILTER PAPER

RETAINER
RING

HEATER

Fig. 3.1   Essential features of the ring oven [58].

allows accurate (90 to 95% in the μg range) quantitative determination of the unknown.

West and co-workers [57, 58] have described the application of ring oven techniques to air pollution analysis. Samples may be collected by filtration on a range of materials, or by impaction or electrostatic precipitation. Dissolution of the sample is then the normal procedure, followed by analysis of an aliquot using the ring oven. Alternatively, samples collected by filtration may be transferred to the ring oven by cutting a triangular portion, about 0.5 in per side from the filter and using it as the sample. This section of filter is fixed at the centre of the ring oven, beneath the normal clean filter paper, and washing out of the sample is performed in the usual manner. In the case of samples collected with an automatic tape sampler, if the sample spot is smaller than the heated ring of the ring oven, the section of tape containing the appropriate spot may be transferred directly to the centre of the oven and analysed as described. West reports that small air samples (1 m^3) collected using an automatic tape sampler allow as little as 2 to 3 μg of a constituent metal of the air sample to be determined with good accuracy [57].

Table 3.9 Detection limits for ring oven procedure [57, 58, 60]

Metal	Detection limit (μg)*	Working range (μg)
Al	0.01	0.03– 0.5
Be	0.01	0.01– 0.2
Cd	0.075	
Co	0.02	0.04– 0.5
Cr	0.15	0.3 – 1.0
Cu	0.04	0.1 – 0.5
Fe	0.01	0.01– 0.5
Mn	2.0 (malonic acid [58]) 0.075 (benzidine [57])	2.0 –10.0
Ni	0.08	0.10– 1.0
Pb	0.015	
Sb	0.08	0.1 – 1.0
Se	0.08	0.1 – 0.5
V	0.01	0.01– 3.0
Zn	0.04 (o-mercaptothenalaniline [58]) 0.075 (ammonium mercury thiocyanate [57])	0.05– 1.0

* In the total ring.

A one-dimensional concentration technique inspired by the ring oven procedure has been used to determine lead at the 10 to 100 ng level collected on air filters, by a colorimetric reaction with sodium rhodizonate [59].

Table 3.9 shows the detection limits of the ring oven method for a number of metals under ideal conditions [57, 58, 60]. Interference effects of other ions may affect the results obtained as many reagents are not entirely specific, and this fact must always be taken into account when performing analysis by this technique.

Experimental procedure for ring oven analysis [57]

Air sampling

Samples are collected by filtration on any filter medium with a low metal blank.

Sample preparation and concentration

Cut out a triangular portion, averaging 0.5 in per side, from the filter. Place the triangle centrally upon a circle of filter paper (e.g. Whatman No. 40) and fix it in place. Membrane filters, made of cellulose acetate, may be fixed by slight moistening of the paper filter with acetone, and pressing onto the membrane filter triangle when the acetone is almost dry. The use of too much acetone may cause clogging of the pores of the filter paper, or occlusion of collected particulate matter by cellulose acetate, and is to be avoided. The filter paper with triangle in place is put in position across the surface of the ring oven, the triangle on the underside. The addition of 8 to 10 portions (5 μl) of 0.1 N HCl to the centre of the filter paper causes transfer and concentration of the sample in a ring on the filter paper.

If paper filters are used for air sampling, a triangular portion is used as above, but is fixed to the receiving filter paper with tiny dots of glue, applied by a fine glass needle. The triangle is placed above the receiving filter paper, and centred directly beneath the guide tube on the oven. Samples collected on glass fibre filters may be treated in a similar manner.

Analysis

The concentrated sample, in the form of a ring, is divided by cutting the receiving filter paper into segments, each of which is analysed for a particular metal.

(a) Al. Spray the sector with a saturated solution of morin in methanol, dry and bathe in 2 N HCl. Observe under u.v. light, while still moist. A yellow—green fluorescent line indicates Al.

(b) Cd. Spray the sector with a solution of cadion in ethanol, containing some NaOH. A red—pink line denotes Cd.

(c) Cr. Oxidize the chromium (III) to chromate with a 1:1 mixture of 10% H_2O_2 and NH_3 solution. (Fill a fine glass capillary with oxidizing mixture and run the capillary rapidly along the circular line on the section; thus only a very narrow moistened line is obtained.) Dry the sector in a drying oven and spray with a freshly prepared 1% solution of diphenylcarbazide in ethanol. Finally dip into 2 N H_2SO_4. A violet line appears if Cr is present.

(d) Co. Fume the sector over NH_3 solution, apply 5% disodium hydrogen phosphate to mask iron, and spray with a 1% solution of 1-nitroso-2-naphthol in acetone. A red—brown line appears if Co is present.

(e) Cu. Fume the sector over NH_3 solution and spray with a 1% solution of dithio-oxamide in ethanol. An olive-green to black line indicates Cu.

(f) Fe. Fume the sector in HCl, and if necessary over bromine water, and finally spray with a 1% potassium ferrocyanide solution. A blue line denotes Fe.

(g) Pb. Treat the sector with freshly prepared 0.2% aqueous solution of sodium rhodizonate and fume over HCl while the yellow colour of the reagent disappears. A violet to red line indicates Pb.

(h) Mn. Treat the sector with 0.05 N KOH solution and spray with a 0.05% solution of benzidine (C a r e :   C a r c i n o g e n i c   R e a g e n t) in dilute acetic acid. A blue line indicates Mn.

(i) Ni. Fume the sector over ammonia solution and spray with a 1% solution of dimethylglyoxime in ethanol. A red line denotes Ni. If the ring zone contains iron, it is advisable to bathe the developed sector in a tartrate solution.

(j) Zn. Moisten the ring zone on the sector with ammonium mercury thiocyanate solution (3.3 g of ammonium thiocyanate + 3 g of mercury (II) chloride are dissolved in 5 ml of water without warming). Bathe and agitate the sector in a 0.02% cobalt solution. A blue line denotes Zn. Blue spots, which might eventually appear on other parts of the paper should be ignored.

Calibration

A standard solution is prepared so as to have a concentration of about 0.01 mg of metal ml^{-1}. Standard rings are prepared by adding 10, 20, 40, 60, 80 and 100 μl of the standard to each of 6 separate filter discs using a 10 μl pipette. The added metal solutions are washed with 5 × 10 μl portions of 0.1 N HCl to transfer the metal to the ring zone. Appropriate reagents are then used to develop a standard series of colours. Unknowns are determined by comparison with the standards, greater precision being achieved by running the sample 3 times, each time using a different sample size.

3.2.3.3 Polarography

A technique which has long been applied to the analysis of metals, polarography is unlikely to be a preferred technique for air pollution analysis. Although it is less sensitive than many methods currently available for the analysis of metals, the application of the technique to the analysis of airborne particulate matter has been described [58, 61].

The technique is electrochemical and the polarographic cell in which the sample solution is placed contains 2 electrodes. A pool of mercury in the bottom of the cell serves as a large non-polarizable reference electrode, whilst a stream of small mercury drops (typically 2 to 6 s per drop), known as a dropping mercury electrode, serves as a polarizable active electrode with a continuously renewed surface [62]. A steadily increasing voltage is applied to the cell and the current passing through the cell is measured. A slow rise in current with applied potential is found, with much faster rises occurring at characteristic potentials corresponding to the electrolysis of specific chemical species. A relatively high concentration of a 'supporting electrolyte' is included in the sample solution and an appropriate choice of this electrolyte can cause separation of the electrolysis steps of the different metal ions in the solution, allowing quantitative determination of 5 or 6 substances from a given polarogram (Fig. 3.2). West [58] reports that although sensitivities as high as 0.1 $\mu g\,ml^{-1}$ may be attained for some metals, in general

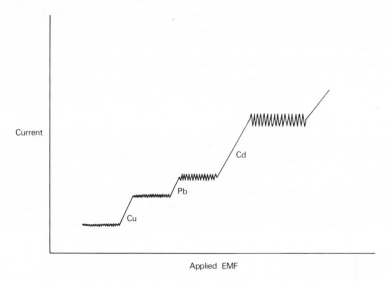

Fig. 3.2   A polarogram [58].

a more realistic figure is $1 \mu g \, ml^{-1}$. Sample volumes of around 1 ml may be used. Some improvement in sensitivity has resulted from refinements applied to conventional polarographic techniques [62, 63].

A major problem in polarographic analysis is the determination of a constituent of low concentration in the presence of large amounts of more easily electrolysable substances, which, in the case of metal analysis, may include organic materials. In general, two techniques are available to overcome this problem. Firstly, selective extraction of the minor consitituent is a valuable method, and this may allow a pre-concentration if extraction is into an organic solvent [64]. Secondly, complexing of the interfering substances to produce less readily electrolysable forms may prove successful.

Application to air pollution analysis has been described by Levine [61]. Samples collected by impingement or electrostatic precipitation were wet ashed, and then dissolved in an appropriate supporting electrolyte. Polarographic analysis of the sample and comparison with a series of standards prepared in a similar manner to the sample, and hence in a comparable matrix, enabled the determination of Pb, ZnO fumes, Cd fumes, chromic acid fumes or mists, and Mn fumes in air to be carried out.

Experimental procedure for polarographic analysis of Pb and Cd [61]

Air sampling

The air is sampled with a modified Greenberg–Smith impinger or an electrostatic precipitator, and in an industrial environment about $1\,m^3$ of air is sampled.

Analysis for Pb fumes

Transfer the collected sample to a casserole, add 6 N HCl (5 ml), 6 N HNO_3 (1 ml) and evaporate to dryness on a steam bath. Take up the residue in exactly 5 ml of 20% citric acid solution. Pour the contents of the casserole into the electrolytic cell, expel oxygen and polarograph at 1/5 the galvanometer sensitivity $(0.015\,\mu A\,mm^{-1})$ in an applied voltage range of -0.4 to -0.8 V. Measure the step height and read the concentration on the calibration curve.

Calibration curve for Pb

Prepare a standard solution of lead nitrate containing $50\,\mu g$ lead ml^{-1}, and add the following quantities of Pb to casseroles containing water (75 ml), 6 N HCl (5 ml) and 6 N HNO_3 (1 ml): 250, 200, 150, 100 and $50\,\mu g$. Evaporate to dryness on the steam bath. Take up in exactly 5 ml of 20% citric acid, pour into the electrolytic cell, and polaragraph as above. A graph plotted of step heights against concentration in $\mu g\,ml^{-1}$ is linear.

Analysis of Cd fumes

Transfer the sample from the impinger or electrostatic precipitator into a casserole, add 6 N HCl (5 ml), and evaporate to dryness on a steam bath. Take up the residue with exactly 5 ml of 20% citric acid solution containing 3 drops of 0.1% methyl red solution per 100 ml. Pour into the cell and polarograph at 1/20 the galvanometer sensitivity $(0.06\,\mu A\,mm^{-1})$ and at a voltage range to include the empirical half-wave potential of about -0.6 V. Measure the step height and read the concentration on the calibration curve.

Calibration curve for Cd

Prepare a standard solution of a Cd salt having a concentration of $25\,\mu g$ Cd ml^{-1}. Using this solution, add the following quantities of Cd to casseroles containing water (75 ml) and 6 N HCl (5 ml): 250, 200, 150, 100, 50 and $25\,\mu g$. Evaporate the solutions to dryness, take up the residues in 5 ml of 20% citric acid solution containing 3 drops of 0.1% methyl red solution per 100 ml, and polarograph as above. A graph is plotted of step height against concentration in $\mu g\,ml^{-1}$.

3.2.3.4 Anodic stripping voltammetry

This technique is closely related to polarography, and samples are prepared in a similar manner. Application of a negative voltage is used to cause electrodeposition of metal ions from the sample solution onto a Hg or solid electrode. Subsequent linear variation of the electrode potential in an anodic direction causes rapid dissolution with a corresponding sharp current peak. Quantitative deposition and stripping of an ion allows calculation of the concentration of that ion from the integrated stripping current. Alternatively, a reproducible fraction of the ion is electrodeposited and subsequently stripped. The stripping current is measured and the concentration of the ion determined by comparison with standards. Since the deposition causes a very substantial concentration of the ions, the sensitivity is increased substantially over that of conventional polarography, at best by several orders of magnitude [65].

Although only relatively recently developed, anodic stripping voltammetry has found application in water analysis [65] and has been comparatively assessed with other instrumental techniques for the analysis of metals in gasoline and other matrices [66]. Air samples collected over 24 h on glass fibre filters have been analysed for Pb, Cd, Cu and bismuth by anodic stripping voltammetry [67], and Colovos et al. [68] used the technique to determine Pb, Cd, Cu and Zn. In the latter study samples were collected on Millipore filters and ashed in a low temperature asher after addition of 0.2 ml 0.1 M potassium sulphate, and prior to analysis were digested with hydrofluoric acid in a PTFE bomb.

MacLeod and Lee [69] have described the analysis by anodic stripping voltammetry of metals collected in a 2 h sample by an AISI tape sampler. After ashing, the spot tape sample was dissolved in acid and diluted prior to analysis. Airborne Cd (7 to 350 ng), Pb (80 ng to 2.4 μg) and Cu (6 ng to 1 μg) were determined and replicate determinations with standard solutions show a maximum relative standard deviation of less than 12% due to instrumental variability or operation error. It was suggested that the method could be easily extended to measure other A.S.V. responsive metals including Zn, Bi, Ag, Tl and Sb.

Experimental procedure for A.S.V. analysis of Pb, Cd *and* Cu [69]

Air sampling

Samples are collected using AISI tape samplers, in which ambient air is drawn through a circular portion 1 inch in diameter of a continuous strip of Whatman No. 4 filter paper 2 inch wide. A sampling interval of 2 h is used, after which the tape is automatically advanced.

Reagents

Buffer: 0.1 M NaCl–sodium acetate solution made with A.C.S. reagent grade chemicals. The buffer is cleaned and stored in an Environmental Science Associates (ESA) reagent cleaning system 2014P at a reduction potential maintained at − 1400 mV.

Water is de-ionized or double distilled in an all-glass still before use and redistilled lead-free grade nitric and perchloric acids are used to prepare 1:1 (v/v) perchloric–nitric acid solution. All glassware is precleaned by soaking for 48 h in a 10% by volume solution of perchloric acid in water.

Procedure

Cut a circle 0.75 in diameter (representing 56.24% of the total exposed area) from the centre of each AISI tape sampler spot using a nickel-burnished steel cork–borer. Place each cut-out in a borosilicate glass sample boat and ash for 30 min in a low-temperature asher (Tracerlab LTA 600) at 200 W with an oxygen flow of 80 cm^3 min^{-1}. After 30 min, only a light residue is left. Carefully add 1:1 perchloric–nitric acid solution (0.1 ml) to the boat and transfer the dissolved sample quantitatively with water to a 10 ml volumetric flask. This sample is stable for 2 weeks.

All instrumental parameters are held constant for the analysis. Keep plating times and stirring times to within 5 s and keep the stripping rate constant throughout. pH is controlled by the addition of identical amounts of buffer and acid solution each time. The mercury coating is regulated by recoating the electrodes overnight with a dilute mercury solution; slight changes are compensated by running standards on each cell several times a day.

Pipette a 5 ml aliquot of buffer into each cell and place the cells on the cell holders. At equal intervals (30 s) switch the cells to the plating mode and turn on the nitrogen stream. 15 s before the end of each plating time turn off the nitrogen stream. With the cell turned to 'strip' turn the sweep rate from 'reset' to 'on' and strip the cell. Do this to each cell in turn. Using an ESA 4-cell multiple anodic stripping unit, MASA 2014, with 4-cell extension unit PMI 1014S the instrumental settings are as follows: hold the unit with a positive sweep rate of 60 mV s^{-1}, auto sweep hold 'on' at 1070 mV, the initial potential at − 1100 mV and the plating potential of the cells at − 1100 mV. Set the current range at 0.5 mA full scale with a recorder sweep of 10 s inch^{-1}.

Add a 1.0 ml aliquot of sample to the buffer in each cell and repeat the plating and stripping sequences. At the end of the stripping sequence, each of the cells is again stripped to ensure that all the metals are removed from the electrode.

Run standards by addition to the sample in the cell, giving a recorder trace of buffer plus sample plus standard, and thus matrix effects are taken into account.

Peak heights of the recorder trace are proportional to the quantity of each metal present in the solution, allowing calculation of the concentration of metal present.

Unused paper-tape is similarly analysed to determine the metal blank.

3.2.3.5 Spark source mass spectrometry

Mass spectrometry is a highly sensitive analytical technique which has, until recently, found little application to inorganic substances. The advent of the spark source, however, has allowed the volatilization and ionization of relatively involatile inorganic substances. The sample is coated on electrodes across which a spark is induced by application of an electrical potential. Three basic types of source are used [70]. In the RF spark source an a.c. potential in the form of pulses of a few microseconds duration is applied to the electrodes causing sparking. The second type of source, the vacuum vibrator, involves 2 electrodes with a low potential difference (10 to 30 V) between them. Mechanical vibration of one electrode causes it to make contact with the other, once in each vibration, and upon parting again a spark is generated across the electrodes. Thirdly, in the direct current hot arc source, a condenser is used to build up a charge between stationary electrodes. A high voltage trigger causes spark formation and discharge of the condenser, after which a new cycle starts.

Conventional mass spectrographic or spectrometric focusing techniques focus the ions formed in the spark on the basis of their mass/charge ratio. Detection may be on a photographic plate [70], or by scanning with an electrostatic and electromagnetic analyser followed by detection by an electron multiplier [71].

Brown and Vossen [71, 72] have described the application of this technique to air pollution analysis. A $10.8 \, m^3$ air sample collected on a nitrocellulose or cellulose acetate filter was pretreated by low temperature ignition of the nitrocellulose or dissolution of the cellulose acetate in acetone after addition of graphite. After removal of the filter material and addition of a $5 \, \mu g$ silver standard, the sample was ground and pressed as a tip on pure graphite electrodes. All elements except hydrogen and helium may be determined using the technique and in an urban air sample of $10.8 \, m^3$, elements may be determined with an estimated precision of plus or minus 30% S.D. A detection limit of 10 pg is attainable [72].

The technique has also been used to determine metals in dustfall [71], in fly-ash and in other matrices and a comparison of these results with those from other instrumental analytical techniques has been reported [66].

Experimental procedure for spark source mass spectrometry [72]

Air sampling

Air is drawn through a nitrocellulose filter pad for 9.5 h at 19 litre min^{-1}.

Sample preparation

Place the filter pad in a clean silica boat with 0.1 g of high purity graphite (Graphite type USP supplied by Ultra Carbon Corporation, Michigan, U.S.A). Add high purity ethanol (1 ml) as a wetting agent and 100 μl of an aqueous solution of silver nitrate containing 5 μg of Ag as internal standard. Exercising extreme caution, ignite the alcohol, causing slow combustion of the nitrocellulose, and finally to ensure complete ashing place the boat in an oven at 450° C for 1 h.

Transfer the resultant mixture to a dessicator to cool, and then grind thoroughly in a clean agate vial with an agate ball pestle. After grinding, compress portions of the sample onto a support substrate of pure carbon. Encapsulation of the sample rods (3/8 inch length; 3/32 inch diameter) in polyethylene during the pressing operation prevents contamination.

Procedure

Mount the sample electrodes in the source sample clamps so that the sample tips form the analysis gap in front of the first slit. A vibrator is used in place of the micromanipulators for manual adjustment of the electrode gap. Record 5 scans at a suitable multiplier gain. Using an A.E.I. MS702 instrument, a multiplier gain of 10^5 allowed a limit of detection of 0.004 μg m^{-3} for Pb. A scan time of 9 min allows examination of all the elements in the periodic table, apart from hydrogen and helium. Spark excitation parameters were as follows: RF voltage, 30 kV; spark pulse repetition rate, 1000 pulses s^{-1}; spark pulse length, 100 μs. The ion extraction voltage was 20 kV, and cryo-absorption pumping was found to be advisable.

Calculation of results

Measure the height of a peak of an isotope of each element and the peak obtained from the m/e 107 isotope for Ag, the internal standard. For an Ag internal standard of 5 μg, and an air sample of 10.8 m^3, element concentrations are calculated from the following expression.

$$5 \times \frac{A_i}{A_s} \times \frac{I_s}{I_i} \times \frac{\text{at wt}_i}{\text{at wt}_s} \div 10.8 = \text{concn. in } \mu\text{g m}^{-3}$$

where A_i = peak height impurity isotope; A_s = peak height $^{107}Ag^+$; I_s = isotope abundance of ^{107}Ag; I_i = isotope abundance of impurity isotope; at wt$_i$ = atomic weight impurity element; at wt$_s$ = atomic weight of silver. Determination of the metal content of a blank filter is advisable.

3.2.3.6 Spectrophotometry and fluorometry

Spectrophotometric methods, other than atomic absorption, have long been used in air pollution analysis, but because of their generally inferior sensitivity, they are steadily being replaced by more modern techniques. Airborne particulate matter is collected by a normal procedure, ashed, and a solution prepared in the usual manner. By addition of a reagent, a complex compound of the metal is formed, which absorbs light in the visible or u.v. region. Spectrophotometric measurement of the light absorption in a narrow band of wavelengths, and comparison with standards, allows calculation of the concentration of the metal complex. Microgram quantities of metals may be determined, but since many reagents form light-absorbing compounds with a range of metals, selectivity must be enhanced by fixing of oxidation states, the use of masking agents and control of pH. Colorimetric, spectrophotometric and related methods are described in detail by Cheng [73].

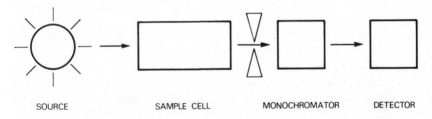

SOURCE   SAMPLE CELL   MONOCHROMATOR   DETECTOR

Fig. 3.3   Essential features of a simple single beam spectrometer.

Spectrometers and spectrophotometers vary considerably in design and sophistication, but Fig 3.3 shows in block form the basic design of a simple single beam spectrometer. Light from the source is passed through a cell containing the sample solution. The beam then passes to a monochromator which selects a narrow band of wavelengths, normally at the absorption maximum of the metal compound, and transmits light in that band to the detector, a photocell which measures the intensity of the beam. The light absorption of a cell containing the reagents, but not added sample, is also measured and used to zero the instrument. The transmittance of the sample, T, is defined as the ratio of the radiant power transmitted by a sample to the radiant power incident upon the sample [74], and the absorbance, $A = \log_{10} 1/T$.

Normally, either transmittance or absorbance is measured by the instrument and this may be used to estimate the concentration of metal in the sample by use of a calibration curve. Beer's law applies to most measurements and this may be stated as $a = A/bc$ where, a, the absorptivity is a constant for the given light absorbing substance; c is the concentration of the substance in solution, and b is the sample length, or internal length of the sample cell. The molar absorptivity, ϵ, also known as the molar extinction coefficient, is a commonly quoted figure and is a constant for a given compound.

$\epsilon = a \times$ Molecular weight.

An important extension of Beer's law is the 'law of additivity' which states that absorption of one molecular species will be unaffected by the presence of other species, absorbing or otherwise. Hence in a mixture of absorbing substances, the absorbances are additive.

Literature relating to spectrophotometric reagents and determination of metals is legion. A valuable list of reagents, molar absorptivities, specificities and experimental methods has been compiled by I.U.P.A.C. [75]. The application of these methods to the analysis of air pollutants has been described in detail for many metals by West [58], the American Public Health Association [76], and by Jacobs [77], and hence will not be reviewed in detail here.

Probably the most frequently described determination by a spectrophotometric method is that of airborne lead with dithizone [58, 76–81], reported to have a threshold sensitivity of $0.2 \mu g$ Pb and a precision on any given result of plus or minus $0.1 \mu g$ Pb [78, 79]. Dithizone may, by adjustment of conditions, also be used for determination of other metals such as Hg, Ag, Bi, Sb, Cd and Zn [58, 82]. Spectrophotometric methods have also been recommended for the determination *inter alia* of As [83], Be [84, 85] and Ni [86] in air. Detection limits normally lie within the region of 0.1 to $1 \mu g$ of the metal [58].

A method reported for determination of chromic acid mists in air involves direct impregnation of the paper used for air filtration with the reagent, and glycerol as a humectant. The Cr is determined by comparison of the colour of the filter with standards after completion of air sampling [24, 87].

Absorption of light in the visible or u.v. region by a molecule involves an electronic excitation. One mechanism by which the molecule may lose energy is first by loss of vibrational energy to return to the ground vibrational state of the excited electronic state, and subsequent emission of light energy, which allows a return to the ground electronic state. The emitted light is known as fluorescence, and is of longer wavelength than the absorbed exciting radiation [73]. Fluorescence measurements, using a fluorometer or spectrofluorometer, usually involve irradiation of the sample by monochromatic light in the u.v. region. Measurement of fluorescent intensity at the emission maximum, via a second monochromator, allows determination of the concentration of the fluorescent compound, upon comparison with standard samples. Be may be determined as the morin $(2', 4', 3, 5, 7$-pentahydroxyflavone) complex by fluorescent measurement [58, 76, 80] and high sensitivity may be attained, allowing measurement of as little as $0.01\,\mu g$ of Be. Selenium (4^+) may also be determined by fluorescence after reaction with 2, 3-diaminonaphthalene [76].

Experimental procedure for determination of Pb *with dithizone* [81]

Air sampling

Approximately 150 to $200\,\text{m}^3$ of air are drawn at $20\,\text{litre min}^{-1}$ through a 47 mm glass fibre filter.

Reagents

Water is distilled or de-ionized before use. Nitric acid—perchloric acid solution is prepared by mixing conc. HNO_3 (300 ml with 72% perchloric acid (200 ml). HNO_3 $(1 + 4)$ is prepared by dilution of conc. HNO_3 (200 ml) with water to 1 litre. Dithizone solution is made by dissolution of diphenyl-thiocarbazone (40 mg) in choloroform (1 litre). It is stored at room temperature in the absence of direct light. Disodium EDTA solution is prepared by dissolution of disodium ethylene-diaminetetracetate (5 g) in water (500 ml). Buffer solution is prepared by solution of dibasic ammonium citrate (400 g), hydroxylamine hydrochloride (10 g) and potassium cyanide (40 g) in water (1 litre), and subsequent mixing with conc. (sp. gr. 0.90) ammonium hydroxide (2 litres).

Procedure

Thallium, stannous tin and trivalent indium can interfere if present in quantities greater than 20, 100 and $200\,\mu g$ respectively, when a modified procedure must be employed. In the absence of such interferences proceed as below.

Remove the filter from the filter holder and place it, exposed side down, in a 150 ml beaker and add nitric—perchloric acid mixture (10 ml). Digest on a hotplate to fumes of perchloric acid (use a fume hood, and exercise extreme care), until all of the dark carbonaceous material has oxidized. Add HNO_3 $(1 + 4)$ (20 ml), mix, crush the filter with a glass rod and allow to cool. Filter the sample through a thin, rapid-filtering paper hardened to great wet strength, directly into a lead-free 100 ml volumetric flask. Rinse the glass fibre with water (3 × 20 ml), make up to 100 ml, stopper and shake the volumetric flask to mix well.

Pipette a suitable aliquot (normally 10 ml) of sample to a 200 ml modified

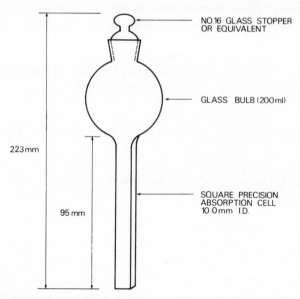

NO.16 GLASS STOPPER
OR EQUIVALENT

GLASS BULB (200ml)

223mm

SQUARE PRECISION
ABSORPTION CELL
10·0mm I.D.

95 mm

Fig. 3.4 Modified absorption cell [81].

absorption cell and add HNO_3 $(1 + 4)$ [20 ml], water (25 ml), buffer (50 ml), mix, and cool to room temperature. The modified absorption cell consists of a normal 10 mm path length spectrometer cell with a 200 ml stoppered glass bulb fused to the top (Fig. 3.4). Add dithizone solution (10 ml) and shake vigorously for 30 s. Insert the cell in the spectrophotometer and measure the absorbance of the lower layer at 510 nm, using air as reference. Add disodium EDTA solution (5 ml), shake vigorously for 90 s and measure the absorbance of the lower layer from 1 to 5 min after the two layers separate. The difference between the two absorbance readings represents the quantity of Pb present in the aliquot. If the initial absorbance reading is greater than 2.0, add additional dithizone solution (10 ml portions) to the sample to dilute the lead dithizonate colour, shake vigorously for 30 s and repeat the absorbance measurements before and after adding EDTA solution.

Calibration

A lead standard is prepared by solution of lead nitrate (0.1599 g; analytical grade) in water (about 200 ml). After addition of conc. HNO_3 (10 ml) the solution is diluted to 1 litre with water. Further dilution of the above solution (20 ml) to 1 litre provides the necessary lead standard (2 μg lead ml^{-1}). Add HNO_3 $(1 + 4)$ [20 ml], water (25 ml) and buffer solution (50 ml) to a 200 ml modified absorption cell, mix and cool to room temperature. Add dithizone solution (10 ml) and shake vigorously for 30 s. Insert the cell into the spectrophotometer and measure the absorbance of the lower layer at 510 nm using air as reference. Add standard lead solution ($2\,\mu g\,ml^{-1}$; 10 ml), shake the mixture vigorously for 30 s and measure the absorbance due to the added lead. Add further standard lead solution (10 ml) and repeat the above procedure.

Calculation

Calculate the calibration factor,

$$F = X/(Y-Z),$$

where $X = \mu g$ lead in calibration sample; $Y =$ absorbance after adding Pb; $Z =$ absorbance before adding Pb.

Calculate the particulate Pb,

$$C_p(\mu g\,m^{-3}) = (A-B)-(C-D) \times F \times (35.3/H) \times (G/10) \times (100/E),$$

where $A =$ sample absorbance before EDTA treatment; $B =$ sample absorbance after EDTA treatment; $C =$ blank absorbance before EDTA treatment; $D =$ blank absorbance after EDTA treatment; $E =$ volume (ml) of aliquot removed from 100 ml volumetric flask; $G =$ volume of dithizone solution (ml); $H =$ volume of air sampled (ft^3).

The blank values are obtained by analysis of an unexposed filter by the above procedure.

Experimental procedure for fluorimetric determination of Be [76]

Sample collection

Ambient air is sampled for 24 h through a suitable filter.

Sample preparation

(Cellulose or Membrane filter sample.) Transfer the filter to a Vycor dish and wet with 1 : 5 H_2SO_4. Heat on a hot-plate until charring occurs, then add a few drops of conc. HNO_3 and evaporate to dryness. Fire with a Meeker burner until carbon-free (a muffle furnace may be used). Allow the dish to cool, wet the ash with mixed acid (50% conc. HNO_3–50% conc. H_2SO_4) and evaporate to fumes on a hot-plate or burner. Repeat if not carbon free. Cool, add a few drops of 1 : 1 H_2SO_4 and heat to strong sulphur trioxide fumes. Remove from hot-plate, cool and add enough water to dissolve salts. Transfer to a 50 ml volumetric flask and make up to volume. (Samples containing beryl ore or high fired beryllium oxide which is difficult to put in solution should be decomposed using a potassium fluoride, sodium pyrosulphate fusion [88].)

Analysis

Pipette sample solution (1 ml) into a 15 ml graduated centrifuge tube. Add 0.05 M aluminium nitrate solution (1 ml), 25% w/v ammonium chloride solution (1 ml) and phenol red (1 drop), and then enough 3 N NH_3 solution to make the solution alkaline. Allow to stand for 10 min, and then centrifuge at 2000 rev min^{-1} for 5 min, rotate the tube 180° in the centrifuge and centrifuge for 5 min further. Pour off and discard the supernatant liquor. Dissolve the precipitate in 4 N NaOH (1 ml), add 25% w/v potassium cyanide (1 drop) and make up to 10 ml with water. Centrifuge for 10 min at 2000 rev min^{-1} and pour the liquid into a Coleman cuvette (19 mm diameter × 100 mm long). Measure the fluorescence in a fluorimeter set at 100% fluorescence with a solution of quinine (4 g litre^{-1}) in 0.1 N H_2SO_4. The fluorimeter should be equipped with a filter transmitting approximately 436 nm between lamp and sample and a filter between sample and photocell which transmits a maximum fluorescent emission of 550 nm. Add morin reagent (1 ml) to the sample, mix and read the fluorescence immediately. The morin reading, minus the sample reading without morin equals the net instrument reading. The morin solution is prepared by dissolution of morin (20 mg) in ethylene glycol (5 ml) and dilution to 100 ml with water. Subsequent dilution of 12 ml of this stock solution to 200 ml with water gives the working solution. The stock solution is stable for 1 month if stored in a dark bottle in the refrigerator, and the working solution, which must stand at least 8 h prior to use, may be stored in a dark bottle in the refrigerator.

Calibration

The equivalent of 0.1 g of pure beryllium powder (the available powder is not pure) is dissolved in 1 : 1 H_2SO_4 (15 ml). After cooling it is diluted to 1 litre with water. This stock solution (100 μg beryllium ml^{-1}) is then diluted ten-fold (10 μg Be ml^{-1}), one hundred-fold (1 μg Be ml^{-1}) and 2000-fold (0.05 μg Be ml^{-1}). The latter solutions are prepared daily before use. Add 2, 5 and 10 ml portions of Be standard solution (0.05 μg ml^{-1}) and a 1 ml portion of (1 μg Be ml^{-1}) to filters similar to those used for sample collection. Run the impregnated filters through the above procedures.

Calculation

$$\text{Total Be in sample} = \frac{\text{ml sample}}{\text{ml aliquot}} \times \frac{\text{net sample I.R.}}{\text{net standard I.R.}} \times \mu g \text{ Be in standard}$$

I.R. = instrument reading; ml sample = volume of dilution of total sample; ml aliquot = portion of sample used for analysis; net sample I.R. = I.R. sample with morin minus I.R. sample without morin; net standard I.R. = I.R. with standard minus I.R. of standard (without morin).

$$\mu g \text{ Be m}^{-3} \text{ of ambient air} = \frac{\text{Total of } \mu g \text{ Be}}{\text{Volume of air sampled (m}^3)}.$$

The procedure is recommended for analysis of 0.01 to 1.0 μg of Be in 10 ml of solution, but greater sensitivity can be achieved. Samples can be stored indefinitely without loss of Be.

3.2.3.7 Atomic absorption spectrometry

An atomic absorption spectrometer bears much similarity to a u.v./visible spectrometer. In place of a cell of sample, however, a flame fills the sample space (Fig. 3.5). Aspiration of the sample solution into the flame causes chemical reduction of metal ions to ground state atoms, the light absorption of which is measured. Atomic absorption spectra are line spectra, unlike molecular spectra which are in the form of broad bands. The atomic absorptions are discrete lines of narrow bandwidth at wavelengths which are characteristic of the given element.

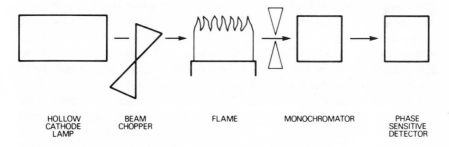

| HOLLOW CATHODE LAMP | BEAM CHOPPER | FLAME | MONOCHROMATOR | PHASE SENSITIVE DETECTOR |

Fig. 3.5 A simple atomic absorption spectrometer.

The light source, a hollow cathode lamp in which the cathode is coated with the metal to be determined, emits light at the discrete absorption wavelengths. Introduction of sample into the flame causes absorption of some of the incident light and after passage through the monochromator, which selects one absorption line, the reduction in beam intensity is measured by the detector. In double beam spectrophotometers, a second reference light beam from the same hollow cathode lamp by-passes the flame and is received at the detector, in alternate pulses to the sample beam and the ratio of beam intensities, after passage through the monochromator, is measured.

The wavelength range of atomic absorption measurements is limited by light absorption by the flame, most usually air–acetylene, at shorter wavelengths, and by the atmosphere at longer wavelengths. This working range, in the u.v. and visible region, includes all metals and semi-metals but excludes most non-metallic elements. It is reported that 65 elements may be determined by atomic absorption methods, although sensitivities vary substantially [76, 89, 90].

The absorption laws applying to spectrophotometric measurements apply to atomic absorption, and most commercial instruments read out in absorbance units. Beer's law of the linear relationship between absorbance and sample concentration applies to most measurements, but deviations occur, particularly at higher concentrations and it is normally advisable to calibrate over the full range of measurements. Additionally, since instrumental parameters are liable to change, it is normal practice to calibrate the instrument at least daily.

Two techniques related to atomic absorption are also currently in use. Some elements, notably the alkali metals and alkaline earths, are electronically excited by the flame, to an extent sufficient for the effect to be analytically useful. The return to the ground electronic state is by emission of light at discrete characteristic wavelengths, and measurement of emission intensities in the absence of a hollow cathode lamp allows determination of metal concentration by comparison with standards. This technique is known as emission spectroscopy or flame photometry. The other related technique, atomic fluorescence spectroscopy does involve use of a light source. Absorption of light by the atoms in the flame occurs and results in electronic excitation. Return to the ground state may be by fluorescent light emission, and the measurement of fluorescent intensity allows determination of metal concentrations. Atomic fluorescence may be more sensitive than conventional atomic absorption spectroscopy [91].

Detection limits for some metals by conventional atomic absorption techniques are shown in Table 3.10. Double beam instruments normally allow a limit of detection several times lower than that attainable with a single beam. In estimating absolute detection limits it must be borne in mind that normally several ml of samples are required to obtain one reading. Recently, the introduction of the 'Sampling Boat' technique, in which an aliquot of sample ($10\,\mu l$–1 ml) is dried in a tantalum boat and then inserted into the flame causing rapid volatilization of the metal [92], and the advent of flameless atomizers has allowed a substantial improvement in detection limits (Table 3.10). Flameless atomizers involve

Table 3.10 Atomic absorption detection limits

Metal	Analytical line (nm)	Flame* (μg ml^{-1})	Sampling boat (ng)	Graphite atomizer (pg)
Ag	328.1	0.002	0.2	0.1
Al	309.3	0.02†		2
As	193.7	0.05‡	20	10
Be	234.9	0.002†		3
Ca	422.7	< 0.0005		
Cd	228.8	0.002	0.1	0.1
Co	240.7	0.01		5
Cr	357.9	0.003		10
Cu	324.7	0.001		2
Fe	248.3	0.005		3
Hg	253.6	0.25	20	
K	766.5	< 0.002		
Mg	285.2	< 0.0001		
Mn	279.5	0.002		0.2
Mo	313.3	0.02†		3
Na	589.0; 589.6	< 0.0002		
Ni	232.0	0.002		10
Pb	283.3	0.01	1	2
Sb	217.6	0.04		20
Se	196.0	0.05‡	10	50
Sn	224.6	0.01‡		100
Sr	460.7	0.002		5
V	318.3; 318.4; 318.5	0.04†		100
Zn	213.9	< 0.001	0.03	0.05

* Double beam instrument; † Nitrous oxide – acetylene flame; ‡ Argon–hydrogen-entrained air flame.
Data cited for Perkin-Elmer spectrophotometer and heater graphite atomizer.

use of a graphite tube, cup, rod or ribbon whose temperature is controlled and may be rapidly varied [93, 94]. Hence an aqueous sample (about 20 μl) may be dried at 100° C, ashed at 550° C and atomized by raising the temperature to 2000° C, the graphite assisting chemical reduction of the metal ions. A sharp peak, corresponding to atomization is recorded on a strip-chart recorder.

Although spectral interferences, involving overlap of absorption lines, are not a problem in atomic absorption measurement, other types of interference may occur. Chemical and ionization interferences occur for some metals, but these are well documented and may be overcome [89]. Matrix effects are common in air pollution measurements by atomic absorption, and must be taken into account. Physical properties of the matrix (e.g. viscosity) affect the rate of sample aspiration into the flame when using conventional techniques. Additionally, other materials present in the matrix affect the rate of atomization of the metal, and hence affect the absorption signal. It is not possible when dealing

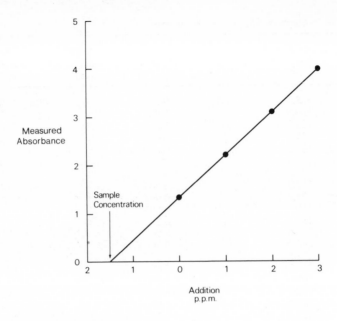

Fig. 3.6 The method of additions.

with air pollution samples to make standard solutions containing identical matrix materials to the those in the sample, and hence use of the method of standard additions [95] is advisable, and its use in air pollution measurements has been described [96–98]. Three, or more aliquots of sample are taken, and the first diluted to a known volume with solvent. The other aliquots are diluted with solvent to the same volume after addition of suitable quantities of standard solution of the metal to be determined. The two, or more, additions are different, and are calculated to approximately double or treble the concentrations of metal in the diluted solutions. A graph of measured absorbance against standard addition may then be plotted after A.A. measurement (Fig. 3.6). Upon extrapolation to zero absorbance, the intercept on the concentration axis corresponds to the concentration of the metal in the diluted solution.

Extraction of the metal using a complexing agent and organic solvent may be beneficial in that interference may be avoided and samples concentrated [64, 99]. Conventional flame techniques are typically two to four times more sensitive when using organic solvents.

Atomic absorption methods for determination of metals in airborne particulate matter have been reviewed by Hwang [9]. Pb in air has been the subject of intensive study, using both conventional [16, 17, 96–98, 100–105] and flameless [106–109] methods. Techniques have also been described for analysis of

Cd [110, 111], Fe [97, 100], V [112] and ranges of metals [16, 17, 64, 97, 101–103, 113, 114] in airborne particulate matter.

Flameless atomization procedures are so sensitive that air samples as small as 200 ml may be used to measure lead [109] at levels as low as $0.1\,\mu g\,m^{-3}$. Filter and reagent blanks, however, tend to become the limiting factor in such measurements. The blanks may be avoided by continuous introduction of sample into the instrument, simply by pumping or drawing polluted air through the flame or flameless atomizer. Such methods have been devised for continuous monitoring of metals in air. These include determination of Pb at levels from $0.16\,\mu g\,m^{-3}$ of air [115, 116], Cd from $0.005\,\mu g\,m^{-3}$ of air [117], Ag from $3\,\mu g\,m^{-3}$ of air [118], and Pb in vehicle exhaust gases [119] and smelter fume [120]. Robinson and Wolcott [121] have described some of the difficulties inherent in the calibration of such techniques. Techniques using a graphite cup from an atomic absorption flameless atomizer as an air filter have a very low blank, and are hence also capable of very high sensitivity [108, 122].

Experimental procedure for atomic absorption analysis of As, Ba, Cd, Cr, Cu, Fe, Mn, Pb, Tl, V, *and* Zn [16, 17]

Air sampling

High-volume air samples (24 h) are taken through 8×10 in glass fibre filters (7×9 in exposed surface) at 50 to 60 ft^3 min^{-1} (about 2200 m^3 total).

Sample preparation

Using a plastic template as a cutting guide, cut a strip from the filter for analysis. The amount of filter taken depends upon the type of sample: for urban samples a 7×1 in strip is cut, for non-urban samples a 7×2 in strip, and for industrial samples the size may be reduced to an extent consistent with the expected levels of airborne metals. Ash the strip in a low-temperature asher at approximately $150°$ C for 1 h at 250 W and 1 mm chamber pressure with an oxygen flow of 50 ml min^{-1}. Alternatively, the filter strip may be ashed in a muffle furnace at 500 to 550° C, but losses of individual metals of up to 55% may occur. Place the ashed filter in a glass thimble which is placed in an extraction tube (Fig. 3.7). A 125 ml Erlenmeyer flask with a 24/40 female joint is charged with constant boiling (about 19%) HCl (8 ml) and 40% HNO$_3$ (32 ml). The flask is attached to the extraction tube, and the tube fitted with an Ahlin condenser. Reflux the acid over the sample for 3 h, during which sample and extraction thimble should remain at the temperature of the boiling acid.

Remove the extraction tube and condenser from the Erlenmeyer flask and fit

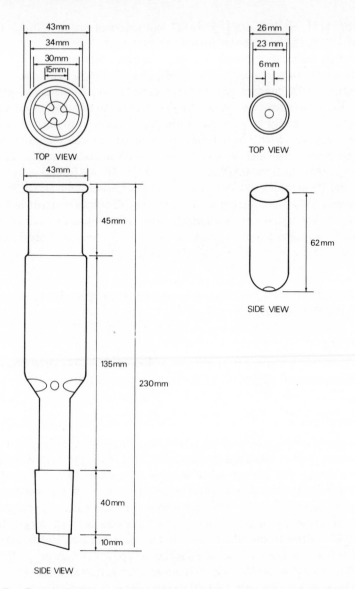

Fig. 3.7   Sample holder and extraction tube for acid extraction of glass fibre filter papers [16, 17]

the flask with a thermometer adapter which serves as a spray retainer. Concentrate the extracted liquid to 1 to 2 ml on a hot-plate and allow to cool and stand overnight. Quantitatively transfer the concentrated material to a graduated 15 ml centrifuge tube with 3 washings of 5 to 10 drops of diluted acid. Dilute urban samples to 4.4 ml per 1 in strip (40 ml per 63 in² of filter) and non-urban samples to 3.0 ml per 2 in strip (13.3 ml per 64 in² of filter). Following dilution, centrifuge the sample at 2000 rev min⁻¹ for 30 min and then decant the supernatant liquor into polypropylene tubes which are capped and stored prior to analysis. 1 ml from each solution is diluted to 10 ml for analysis by atomic absorption.

Analysis

Determine the metal concentration of the solution by comparison of the absorbance of the sample solution to the absorbance of the standard metal solutions. General instrumental operating parameters for a Perkin-Elmer Model 303 are shown in Table 3.11. Working standards are prepared daily from stock solutions of 500 to 1000 μg ml⁻¹ of each metal. Working standard concentrations will vary according to the sensitivity and detection limits of each metal. Results are calculated by multiplying the number of μg of each metal per ml of sample extract by the appropriate dilution factor and dividing by the number of m³ of air represented by the sample.

Table 3.11   Operating and analytical parameters [16, 17]

Element	Wavelength (nm)	Range (μg ml⁻¹)	Minimal detect. conc. (μg m⁻³)*
As	193.7	1.00– 50	0.02
Ba†	553.6	1.00–100	0.02
Ca	422.7	0.01– 8	0.0002
Cd	228.8	0.10– 20	0.0002
Cr	357.9	0.10– 20	0.002
Cu	324.7	0.05– 20	0.001
Fe	372.0	0.50– 40	0.01
Mn	279.8	0.05– 20	0.001
Ni	232.0	0.20– 20	0.004
Pb	217.0	0.10– 40	0.002
Tl	276.8	0.50– 50	0.01
V	318.4	0.50– 50	0.01
Zn	213.8	0.01– 2	0.0002

* Based on a 2000 m³ air sample.
† Nitrous oxide–acetylene burner required.

Note

(i) The ten-fold dilution prior to analysis is necessary to reduce the dissolved solids content to less than 0.5%, and thereby eliminate matrix effects.

(ii) Silica extracted from glass-fibre filters and from the particulate matter itself can also cause significant interference with Fe, Ca, Mn and Zn. This interference may be overcome by allowing the acid extracts to stand for 12 to 24 h followed by centrifugation. The acid extract is then separated from the precipitated silica by decantation into a polypropylene tube.

(iii) All glassware is soaked in 20% HNO_3 for 2 to 6 h and washed with distilled water prior to use. HNO_3 and HCl acids are distilled in all-glass apparatus to remove metal contaminants.

(iv) Blank glass-fibre filters are analysed to determine the metal background. The use of membrane filters permits the analysis of the extremely low levels of elements such as Fe, Ba and Zn that are found in high concentrations in glass fibre filters.

Experimental procedure for atomic absorption analysis of Al, Be, Bi, Ca, Co, Cr, Cd, Cu, Fe, K, Li, Mg, Mn, Na, Ni, Pb, Rb, Si, Sr, Ti, V *and* Zn [123]

Air sampling

Air is sampled through polystyrene filters (microsorban Type 99/97, Delbag-Luftfilter GMBH, Berlin, Germany) (25 × 20 cm sheet) for 24 h using a high-volume sampler (about 2000 m^3 sample).

Sample preparation

The filter is folded, placed in a 20 ml beaker, covered with a watch glass and ashed overnight (12 to 16 h) at 400 to 425° C in a muffle furnace. After ashing, transfer the residue to the PTFE cup of an acid digestion bomb (Parr Instrument Co., Moline, Ill.) Scrape the residue out of the beaker using a PTFE-coated spatula and add 65% HNO_3 (2 ml) and 30% HCl (0.5 ml) to the beaker. Warm the beaker on a hot-plate (60° C) and cool before pouring the solution into the PTFE cup. Add 65% HNO_3 (2 ml) to the beaker and warm on the hot-plate (60° C) for 5 min. Cool the beaker and pour the acid solution into the PTFE cup. The procedure is repeated with 65% HNO_3 (2 ml) and 30% HCl (1 ml) and finally 40% HF (3 ml) is added directly to the PTFE cup (Extreme caution). Place the cover on the PTFE cup and seal the cup in the acid digestion bomb. Heat the bomb in an oven at 125° C for 6 h, and after digestion cool in a freezer to − 5° C for 2 h.

Open the bomb, remove the PTFE cup and allow to reach room temperature. If necessary, the volume of solution is adjusted to a known volume in polypropylene labware (normally the final volume of solution on opening the bomb is 10.0 ml plus or minus 0.2 ml, in which case this step may be omitted). An aliquot (2 ml) is taken and placed in a polypropylene bottle (0.5 oz) containing

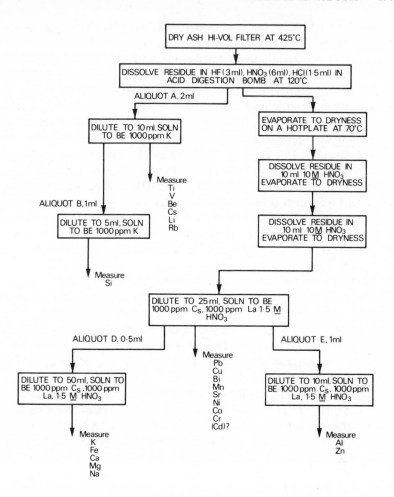

Fig. 3.8 Flow chart for the atomic absorption analysis of Hi-vol atmospheric particulate samples collected on polystyrene filters [123]

10 000 ppm potassium solution (1 ml) and de-ionized water (7 ml). This solution, designated aliquot A (1.8 M hydrofluoric acid; 1.9 M HNO_3; 0.4 M HCl; and 1000 ppm K approximately) is used to measure Be, Cs, Li, Rb, Ti and V.

Aliquot A (1 ml) is added to another polypropylene bottle (0.5 oz) containing 5000 ppm potassium solution (1 ml) and de-ionized water (3 ml). The resultant solution, designated Aliquot B (0.4 M hydrofluoric acid; 0.4 M HNO_3; 0.1 M HCl; and 1000 ppm K) is used to measure Si.

Evaporate the sample solution left in the PTFE cup to dryness on a hot-plate (about 70° C). Dissolve the residue in 10 M HNO_3 (10 ml) and evaporate to dryness. Redissolve the residue in 10 M HNO_3 (10 ml) and evaporate to half its original volume. Quantitatively transfer the sample to a 25 ml volumetric flask containing 6250 ppm lanthanum in 8 M HNO_3 (4 ml) and 10 000 ppm Cs (2.5 ml). Wash the PTFE cup several times with de-ionized water, and add the wash to the volumetric flask. Dilute the sample to 25 ml with de-ionized water. This solution (1.5 M HNO_3; 1000 ppm Cs and 1000 ppm La), designated Aliquot C is used to determine Bi, Cd, Co, Cr, Cu, Mn, Ni, Pb and Sr.

Add Aliquot C (0.5 ml) to a 50 ml volumetric flask containing 6250 ppm La in 10 M HNO_3 (8 ml), and 10 000 ppm Cs (5 ml). Dilute the sample to 50 ml with de-ionized water and the solution (1.5 M HNO_3, 1000 ppm La, 1000 ppm Cs), designated Aliquot D is used to determine Ca, Fe, Mg, Na and K. Add Aliquot C (1 ml) to a (10 ml) volumetric flask containing 6250 ppm La in 10 M HNO_3 (1.6 ml) and 10 000 ppm Cs (1 ml). Dilute to 10 ml, and the solution (1.5 M HNO_3; 1000 ppm La; and 1000 ppm Cs, designated Aliquot E is used to measure Al and Zn.

All standards used in the analysis are prepared so as to have as nearly as possible the same matrix as the samples. Interferences may be checked by use of the standard additions method. Standard stock solutions (1000 ppm) of the metals are prepared from reagent grade chemicals, and diluted working standards prepared fresh daily. High purity HCl, HNO_3 and hydrofluoric acid are used and lanthanum nitrate hexahydrate and caesium nitrate solutions are prepared from the purest grades available.

The elements Cs, Li and Rb are best determined by atomic emission and Al, Be, Ca, Li, Mg, Si, Sr, Ti and V require a nitrous oxide—acetylene flame for atomic absorption measurement. All other metals are determined by conventional A.A. techniques. Fig. 3.8 shows the analytical scheme and Table 3.12 the practical detection limits for the procedure.

Experimental procedure for determination of Pb *by flameless atomic absorption* [106]

Air sampling

Air is sampled through Whatman No. 41 cellulose filter paper using either a low vacuum high-volume sampler (20 × 25 cm filter holder; initial rate 70 $m^3 h^{-1}$) or a high vacuum pump (10 cm diameter filter holder; 20 $m^3 h^{-1}$) for 24 h, equivalent to 1200 m^3 and 400 m^3 of air respectively. For Pb determination a 10 cm^2 piece is cut from a 20 × 25 cm filter and 1/8 segment from the 10 cm diameter filter.

Table 3.12 Practical detection limits for atomic absorption analysis of samples collected on polystyrene filters [123]

Element	$\mu g\,m^{-3}$ in air sample*
Al	0.4
Be	0.000 04
Bi	0.002
Ca	0.03
Cd	0.0003
Co	0.000 8
Cr	0.000 3
Cs	0.000 04
Cu	0.003
Fe	0.07
K	0.04
Li	0.000 3
Mg	0.01
Mn	0.000 3
Na	0.03
Ni	0.002
Pb	0.007
Rb	0.000 5
Si	0.8
Sr	0.003
Ti	0.07
V	0.006
Zn	0.03

* The amount of metal required to provide a signal which is equal to twice the S.D. of the blank signal for the same element when $2000\,m^3$ of air is sampled.

Sample preparation

Transfer the filter sector into 0.1 M HNO_3 (about 50 ml) and expose it to ultrasonic vibration (Branson ultrasonic cleaner, consisting of an ultrasonic generator Model LG-150 and an ultrasonic tank-type transducer LTH-60) for 5 min. Decant the solution and repeat ultrasonic treatment in 0.1 M HNO_3 (30 ml) for 5 min. Wash the beaker several times with 0.1 M HNO_3, combine all extracts and washings and dilute to 100 ml in a volumetric flask. Treat a blank filter in the same way. Use only glassware pretreated with HNO_3 for 24 h.

Determine Pb by injecting 10 to 50 μl of the extract with an Eppendorf micropipette into the graphite cell (Perkin-Elmer HGA-70) (Programme 5—drying at 100° C and charring at 490° C; atomization voltage, 9 V, 2400° C) fitted to a Perkin-Elmer Model 303 double beam A.A. spectrophotometer equipped with a deuterium background corrector. Measure the absorbance due to lead at the 283.3 resonance line, running both sample and blank at least twice. Subtract the blank absorbance value from that of the sample filter and determine the concentration from a calibration curve.

Calibration

Prepare a standard solution of Pb (1000 ppm) from lead nitrate (1.599 g) dissolved in 0.1 M HNO_3 and diluted with 0.1 M HNO_3 to 1 litre. Dilute the stock solution with 0.1 M HNO_3 to concentrations from 20 to 250 μg Pb litre^{-1}. The dilutions are prepared weekly. Inject aliquots of standards, of the same fixed volume as that used for the sample, into the graphite furnace. Plot the absorbance against the quantity of lead present in the furnace, a graph which has been found to be linear up to 5 ng lead. 1% absorption was found for 0.1 ng Pb, and the method may be used to determine 0.01 μg m^{-3} of Pb, as described.

Application to Cd [111]

Air samples are collected and prepared exactly as for Pb, with the first ultrasonic treatment being lengthened to 10 min. In this instance, the instrumental parameters are Programme 4 and atomization voltage of 9 V: drying at 100° C; charring at 330° C and atomization at 2400° C. Absorbance peak heights are measured at the 228.8 nm Cd line. If one half of a 10 cm diameter filter is taken, the dilution is made to 50 ml and 50 μl is injected, Cd levels as low as 0.2 ng m^{-3} may be determined with a S.D. of 10%.

3.2.3.8 Other analytical methods

Titrimetric procedures have long been applied in air pollution analysis. Because of their generally low sensitivity and selectivity they are of limited application, but may be of value in the industrial hygiene field where levels of airborne metals are generally higher than in ambient air in other locations and a limited range of metals is present at an appreciable level. Thus, for instance, Fe (100 μg to 400 mg) may be determined by titration with ceric sulphate or potassium dichromate [76], and Jacobs [77] describes the determination of a range of metals by direct or indirect titration. An iodine microtitration method using an amperometric titrator for the determination of airborne arsenic has also been recommended [83].

Other procedures applicable to the analysis of metals may also be applied to air pollution analysis. Be, collected on glass fibre filters may be converted to the bis-trifluoroacetylacetone complex and determined by gas chromatography. The method has a limit of detection of 4 × 10^{-14} g Be and readily determines levels of Be in urban air (about $3 \times 10^{-4} \mu g \, m^{-3}$) from a portion of a filter used to sample 2200 m^3 of air over a 24 h period [124].

Other more specialized observations and techniques have included electron spin resonance (e.s.r.) spectra of ferromagnetic particles of airborne dust collected on filters. A clear relationship was found between the e.s.r. signal and the amount of Fe on the filter, and this allowed calculation of airborne Fe in a 100 m^3 air sample [125].

Anodic deposition of Pb, followed by isotope ratio mass spectrometry is reported to be applicable to the analysis of 10 ng to 10 μg of Pb and has been used for determination of Pb collected on filters [126]. The method is particularly valuable where isotope ratio data are required.

Experimental procedure for gas chromatographic determination of Be [124]

Air sampling and sample preparation

Air samples (about 2200 m^3) are collected on glass fibre filters in a manner identical to that of Thompson et al. [16, 17], as described in detail under experimental procedures for atomic absorption analysis. Ashing and extraction of the filter are performed in an identical manner. After dilution to 4.4 ml with distilled water and centrifugation at 2000 rev min^{-1} for 30 min, the supernatant is transferred to a polypropylene tube, ready for analysis.

Reagents

0.164 M trifluoroacetylacetone [H(tfa)] is prepared by diluting freshly distilled H(tfa) (2 ml) in high purity benzene (100 ml). Store the solution in a silanized 100 ml borosilicate glass volumetric flask. H(tfa), both neat and in benzene solution is subject to slow decomposition on standing. Ethylenediaminetetraacetic acid (EDTA) – buffer solution is prepared by weighing disodium EDTA monohydrate (5.15 g), sodium acetate trihydrate (85 g) and glacial acetic acid (6.25 ml) and dissolving the reagents in de-ionized water (500 ml). Standard Be(tfa)$_2$ solutions are prepared by dissolving Be(tfa)$_2$, purified by sublimation, in high purity benzene. Further dilutions are made to a resultant concentration of 1.0×10^{-11} g beryllium μl^{-1}. Prepare this solution every 2 weeks.

Standard solutions are stored in borosilicate glass volumetric flasks, cleaned in Chromorge cleaning solution and silanized to reduce any active sites. Silanization is achieved by filling the flasks with 20% hexamethyldisilazane (HMDS) in benzene and standing overnight. The reaction vessels (5 ml culture tubes) are treated similarly.

Analysis

1 ml of the *aqua regia* filter digest is pipetted into a 5 ml borosilicate glass culture tube fitted with a screw cap. EDTA − sodium acetate buffer solution (1 ml) is added to the culture tube. Then add 3 N NaOH solution to a final pH of 5.5 to 6.0; normally from 0.8 to 1.3 ml is required. The tube, containing a small PTFE-coated magnetic stirrer is capped and inserted into an oil bath maintained at 93° C on a Corning stirring hot-plate. Heat the sample and stir vigorously for 10 min. Cool the tube and add 0.164 M H(tfa) in benzene (1 ml). Cap the tube and place above the stirring mechanism of the hot-plate and stir for another 15 min at ambient temperatures.

Stand the tube until aqueous and organic layers separate. Transfer the organic layer to a 2-dram vial with a medicine dropper and add 0.1 N NaOH solution (2 ml). Shake the mixture quickly by hand for 5 s to remove excess H(tfa) from the organic layer. Separate the phases immediately by withdrawing the organic layer with a medicine dropper and transferring it to another 2-dram vial which is then capped. This step must be performed rapidly and reproducibly or intolerably large amounts of Be will be lost and both precision and accuracy will suffer.

Repetitively inject $5 \times 1\,\mu l$ aliquots into the GC. The mean of peak heights from the $Be(tfa)_2$ peaks in the unknown solution are compared with the average from 5 analyses of a standard solution of $Be(tfa)_2$. Over the concentration range used the response is essentially linear so peak height ratios can be used for calculation of Be found.

Instrumental parameters

Ross and Sievers [124] used a Hewlett-Packard Model 402 high efficiency GC equipped with a ^{3}H source electron-capture detector. Column: 2 m × 3 mm internal diameter borosilicate glass column packed with 2.8% W-98 silicone (Union Carbide) on Diataport S (Hewlett-Packard). Carrier gas: methane, 10%; argon, 90%; 54.5 ml min^{-1}. Column temp: 110° C; detector temp: 200° C; on-column injection (no additional heat at site of injection).

The blank from reagents and filter was found to be 4% of the Be in typical urban air samples. The analytical method is sensitive to as little as 4×10^{-14} g Be, corresponding to less than $10^{-6}\,\mu$g Be m^{-3} of air.

3.3 Gases and vapours

3.3.1 General sampling considerations

Few metallic elements and compounds are of significant volatility at ambient temperatures. Those that are present in air in the vapour phase, however, provide a totally different sampling problem from that of the particulate forms, although analytical techniques may be similar. Some volatile metal compounds may be determined by continuous monitoring at certain concentrations (e.g. nickel carbonyl, tetraethyl lead), whilst others require a pre-concentration stage. Pre-concentration may be by adsorption tube, liquid scrubber, thermal decomposition to an involatile form, or freeze-out. No technique can be assumed to give a quantitative collection, and in all cases collection efficiency must be established by calibration. The effect of temperature upon such procedures should also be investigated. It must also be recognized that the efficiency of recovery of collected compounds from adsorbents and freeze-out traps may be less than quantitative.

Choice of unsuitable materials for the construction of apparatus and sample lines can lead to considerable error. Some types of tubing, including glass, may strongly absorb gaseous compounds. Hence, initially low results may be obtained, whilst subsequent desorption into a gas stream containing low levels of the compound may cause false positive readings. Adsorption on surfaces and tubing may also catalyse reactions with other adsorbed compounds and lead to error. Thin tubing may be significantly permeable to gaseous pollutants, and both outward diffusion of sample and inward diffusion of the surrounding atmosphere may occur. Greases and oils used in seals may strongly absorb gaseous materials; PTFE which is self-lubricating, is recommended for making such seals. Surface effects and diffusion effects will be most pronounced if samples are stored, and where possible rapid analysis of grab samples is desirable. Mechanical leakage may be a large source of error, and corrections for temperature and pressure are important when calculating pollutant levels.

Particulate matter in an air sample may foul rate meters and total volume meters, as well as blocking orifices in impinger-type collectors. It is frequently necessary to use a pre-filter to avoid such problems, although the possibility of adsorption of gaseous pollutants upon particulate matter must be borne in mind. In some cases, such as the collection of lead alkyls with iodine monochloride solution, efficient prior filtration of lead particulate matter is essential if analysis of particulate Pb with the volatile lead alkyls is to be avoided.

3.3.2.1 Metal carbonyls

Metal carbonyls are formed in industrial processes, both by intent and as undesired by-products of catalyst decomposition. Hence, analytical procedures for Ni and Fe carbonyls have been devised for use in the context of industrial hygiene. These compounds also occur in town gas, and Densham et al. [127] have reviewed the analytical methods available for their determination. Jacobs [77] describes a method for determination of iron and nickel carbonyls in air which involves thermal decomposition to the metal in a silica or glass tube, and subsequent analysis of the deposited metal. This method, however, is reported not to be quantitative at the thermal decomposition stage [127].

Numerous liquid and solid adsorbents have been used for collection of metal carbonyls from air [128]. An acidified solution of chloramine-B in alcohol has been used [129], as well as alcoholic I [130, 131]. These methods involve a subsequent spectrophotometric determination of the metal as a complex, and this limits their sensitivity. Analysis by atomic absorption spectrometry might be expected to allow a substantial improvement in sensitivity. Filtration of the air sample initially is necessary to ensure elimination of particulate metal.

Brief *et al.* [132] describe the collection of nickel carbonyl in 10–15 ml of 3% HCl in a midget impinger. Reaction with α-furildioxime in the presence of chloroform forms a yellow complex which is determined spectrophotometrically. Collection efficiency is greater than 90%, but Cr, Fe and V interfere with the analysis. Sampling for 20 to 30 min at 0.1 ft^3 min^{-1} gives a detection limit of 0.0008 ppm of nickel carbonyl. For collection of iron pentacarbonyl Brief *et al.* [133] use 3% HCl mixed with an I/KI solution immediately prior to use. Air is sampled at 2–3 litre min^{-1} through 10 ml of this solution in a fritted bubbler, after prior removal of particulate matter by filtration. Collection was found to be virtually quantitative. After reduction to the ferrous state, Fe is determined spectrophotometrically as its 1,10-phenanthroline complex, and interferences may be overcome by extraction of Fe, as ferric chloride, into isopropyl ether. In a 50 litre air sample 0.009 ppm of iron carbonyl may be detected.

Densham *et al.* [127], using a sintered bubbler containing iodine monochloride in glacial acetic acid, were able to collect both iron and nickel carbonyls. After evaporation of the reagent and addition of the complexing agent, determination of the metal by visual colorimetric means is carried out. Sensitivities were 0.006 ppm for nickel carbonyl and 0.01 to 0.03 ppm for iron carbonyl.

A continuous procedure for measurement of metals carbonyls in town gas is also described by Densham *et al.* [127]. Polluted town gas was used to supply the combustion gas stream of an atomic absorption instrument. In order to obtain blank readings, the gas stream was purified by prior passage through activated carbon, and the absorbance measured. The detection limits were 0.002 ppm of nickel carbonyl in the gas, and 0.01 ppm of iron carbonyl. Subsequent improvements in atomic absorption instrumentation should allow a substantial lowering of these detection limits.

A continuous monitor for metal carbonyls in air, unable to discriminate between Ni and Fe, has been described by McCarley *et al.* [128]. Polluted air impinges upon a hot borosilicate glass plate causing thermal decomposition of metal carbonyls and deposition of the metal. An ingenious optical system produces a collimated beam of plane polarized light which is incident upon the surface of the glass at the Brewsterian angle. In this arrangement, clean glass reflects no light, and reflectance is due solely to deposited metal. Concentrations of 0.05 to 4 ppm of nickel carbonyl in air may be continuously determined.

It is also reported that a direct field instrument which is commercially available is suitable for continuous monitoring of nickel carbonyl in the 10 to 1500 ppb range [130]. Kincaid *et al.* [131] give a brief description of a continuous monitoring instrument in which nickel carbonyl is reacted with Cl_2 or Br_2 in the gas phase forming a smoke of nickel halide. The light scattering caused by the smoke is measured, and the intensity of scattered light is related to the concentration of nickel carbonyl in the contaminated air. Levels of a fraction of 1 ppm may be determined.

A commercially available i.r. monitor with a 40 m path length absorption cell is capable of determining nickel carbonyl with a detection limit of 0.2 ppb, and by automatically measuring absorbance at three different wavelengths avoids interference by considerable excess concentrations of CO.

Experimental procedure for determination of nickel carbonyl [132]

Air sampling

Air or process gas is sampled by bubbling through 10 to 15 ml of 3% HCl in a midget impinger for 30 to 60 min at $0.1 \, \text{ft}^3 \, \text{min}^{-1}$. A filter assembly precedes the bubbler to ensure that no nickel solids enter the bubbler.

Analysis

Transfer the liquid to a 60 ml separatory flask and add in sequence:
   (a) 2 drops of phenolphthalein;
   (b) 4 drops of ammonium hydroxide (about 20 to 30% aqueous solution);
   (c) 20% aqueous NaOH to the phenolphthalein endpoint, plus 3 drops excess;
   (d) 3.0 ml α-furildioxime solution (1% α-furildioxime in 1:1 alcohol:water);
   (e) 5 ml chloroform.

After shaking for at least 1 min, allow the chloroform layer to separate from the aqueous phase and draw the chloroform layer into a test tube. The colour may be compared with pre-established standards to determine µg of Ni in the chloroform. For greater accuracy the final colour is compared against reagent blanks in 1 cm cells in a spectrophotometer set at 435 nm. The volume of air sampled and the weight of Ni collected are then used to calculate the Ni concentration in the gas.

Sensitivity

The method is visually sensitive to at least 0.002 ppm, and using a spectrophotometer a sensitivity of 0.0008 ppm has been reported for 20 to 30 min sampling periods. No serious interferences are known, although Cu gives a brown precipitate which can be removed by washing the chloroform extract twice with 10% ammonium hydroxide. Cr can be tolerated up to about 0.1% and Fe and V up to about 0.5%.

Iron carbonyl [133]

Air sampling

The absorption solution is made up in 2 parts and mixed just prior to use. One part consists of 3% v/v HCl and the other I_2–KI solution. The latter is prepared by dissolving I (4 g) and KI (10 g) in water, filtering the solution through glass wool, diluting the filtrate to 100 ml and storing it in a brown bottle. When required the absorption solution is prepared by mixing 1 ml of the I_2–KI solution to 10 ml of the HCl solution.

Prior to use all glassware used in sampling and analysis must be rinsed with 10% HCl and then with distilled water. The air sample is bubbled through 10 ml of the absorption solution in a fritted bubbler, designed for this volume of solution, at a rate of 2 to 3 litres min^{-1} and about 50 litres of air are sampled. A high efficiency filter precedes the bubbler to collect particulate iron before it reaches the sampler.

Analysis

Transfer the sample to a 50 ml volumetric flask, using a minimum of wash water. Then, add in order:

(a) 1% w/v sodium sulphite solution dropwise until the I is just decolorized (this solution is prepared fresh daily);

(b) 20% w/v hydroxylamine hydrochloride (1 ml) reduces Fe^{3+} to Fe^{2+};

(c) 30% w/v sodium acetate solution – the volume added is that found necessary to adjust the pH of a reagent blank to 5.0;

(d) 1% w/v 1,10-phenanthroline solution (2 ml).

Make up the liquid volume to 50 ml with water. (If the 1,10-phenanthroline solution changes colour, discard it and prepare a fresh solution.) Allow at least 10 min for colour development and determine the absorbance at 508 nm spectrophotometrically.

Standardization

Electrolytic iron wire is cleaned with fine sandpaper and dissolved in 15% v/v H_2SO_4 to make a stock solution of $100\,mg\,litre^{-1}$, which is diluted to create standards of 1 to $10\,\mu g\,ml^{-1}$. The method is sensitive to $1\,\mu g$ Fe, and will detect 0.009 ppm of iron carbonyl in a 50 litre air sample. Interferences include Cr, Cu, Ni, Co, Zn, Cd and Hg. If the presence of these metals is suspected, make up the sample solution to 7 to 8 N in HCl, and extract the ferric chloride into isopropyl ether. After re-extraction of the iron into water, reduce with hydroxylamine, and form the colour complex in acetate-buffered solution as described previously. A new calibration curve should be developed.

3.3.2.2 Hg and its compounds

Hg may be found in air in several distinct chemical and physical forms. Elemental mercury is a common air pollutant, present as the vapour, or as an aerosol of liquid droplets. Organically bound mercury can exist in air as monoalkyl mercury compounds such as methylmercury chloride [CH_3HgCl] or dialkyl mercury compounds such as dimethylmercury [$(CH_3)_2Hg$], both present predominantly in the vapour phase. Inorganic mercury compounds are relatively involatile and are present in air as particles, although some, notably mercuric chloride, may be present substantially in the vapour phase.

The more sophisticated analytical procedures discriminate between the various forms of Hg in air; an important facet in view of the differing toxicology associated with the different chemical forms. It must be emphasized, however, that many sampling and analytical procedures either measure only one form of the metal or measure the sum of the concentrations of all forms, and this may be acceptable only in industrial situations where adequate knowledge of the predominant form of airborne mercury is available.

Impingers and scrubbers have been used for sampling Hg in air. Jacobs [77] describes a method utilizing I/KI for collection of Hg from air, followed by a colorimetric determination. Panek used acidified 6% potassium permanganate for collection, following with a colorimetric determination with dithizone [134]. Linch et al. [135], however, found poor collection of diethylmercury by acid permanganate and very inefficient collection of dimethyl- and diethylmercury by I/KI. Crystalline iodine was found to retain little diethylmercury, but collected elemental Hg. They did, however, find that 0.1 N iodine monochloride in 0.5 M HCl gave highly efficient recovery of alkyl, dialkyl, inorganic and elemental mercury including particulate forms. Determination of collected Hg by a dithizone procedure was performed. This method has, however, been criticized for high and irreproducible blanks, instability of collected Hg compounds and interference by high levels of SO_2.

Elemental Hg in the air may be directly determined by continuous atomic absorption measurement at 253.7 nm [77]. Commercially available instruments are fairly sensitive but respond to various organic vapours also present in air. Although sensitivity to molecular absorption by organic compounds is not as high as for atomic absorption by mercury, under some circumstances, errors may arise. A typical instrument will measure mercury levels of 5 μg to 1 mg m^{-3} of air.

Hg vapour in air may be collected by amalgamation on Ag wool [136] or gauze [137, 138]. Subsequent release by thermal de-amalgamation and elution into an atomic absorption or u.v. spectrophotometer allows determination of the Hg without interference from organic air pollutants. For 24 h sampling times ambient Hg levels from 15 ng m^{-3} to 10 μg m^{-3} may be determined using Ag wool collection [136] or from 5 ng m^{-3} to 100 μg m^{-3} using sampling times of 2 h or less and an Ag gauze collector [137, 138]. Scaringelli *et al.* [139] describe collection of all forms of Hg by adsorption on charcoal, backed by a fibre filter to ensure efficient collection of particulate matter. Subsequent heating of the sampler and nitrogen elution frees the collected compounds which are pyrolysed to form elemental Hg, and collected on Ag wool. Heating of the silver wool causes de-amalgamation of the mercury which is determined by atomic absorption at 253.7 nm. A porous graphite tube plated with gold has been shown to be an efficient collector of particulate and elemental Hg [140]. Subsequent insertion of the tube in the carbon-rod atomizer of an atomic absorption instrument allows determination of the collected Hg.

Conventional flame atomic absorption of Hg is of poor sensitivity. A flameless technique has been developed, however, involving reduction of Hg compounds in solution using stannous chloride or hydrazine hydrate and introduction of the resultant elemental mercury as vapour into a cell in the sample beam of an atomic absorption instrument [141, 142]. Alternatively, the Hg may be electrolytically amalgamated onto Cu wire and released into a cell in the spectrophotometer light beam by heating the wire [143]. The detection limit is 0.2 ng of Hg. Methods involving the chemical reduction of Hg compounds and the flushing of the element from solution may also introduce volatile organic materials into the light path thus causing erroneous results. This may be overcome by prior ashing of the sample [137] or the use of a background correction device [142], commonly available as an accessory on double beam atomic absorption spectrophotometers. After suitable chemical treatment, such analytical methods may be applied to particulate mercury collected by filtration or impingement, or samples of mercury and its compounds collected by absorption in a reagent. The method is of high sensitivity and is specific. A valuable critical report of sampling and analytical techniques for mercury in stationary sources has been compiled by Driscoll [144].

Elegant techniques for separate collection and analysis of different forms of

Hg have been described. Henriques *et al.* [145] separated particulate Hg initially by filtration through a Millipore filter. A gold filter, coated with a gold–silicon alloy collected elemental Hg vapour, a scrubber containing acid permanganate solution trapped methylmercury (and other readily oxidized mercury compounds) and, last in the train, a pure gold filter collected dimethylmercury and other gold-soluble mercury compounds. Analysis was by vapour atomic absorption.

Braman and Johnson [146] described a technique suitable for determination of several forms of Hg at levels down to 0.1 to $1 \, \text{ng m}^{-3}$. A glass wool filter removed particulate matter above $0.3 \, \mu\text{m}$ from an air sample. This was followed by adsorption tubes containing respectively siliconized Chromosorb W treated with hydrogen chloride (collects mercuric chloride type compounds and particulate matter which pass through the filter); Chromosorb W treated with caustic soda (methylmercury chloride type compounds); silvered glass beads (elemental mercury); and finally gold coated glass beads (dimethylmercury). High collection efficiencies were found. Hg was eluted from the adsorption tubes by heating with helium carrier gas, and passage through a d.c. discharge chamber caused excitation of the Hg and emission at $257.3 \, \text{nm}$ was measured. The detection limit of the analytical procedure was about $0.01 \, \text{ng}$.

Experimental procedure for determination of elemental Hg [136]

Air sampling

Hg in ambient air is sampled by drawing air at a known rate for a fixed period of time through cleaned collectors, each containing 1 to 2 g of Ag wool and connected in series. For field sampling purposes a flow rate of $100 \, \text{ml min}^{-1}$ for 24 h is anticipated. Flow rate, collection time and number of collectors is varied to suit the expected Hg level. After use the collector is disengaged from the pump and capped to prevent contamination. Ag wool in an 8 mm internal diameter tube has been shown to retain 3 to $4 \, \mu\text{g}$ of Hg per gram of Ag wool before breakthrough.

Collectors are 100 mm long × 5 mm internal diameter borosilicate glass tube equipped with ball joints on the ends and packed snugly with 1 to 2 g of cleaned Ag wool. The Ag wool is Fisher micro-analysis grade and is initially cleaned by placing it in a furnace at 800° C for 2 h. All collectors are permanently wrapped with exactly 100 cm of 22 gauge Nichrome heating wire.

Analysis

After calibration of the analytical system (Fig. 3.9) using a similar collector, the collector to be analysed is clamped into the system by ball-joint clamps, the

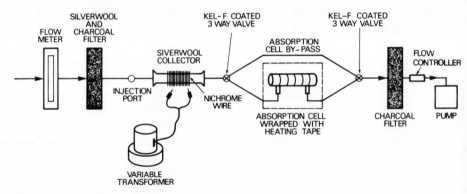

Fig. 3.9   Analytical system for mercury analysis [136].

carrier gas flow is adjusted to 200 ml min^{-1} and the collector is heated for 30 s at 24 V producing a temperature of 400° C (after one 30 s heating period the collector is Hg free, and ready for future use). The gas train, constructed of 0.25 in Tygon tubing and Kel-F coated valves, carries the Hg vapour to a 20 cm, 3.3 cm, internal diameter, fused silica absorption cell, heated to 90° C with heating tape to prevent condensation of Hg, placed in the light beam of an atomic absorption spectrophotometer set up for mercury measurement at 253.65 nm (Long *et al.* [136] used a Perkin-Elmer 403 instrument). Very short lengths of Tygon tubing are used for connections owing to Hg absorption problems, which are far more acute with rubber or PTFE. Air entering the analytical train is pre-filtered through Ag wool and activated charcoal. The atomic absorption signal is recorded on a pen-recorder and integrated by an Automatic Digital Integrator.

Calibration

The injection of known amounts of air saturated with Hg vapour onto a collector in the analytical train at a carrier flow of 200 ml min^{-1} is used to standardize the detection system. The Hg is subsequently released by heating, and analysed in the usual manner (direct injection gives lower results). Three 1 litre borosilicate glass bottles containing sufficient Hg to cover the bottoms and equipped with serum caps are used as standardization reservoirs. They are maintained at 20.0 plus or minus 0.1° C in a constant temperature bath. At 20° C and 1 atm pressure 1.0 cm^3 of air contains 13.19 ng Hg. By using gas-tight syringes of volume 0.02 to 100 cm^3, 0.26 to 1319 ng can be introduced into the analytical system through the injection port.

Interferences

SO_2 and NO_2 do not interfere significantly. H_2S gives a negative interference of about 6% at levels of 13 to $650\,\mu g\,m^{-3}$ (significantly above normal ambient levels of H_2S). Dimethylmercury is not collected significantly (less than 1%). Very high levels ($20\,\mu g$ injections) of chlorine attack the Ag wool, but a basic scrubber before the collector will eliminate this problem.

Sensitivity

A detection limit of $0.3\,ng$ is found and calibration curves are reproducible to within 11% at $0.5\,ng$, and to within 3%, relative S.D., beyond $6\,ng$ of Hg.

3.3.2.3 Volatile Pb compounds

Tetra-alkyl lead compounds, used worldwide as motor fuel additives are volatile and occur in air in the vapour phase. Filtration of the air sample allows separation from lead particulates.

Several methods are available for trapping lead alkyl compounds. Crystalline iodine, used initially [79] proved inefficient [147] and has been superseded by iodine monochloride in HCl solution [105, 147–149]. Collection efficiency in the latter solution is near 100% [147] and subsequent analysis of trapped organic Pb may be performed spectrophotometrically with dithizone [147–148] (Pb concentrations down to $10\,\mu g\,m^{-3}$; 8 h sampling time [147]), or by atomic absorption measurements [105, 149] (organic Pb concentrations down to $0.2\,\mu g\,m^{-3}$; 24 h sampling time [105]).

An activated carbon scrubber is reported to give efficient collection of organic Pb from air [150]. After sampling 100 to $200\,m^3$ of air, the carbon is digested with nitric and perchloric acids, and the lead determined spectrophotometrically with dithizone, allowing detection of extremely low levels of organic lead [81, 150]. Recent work, however, brings into question the efficiency of conventional filters for collection of particulate Pb [151] and consequently leakage of Pb particulates past the filter may be a cause of erroneous elevated readings for organic Pb determined after collection by activated charcoal or iodine monochloride. In a recent refinement of the iodine monochloride procedure, this problem is overcome by selective complexation of dialkyl lead derived from tetra-alkyl lead compounds. If used with flameless atomic absorption analysis a very high sensitivity can be attained [149].

Adsorption tubes packed with gas chromatography column-packing materials have been used for collection of organic Pb compounds from air [152–155]. Cantuti and Cartoni [152] allowed equilibration of organic lead in the gas phase with that adsorbed on the packing material by sampling at ambient temperatures or $0°$ C. Laveskog, [153] by cooling to $-80°$ C, achieved total collection from a 1 litre air sample, whilst Harrison and co-workers [154, 155] used liquid nitrogen cooling and demonstrated quantitative collection of lead alkyl compounds from air samples of up to $0.07 \, m^3$. Sampling of greater volumes of ambient air is generally precluded by ice formation in the sample tube. Desorption of the adsorbed compounds is effected by heating and nitrogen or helium elution. If the eluted material is passed into a gas chromatograph, separation of individual lead alkyl compounds is possible, with detection by electron capture [152] or mass spectrometry (GC/MS) [153]. Alternatively, elution may be into the combustion air stream of an atomic absorption spectrophotometer and this allows determination of total volatile Pb with a high sensitivity (down to $0.003 \, \mu g \, m^{-3}$) with short sampling times [154, 155].

A continuous atomic absorption procedure designed for measurement of airborne lead in industrial plants manufacturing lead alkyl compounds has been described [156]. The polluted air is used as the supply for a specially designed burner, and measurement of atomic absorption allows determination of organic Pb in air at levels down to $1 \, \mu g \, m^{-3}$, although particulate lead is not discriminated.

Gas chromatographic determination of tetraethyl-lead (T.E.L.) [152]

Air sampling

Sampling tubes are prepared from 1 ml hypodermic syringes, about 8 cm long and 0.5 cm internal diameter and packed with 10% SE52 on 80/100 mesh Chromosorb P. The air is drawn through the tube at $1.5 \, litre \, min^{-1}$ for 10 to 15 min, the tube being maintained at ambient temperatures or at $0°$ C (in ice), as shown in Fig. 3.10. Before entering the sample tube the air passes through an empty U-tube where the excess of atmospheric water is condensed.

Analysis

The sample tube is transferred to the injection system shown in Fig. 3.11. The carrier gas (1) can pass either directly from the stopcock (2) through the chromatographic column, or through the sample tube. For introduction of the sample, the sample tube (4) is flash heated to about $130°$ C by a small electric furnace (3) positioned immediately above the injection port (5). After heating, the syringe is pushed so that the needle passes through the silicone rubber septum; by rotating the stopcock (2) the carrier gas flushes the sample into the chromatographic column (10% SE52 on 80/100 mesh Chromosorb P; 1 m × 0.3 mm internal diameter glass; oven temperature, $80°$ C; injector and detector

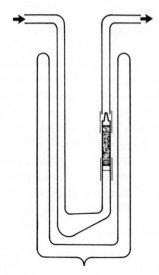

Fig. 3.10 Sampling apparatus for collecting T.E.L. from polluted air [152].

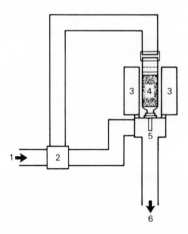

Fig. 3.11 Injection system for the introduction of the collected samples into the gas chromatograph [152].

temperature, 130° C). Carrier gas is pure nitrogen with an inlet pressure of 1 atm and a flow of 30 ml min^{-1}. Detection is by an electron capture detector, purged with nitrogen at 150 ml min^{-1} and with the applied potential adjusted for maximum sensitivity.

Calculation and calibration

For calculation of the concentration of T.E.L. in air the specific retention volume of the adsorption tube, V_g^0 must be known. This may be measured by syringe injection of a dilute T.E.L. solution into carrier gas at the inlet end of an adsorption tube maintained at a known temperature. Passage of the carrier gas at a known rate through the tube directly into the GC detector allows determination of the volume of gas which must be passed to cause breakthrough of T.E.L. Correction of volumes to atmospheric pressure and $0°$ C allows calculation of the specific retention volume, or retention volume per g of liquid stationary phase. A plot of log V_g^0 against $1/T$, the inverse of the absolute temperature of the adsorption tube during sampling should be linear and interpolation allows determination of V_g^0 for any sampling temperature. The GC detector may be calibrated by direct injection of dilute solutions of T.E.L., and then

$$\text{Concn. of T.E.L. in air}, C_g = m273/wV_g^0T$$

where m = weight of T.E.L. detected by GC; w = weight of liquid phase in the sample tube (g).

In order to verify the overall procedure, a known concentration of T.E.L. in air may be prepared with a dynamic dilution apparatus. A slow flow of nitrogen saturated with T.E.L. at $0°$ C by passage through a Dreschel type bottle is diluted with a large volume of pure air from a cylinder. The sample tube is placed at the exit of the mixing chamber and the air, containing a known concentration of T.E.L. at $0°$ C, is sampled by suction.

The minimum volume of air drawn through the sample tube required to saturate the tube and reach equilibrium conditions is approximately twice the retention volume. Concentrations of T.E.L. in air may be determined to below $100\,\mu g\,Pb\,m^{-3}$, and the method may be extended to the analysis of tetramethyl lead (T.M.L.) and other lead alkyls.

Experimental procedure for determination of very low levels of volatile Pb in air [81]

This procedure is intended for use in conjunction with the determination of inorganic Pb, as described in the section on spectrophotometric procedures. Reagents described are prepared in an identical manner.

Air sampling

Air is sampled through a glass fibre or $0.45\,\mu m$ Millipore filter to remove particulate Pb, and then through the sample tube containing activated carbon shown in Fig. 3.12. The activated carbon is 20 to 50 mesh or equivalent, should be low in

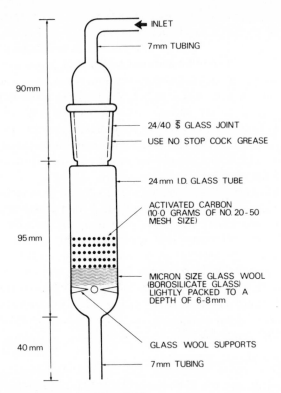

Fig. 3.12 Activated carbon scrubber [81].

lead, and stored in a tightly covered container. Sample approximately 150 to 200 m^3 of air at about 20 litre min^{-1}. If membrane filters are used two must be connected in parallel to achieve this flow rate.

Preparation of sample and analysis

After sampling, pour the activated carbon into a 500 ml Erlenmeyer flask. Add conc. HCl (25 ml), swirl and add conc. HNO$_3$ (75 ml) and swirl to mix. Digest overnight (about 16 h) on a hot-plate or steam bath at 90 to 100° C. Add conc. HNO$_3$ (30 ml) to the low volume or nearly dry residue. Heat (80 to 90° C) for 1 h, add water (about 100 ml) and mix well. Allow the mixture to stand at room temperature for about 2 h. Decant the supernatant liquid through a thin, rapid filter paper hardened to great wet strength, into another 500 ml Erlenmeyer flask. Rinse the residue carbon on the filter with three portions of water. Discard the filter paper and carbon.

Add nitric–perchloric acid solution (10 ml) to the acid extract and heat on a hot-plate (175 to 200° C) to fumes of perchloric acid. If all carbon is not oxidized and the solution has gone to dryness, add an additional portion (5 ml) of nitric–perchloric acid solution and heat to fumes of perchloric acid. (Caution: Oxidation with perchloric acid can be hazardous if perchloric acid hoods are not available; oxidation with HNO_3 and H_2SO_4 (2.0 ml) is also satisfactory but is somewhat slower.) To the slightly cooled perchloric acid sample, add water (about 25 ml) from a wash bottle whilst washing down the sides of the Erlenmeyer flask, and add HNO_3 (1 + 4) (20 ml). Allow about 30 min for complete solution of the sample and transfer the mixture to the 200 ml modified absorption cell. To the sample in the modified cell, add buffer solution (50 ml) and reducing solution (10 ml), mix and allow 15 to 20 min for complete reduction of the sample. (The reducing solution is prepared from potassium cyanide (20 g), dibasic ammonium citrate (40 g) and anhydrous sodium sulphite (200 g) and diluted to 1 litre with water. Add concentrated NH_3 solution (sp. gr. 0.90; 600 ml). Prepare this solution in a well-ventilated hood.) Add 10.0 ml of dithizone solution and shake vigorously for 30 s. Proceed as in the determination of particulate Pb from this point, and calculate the Pb concentration as for particulate Pb. It is imperative to run a blank on the activated carbon and reagents for subtraction from the sample reading, and avoidance of leakage of particulate Pb past the filter is essential.

References

[1]   Lee, R.E. and von Lehmden, D.J. (1973). *J. Air Pollut. Control Assoc.* 23(10) 853–7.
[2]   U.S. Dept. of Health, Education and Welfare (1966). Air Quality Data from the National Air Sampling Networks and Contributing State and Local Networks (1964–65). Public Health Service, Division of Air Quality, Cincinnati, Ohio.
[3]   Spurny, K.R., Lodge, J.P., Frank, E.R. and Sheesley, C.D. (1969). *Environ. Sci. Technol.* 3(5) 453–64.
[4]   Spurny, K. and Lodge, J.P. (1968). *Staub-Reinhalt. Luft.* 28(5) 1–10.
[5]   Spurny, K. and Pich, J. (1965). *Coll. Czech. Chem. Commun.* 30, 2276–86.
[6]   U.S. Dept. of Health, Education and Welfare (1962). Air Pollution Measurements of the National Air Sampling Network – Analyses of suspended Particulates (1957–61) Public Health Service Publication No. 978, U.S. Government Printing Office, Washington, D.C.
[7]   Cohen, A.L. (1973). *Environ. Sci. Technol.* 7(1) 60–1.
[8]   Pate, J.B., and Tabor, E.C. (1962). *Am. Ind. Hyg. Ass. J.* 23, 145–50.
[9]   Hwang, J.Y. (1972). *Analyt. Chem.* 44, 20A–27A.
[10] Dams, R., Rahn, K.A. and Winchester, J.W. (1972). *Environ. Sci. Technol.* 6(5) 441–8.
[11] Zoller, W.H. and Gordon, G.E. (1970). *Analyt. Chem.* 42(2) 257–65.

[12]   Vogg, H. and Haertel, R. (1973). *Fresenius' Z. Anal. Chem.* **267**(4) 257–60.

[13]   Gandrud, B.W. and Lazrus, A.L. (1972). *Environ. Sci. Technol.* **6**(5) 455–8.

[14]   Kometani, T.Y., Bove, J.L., Nathanson, B., Siebenberg, S. and Magyar, M. (1972). *Environ. Sci. Technol.* **6**(7) 617–20.

[15]   Luke, C.L., Kometani, T.Y., Kessler, J.E., Loomis, T.C., Bove, J.L. and Nathanson, B. (1972). *Environ. Sci. Technol.* **6**(13) 1105–9.

[16]   Thompson, R.J., Morgan, G.B. and Purdue, L.J. (1969). *Analyt. Instr.* **7**, 9–17.

[17]   Thompson, R.J., Morgan, G.B. and Purdue, L.J. (1970). *At. Absorp. Newsl.* **9**(3) 53–7.

[18]   Struempler, A.W. (1973). *Analyt. Chem.* **45**(13) 2251–4.

[19]   King, W.G., Rodriguez, J.M. and Wai, C.M. (1974). *Analyt. Chem.* **46**(6) 771–3.

[20]   Feldman, C. (1974). *Analyt. Chem.* **46**(1) 99–102.

[21]   Newton, D.W. and Ellis, R. (1974). *J. Environ. Qual.* **3**(1) 20–3.

[22]   Dyck, W. (1968). *Analyt. Chem.* **40**(2) 454–5.

[23]   West, F.K., West, P.W. and Iddings, F.A. (1966). *Analyt. Chem.* **38**(11) 1566–70.

[24]   Silverman, L. and Ege, J.F. (1947). *J. Ind. Hyg. Toxicol.* **29**(2) 136–9.

[25]   Macdonald, G.L. (1971). *Comprehensive Analytical Chemistry*, Vol. IIC (ed. Wilson, C.L. and Wilson, D.W.) Elsevier, Amsterdam.

[26]   Gilfrich, J.V., Burkhalter, P.G. and Birks, L.S. (1973). *Analyt. Chem.* **45**(12) 2002–9.

[27]   Hammerle, R.H., Marsh, R.H., Rengan, K., Giauque, R.D., and Jaklevic, J.M. (1973). *Analyt. Chem.* **45**(11) 1939–40.

[28]   Rhodes, J.R., Pradzynski, A.H., Hunter, C.B., Payne, J.S. and Lindgren, J.L. (1972). *Environ. Sci. Technol.* **6**(10) 922–7.

[29]   Hwang, J.Y. (1970). *Talanta* **17**, 118–21.

[30]   Beitz, L., Haase, J. and Weichert, N. (1974). *International Symposium – Environment and Health*, Paris.

[31]   Bowman, H.R., Conway, J.G. and Asaro, F. (1972). *Environ. Sci. Technol.* **6**(6) 558–60.

[32]   Epler, R.J. (1974). *Environ. Sci. Technol.* **8**(1) 28–30.

[33]   Ter Haar, G.L. and Bayard, M.A. (1971). *Nature* **232**, 553–4.

[34]   Heichel, G.H. and Hankin, L. (1972). *Environ. Sci. Technol.* **6**(13) 1121–2.

[35]   Yakowitz, H., Jacobs, M.H. and Hunneyball, P.D. (1972). *Micron* **3**, 498–505.

[36]   Altshuller, A.P. (1972). *Analytical Chemistry: Key to Progress on National Problems* (ed. Meinke, W.W. and Taylor, J.K.) N.B.S. Special Publ. No. 351.

[37]   Gibbons, D. and Lambie, D.A. (1971). *Comprehensive Analytical Chemistry*, Vol. IIC, (ed. Wilson, C.L. and Wilson, D.W.) Elsevier, Amsterdam.

[38]   Iddings, F.A. (1969). *Environ. Sci. Technol.* **3**(2) 132–140.

[39]   Dams, R., Robbins, J.A., Rahn, K.A. and Winchester, J.W. (1970). *Analyt. Chem.* **42**(8) 861–7.

[40]   Gordon, C.M. and Larson, R.E. (1964). *N.R.L. Quarterly on Nuclear Science and Technology*, pp. 17–22. Naval Research Laboratory, Washington, D.C.

[41]   Brar, S.S., Nelson, D.M., Kline, J.R., Gustafson, P.F., Kanabrocki, E.L., Moore, C.E. and Hattori, D.M. (1970). *J. Geophys. Res.* 75(15) 2939–45.

[42]   Swindle, D.L. and Schweikert, E.A. (1973). *Analyt. Chem.* 45(12) 211–5.

[43]   Riddle, D.C. and Schweikert, E.A. (1974). *Analyt. Chem.* 46(3) 395–8.

[44]   Parsa, B. and Markowitz, S.S. (1974). *Analyt. Chem.* 46(2) 186–9.

[45]   Mitteldorf, A.J. (1965). *Trace Analysis – Physical Methods* (ed. Morrison, G.H.) Interscience, New York.

[46]   Homan, R.E. and Morgan, G.B. (1968). 19th Pittsburgh Conference on Analytical Chemistry and Applied Spectroscopy.

[47]   Morgan, G.B., Ozolins, G. and Tabor, E.C. (1970). *Science* 170, 289–96.

[48]   Tabor, E.C. and Warren, W.V. (1958). *A.M.A. Arch. Ind. Health*, 17, 145–51.

[49]   National Air Pollution Control Administration (1968). Air Quality Data from the National Air Surveillance Networks and Contributing State and Local Networks, 1966 Edn, APTD68–9, 167P.

[50]   American Industrial Hygiene Association (1969). Analytical Guide: Beryllium, *Am. Ind. Hyg. Assoc. J.* 103–5.

[51]   Cholak, J. and Hubbard, D.M. (1948). *Analyt. Chem.* 20(1) 73–6.

[52]   Smith, R.G., Boyle, A.J., Fredrick, W.G. and Zak, B. (1952). *Analyt. Chem.* 24(2) 406–9.

[53]   Fitzgerald, J.J. (1957). *A.M.A. Arch. Ind. Health*, 15, 68–73.

[54]   Churchill, W.L. and Gillieson, A.H.C.P. (1952). *Spectrochim. Acta*, 5, 238–50.

[55]   Seeley, J.L. and Skogerboe, R.K. (1974). *Analyt. Chem.* 46(3) 415–21.

[56]   Weisz, H. (1970). *Microanalysis by the Ring Oven Technique*, 2nd Edn., Pergamon Press, Oxford.

[57]   West, P.W., Weisz, H., Gaeke, G.C. and Lyles, G. (1960). *Analyt. Chem.* 32(8) 943–6.

[58]   West, P.W., *Air Pollution*, 2nd Edn. Vol. II (ed. Stern, A.C.) pp. 147–85. Academic Press, New York.

[59]   Ronneau, C.J.-M., Jacob, N.M. and Apers, D.J. (1973). *Analyt. Chem.* 45(12) 2152.

[60]   West, P.W. and Thabet, S.K. (1967). *Analyt. Chim. Acta.* 37, 246–52.

[61]   Levine, L. (1945). *J. Ind. Hyg. Toxicol.* 27(6) 171–7.

[62]   Taylor, J.K., Maienthal, E.J. and Marinenko, G. (1965). *Trace Analysis – Physical Methods*, (ed. Morrison, G.H.) Interscience, New York.

[63]   Ferrett, D.J., Milner, G.W.C., Shalgosky, H.I. and Slee, L.J. (1956). *Analyst* 81, 506–12.

[64]   Sachdev, S.L. and West, P.W. (1970). *Environ. Sci. Technol.* 4(9) 749–51.

[65]   Mancy, K.H. (1972). *Analytical Chemistry: Key to Progress on National Problems* (ed. Meinke, W.W. and Taylor, J.K.) N.B.S. Special Publ. No. 351.

[66]   von Lehmden, D.J., Jungers, R.H. and Lee, R.E. (1974). *Analyt. Chem.* **46**(2) 239–45.

[67]   Harrison, P.R. and Winchester, J.W. (1971). *Atmos. Environ.* **5**(10) 863–80.

[68]   Colovos, G., Wilson, G.S. and Moyers, J. (1973). *Analyt. Chim. Acta.* **64**, 457–64.

[69]   MacLeod, K.E. and Lee, R.E. (1973). *Analyt. Chem.* **45**(14) 2380–3.

[70]   Roboz, J. (1965). *Trace Analysis – Physical Methods* (ed. Morrison, G.H.) Interscience, New York.

[71]   Brown, R. and Vossen, P.G.T. (1971). *Int. Symp. Ident. Meas. Environ. Pollut. (Proc)*, (ed. Westley, B.) pp. 427–31.

[72]   Brown, R. and Vossen, P.G.T. (1970). *Analyt. Chem.* **42**(14) 1820–2.

[73]   Cheng, K.L. (1965). *Trace Analysis – Physical Methods* (ed. Morrison, G.H.) Interscience, New York.

[74]   Spectrometry Nomenclature (1973). *Analyt. Chem.* **45**(14) 2449.

[75]   I.U.P.A.C. (1963). *Tables of Spectrophotometric Absorption Data of Compounds used for the Colorimetric Determination of Elements*, Butterworth, London.

[76]   American Public Health Association (1972). *Methods of Air Sampling and Analysis*, Washington, D.C.

[77]   Jacobs, M.B. (1967). *The Analytical Toxicology of Industrial Inorganic Poisons*, Interscience, New York.

[78]   Robinson, E. and Ludwig, F.L. (1967). *J. Air Pollut. Contr. Ass.* **17**(10) 664–9.

[79]   Cholak, J. (1964). *Arch. Environ. Health* **8**(2) 222–31.

[80]   American Industrial Hygiene Association (1969). Community Air Quality Guides: Lead. *Am. Ind. Hyg. Ass. J.*, 95–97; Analytical Guides: Lead, 102–3.

[81]   American Society for Testing and Materials, A.S.T.M. Method D3112–72T.

[82]   Saltzman, B.E. (1953). *Analyt. Chem.* **25**(3) 493–6.

[83]   American Industrial Hygiene Association (1964). Hygienic Guide Series: Arsenic and Its Compounds. *Am. Ind. Hyg. Ass. J.*, 610–3.

[84]   Hiser, R.A., Donaldson, H.M. and Schwenzfeier, C.W. (1961). *Am. Ind. Hyg. Assoc. J.* 280–5.

[85]   McCloskey, J. (1967). *Microchem. J.* **12**, 32–45.

[86]   American Industrial Hygiene Association (1966). Hygienic Guide Series: Nickel. *Am. Ind. Hyg. Ass. J.*, 202–5.

[87]   Ege, J.F. and Silverman, L. (1947). *Analyt. Chem.* **19**(9) 693–4.

[88]   Sill, C.W. (1961). *Analyt. Chem.* **33**, 1684–6.

[89]   Kahn, H.L. (1968). *Adv. Chem. Ser.* No. **73**, American Chemical Society, pp. 183–229.

[90]   Weberling, R.P. and Cosgrove, J.F. (1965). *Trace Analysis – Physical Methods* (ed. Morrison, G.H.) Interscience, New York.

[91]   Browner, R.F., Dagnall, R.M. and West, T.S. (1970). *Analyt. Chim. Acta.* **50**, 375–81.

[92]   Kahn, H.L., Peterson, G.E. and Schallis, J.E. (1968). *At. Absorp. Newsl.* **7**(2) 35–9.

[93]   Woodriff, R. (1969). *Trace Subst. Environ. Health -3*, Proc. Univ. Mo. Ann. Conf. 3rd, pp. 297–303.

[94]   Manning, D.C. and Fernandez, F. (1970). *At. Absorp. Newsl.* **9**(3) 65–70

[95]   Beukelman, T.E. and Lord, S.S. (1960). *Appl. Spectros.* **14**(1) 12–17.

[96]   Burnham, C.D., Moore, C.E., Kanabrocki, E. and Hattori, D.M. (1969). *Environ. Sci. Technol.* **3**(5) 472–5.

[97]   Jackson, G.B. and Myrick, H.N. (1971). *Internat. Lab.*, 41–7.

[98]   Hwang, J.Y. (1971). *Can. Spectrosc.* **16**, 43–45; 53.

[99]   Koirtyohann, S.R. and Wen, J.W. (1973). *Analyt. Chem.* **45**(12) 1986–9.

[100] Lundgren, D.A. (1970). *J. Air Pollut. Control Ass.* **20**(9) 603–8.

[101] Lee, R.E., Patterson, R.K., and Wagman, J. (1968). *Environ. Sci. Technol.* **2**(4) 288–90.

[102] Morgan, G.B. and Homan, R.E. (1967). 18th Pittsburgh Conference on Analytical Chemistry and Applied Spectroscopy.

[103] Kneip, T.J., Eisenbud, M., Strehlow, C.D. and Freudenthal, P.C. (1970). *J. Air Pollut. Control Ass.* **20**(3) 144–9.

[104] Burnham, C.D., Moore, C.E., Kowalski, T. and Krasniewski, J. (1970). *Appl. Spectrosc.* **24**(4) 411–4.

[105] Purdue, L.J., Enrione, R.E., Thompson, R.J. and Bonfield, B.A. (1973). *Analyt. Chem.* **45**(3) 527–30.

[106] Janssens, M. and Dams, R. (1973). *Analyt. Chim. Acta* **65**, 41–7.

[107] Begnoche, B.C. and Risby, T.H. (1975). *Analyt. Chem.* **47**, 1041–5.

[108] Woodriff, R. and Lech, J.F. (1972). *Analyt. Chem.* **44**(7) 1323–5.

[109] Matousek, J.P. and Brodie, K.G. (1973). *Analyt. Chem.* **45**(9) 1606–9.

[110] Zdrojewski, A., Quickert, N. and Dubois, L. (1973). *Int. J. Environ. Analyt. Chem.* **2**, 331–41.

[111] Janssens, M. and Dams, R. (1974). *Analyt. Chim. Acta.* **70**, 25–33.

[112] Sachdev, S.L., Robinson, J.W. and West, P.W. (1967). *Analyt. Chim. Acta* **37**, 12–19.

[113] Sachdev, S.L., Robinson, J.W. and West, P.W. (1967). *Analyt. Chim. Acta* **38**, 499–506.

[114] Hwang, J.Y. and Feldman, F.J. (1970). *Appl. Spectrosc.* **24**(3) 371–4.

[115] Loftin, H.P., Christian, C.M. and Robinson, J.W. (1970). *Spectrosc. Lett.* **3**(7) 161–74.

[116] Robinson, J.W. (1972). Proc. 'International Symposium: Environmental Health Aspects of Lead', pp. 1099–105, Amsterdam.

[117] Robinson, J.W., Wolcott, D.K., Slevin, P.J. and Hindman, G.D. (1973). *Analyt. Chim. Acta* **66**, 13–21.

[118] Edwards, H.W. (1969). *Analyt. Chem.* **41**(10) 1172–5.

[119] Clayton, P. and Wallin, S.C. (1973). C.C.M.S./C.P.P.S.D. Conference: Ann Arbor, Michigan.

[120] White, R.A. (1967). *J. Sci. Instr.* **44**, 678–80.

[121] Robinson, J.W. and Wolcott, D.K. (1973). *Analyt. Chim. Acta* **66**, 333–42.

[122] Siemer, D., Lech, J.F. and Woodriff, R. (1973). *Spectrochim. Acta* **28B**, 469–71.

[123] Ranweiler, L.E. and Moyers, J.L. (1974). *Environ. Sci. Technol.* **8**(2) 152–6.

[124] Ross, W.D. and Sievers, R.E. (1972). *Environ. Sci. Technol.* **6**(2) 155–8.

[125]   Strackee, L. (1968). *Nature* **218**, 497–8.
[126]   Barnes, I.L., Murphy, T.J., Gramlich, J.W., and Shields, W.R. (1973). *Analyt. Chem.* **45**(11) 1881–4.
[127]   Densham, A.B., Beale, P.A.A., Palmer, R. (1963). *J. Appl. Chem.* **13**, 576–80.
[128]   McCarley, J.E., Saltzman, R.S. and Osborn, R.H. (1956). *Analyt. Chem.* **28**(5) 880–2.
[129]   Belyakov, A.A. (1960). *Zavodskaya Laboratoriya* **26**, 158–9.
[130]   American Industrial Hygiene Association (1968). Hygienic Guide Series: Nickel Carbonyl, *Am. Ind. Hyg. Ass. J.* 304–7.
[131]   Kincaid, J.F., Stanley, E.L., Beckworth, C.H. and Sunderman, F.W. (1956). *Am. J. clin. Pathol.* **26**, 107–119
[132]   Brief, R.S., Venable, F.S. and Ajemian, R.S. (1965). *Am. Ind. Hyg. Ass. J.* 72–6.
[133]   Brief, R.S., Ajemian, R.S. and Confer, R.G. (1967). *Am. Ind. Hyg. Ass. J.* 21–30.
[134]   Panek, J. (1973). *Cesk. Hyg.* **18**(5) 244–9.
[135]   Linch, A.L., Stalzer, R.F. and Lefferts, D.T. (1968). *Am. Ind. Hyg. Ass. J.* **29**, 79–86.
[136]   Long, S.J., Scott, D.R. and Thompson, R.J. (1973). *Analyt. Chem.* **45**(13) 2227–33.
[137]   Corte, G., Dubois, L. and Monkman, J.L. (1973). *Sci. Tot. Environ.* **2**(1) 89–96.
[138]   Corte, G.L., Thomas, R.S., Dubois, L. and Monkman, J.L. (1973). *Sci. Tot. Environ.* **2**, 251–8.
[139]   Scaringelli, F.P., Puzak, J.C., Bennett, B.I. and Denny, R.L. (1974). *Analyt. Chem.* **46**(2) 278–83.
[140]   Siemer, D., Lech, J. and Woodriff, R. (1974). *Appl. Spectrosc.* **28**(1) 68–71.
[141]   Hatch, W.R. and Ott, W.L. (1968). *Analyt. Chem.* **40**(14) 2085–7.
[142]   Hwang, J.Y., Ullucci, P.A. and Malenfant, A.L. (1971). *Can. Spectrosc.* **16**, 100–6.
[143]   Brandenberger, H. and Bader, H. (1968). *At. Absorp. Newsl.* **7**(3) 53–4.
[144]   Driscoll, J.N. (1974). *Health Lab. Sci.* **11**, 348–53.
[145]   Henriques, A., Isberg, J. and Kjellgren, D. (1973). *Chemica Scripta* **4**, 139–42.
[146]   Braman, R.S. and Johnson, D.L. (1974). *Environ. Sci. Technol.* **8**, 996–1003.
[147]   Moss, R. and Browett, E.V. (1965). *Analyst* **91**, 428–38.
[148]   Linch, A.L., Weist, E.G. and Carter, M.D. (1970). *Am. Ind. Hyg. Ass. J.* 170–9.
[149]   Hancock, S. and Slater, A. (1975). *Analyst* **100**, 422–9.
[150]   Snyder, L.J. (1967). *Analyt. Chem.* **39**(6) 591–5.
[151]   Edwards, H.W. (1974). International Symposium – Environmental and Health, Paris.
[152]   Cantuti, V. and Cartoni, G.P. (1968). *J. Chromatog.* **32**, 641–7.
[153]   Laveskog, A. (1970). Second Int. Clean Air Congress, Washington, D.C.

[154] Harrison, R.M., Perry, R. and Slater, D.H. (1974). *Atmos. Environ.* **8,** 1187–94.
[155] Harrison, R.M., Perry, R. and Slater, D.H. (1974). International Symposium - Environmental and Health, Paris.
[156] Thilliez, G. (1967). *Analyt. Chem.* **39**(4) 427–32.

CHAPTER FOUR

Nitrogen and sulphur compounds

4.1 Introduction

The gaseous compounds of sulphur and nitrogen which are of interest in atmospheric pollution studies fall into three main chemical groups – oxides, hydrides and organic compounds of sulphur and nitrogen.

Of the oxides of sulphur only SO_2 and SO_3 are important air pollutants. SO_2 is a major pollutant causing widespread concern. The main source is the combustion of fossil fuels, when most of the sulphur present in the fuel is oxidized to SO_2. Other major sources include the metallurgical, cement, petroleum refining and miscellaneous chemical process industries. Motor vehicles are a relatively minor source of SO_2 since refined motor fuel normally has a low sulphur content.

There are a number of well documented deleterious effects of atmospheric SO_2, such as damage to vegetation of all kinds, deterioration of textiles and corrosion of metals and building materials [1]. Also sulphate aerosols, produced as a result of oxidation of atmospheric SO_2, contribute significantly to the aerosol burden of the atmosphere, giving rise to loss of visibility and acidic precipitation. Health hazards of SO_2 are less easily defined. The gas is toxic at high concentrations (TLV = 5 ppm) but there is no clear evidence that any injurious effects on the health of city dwellers are directly attributable to SO_2 itself. It is believed, however, that SO_2 in combination with other air pollutants e.g. smoke, can be injurious to health.

SO_3 is also produced during the combustion of fossil fuels but to a much lesser extent than SO_2. Chemical installations, such as those manufacturing H_2SO_4, may also constitute sources of SO_3. In view of the extreme reactivity of SO_3, emissions are closely controlled to prevent damage to plant, personnel etc. On contact with water vapour in the atmosphere SO_3 is rapidly converted into sulphuric acid aerosol. The presence of free SO_3 in the atmosphere has never been demonstrated and is, in fact, very unlikely. Analysis for SO_3 is therefore of concern only at the source.

The oxides of nitrogen which are of major concern in atmospheric pollution studies are nitric oxide (NO) and nitrogen dioxide (NO_2). The higher oxides of nitrogen dinitrogen trioxide (N_2O_3) and dinitrogen tetroxide (N_2O_4) exist in equilibrium with NO and NO_2 but at atmospheric concentrations of the latter the N_2O_3 and N_2O_4 components are negligible. Similar remarks apply to the other higher oxides nitrogen trioxide (NO_3) and dinitrogen pentoxide (N_2O_5) which are believed to be important intermediates in the photochemical smog-forming reactions. The major source of nitrogen oxides is combustion when fixation of atmospheric nitrogen occurs at the high flame temperature. The oxides are emitted mainly as NO which is normally rapidly oxidized to NO_2 by atmospheric O_2 and O_3. Motor vehicle exhaust contributes a sizable fraction of the total emissions of oxides of nitrogen. As well as stationary combustion sources, the manufacture of nitric acid and nitrate fertilizer are sources of NO and NO_2.

As with sulphur oxides, nitrogen oxides may have many deleterious effects [2]. NO is non-toxic but NO_2 is a powerful lung irritant. Adverse human health effects have been observed as a result of long term exposure to concentrations as low as 0.1 ppm NO_2. Concentrations greater than 100 ppm are lethal to most species. NO_2 assists corrosion of metals, deterioration of textiles and can damage vegetation. The oxides of nitrogen are also precursors in the formation of photochemical smog and their oxidation product, nitric acid, contributes to the aerosol burden in the atmosphere. The chemical reactivity of the oxides of nitrogen in the atmosphere is a major factor warranting their control.

The gaseous hydrides of sulphur and nitrogen, namely hydrogen sulphide (H_2S) and ammonia (NH_3) are pollutants of secondary importance but may present considerable problems in specific locations. The major sources of H_2S include pulp and paper manufactures and refining and coking operations; ammonia is emitted during fertilizer manufacture and sewage treatment. The toxic, odorous and corrosive properties of H_2S are well known; also, H_2S, in the atmosphere, is rapidly oxidized to SO_2 with its associated effects. Deleterious effects of ammonia are mainly associated with its role in the formation of atmospheric particulate matter.

Emissions of organic sulphur and nitrogen compounds are, in volume terms, comparatively minor. They are, however, of great importance from the point of view of odour nuisance, a field that is receiving a growing amount of attention. The main sources are industries which process natural products such as wood pulp, paper and animal offal, as well as miscellaneous chemical processes. Few odorous compounds have been definitely identified but mercaptans and other organosulphur compounds have been detected as pollutants from paper manufacture. The organic nitrogen containing esters e.g. peroxyacetylnitrate (PAN) are characteristic products of photochemical smog and will be considered in Chapter 5. Miscellaneous compounds such as HCN, amines, carbon disulphide are pollutants associated with certain chemical process industries but are not normally encountered in ambient air.

It should be noted that many of the gases classified above as pollutants are also emitted from natural sources e.g. H_2S from swamps, NH_3 from animal urine and amines from chicken dung. Interest in the behaviour and status of trace gases in the natural atmosphere has further stimulated research into sensitive analytical techniques for the reliable measurement of low concentrations of gases in air.

The optimum analytical technique to be employed in a particular monitoring exercise will depend on the concentration range which is likely to be encountered and the time-variation of the pollutant concentration. These factors will largely depend on where the measurements are to be made. Locations for air pollution measurements may be broadly divided into five categories, i.e. source, source vicinity, urban and industrial regions, rural regions and remote regions. The range of concentrations of several gaseous nitrogen and sulphur pollutants which may be expected in these regions are given in Table 4.1. Concentrations in the vicinity of a particular source will depend greatly on the source strength and wind direction and large short term variations in concentration may be observed. At locations increasingly removed from the sources the short term variations become progressively less. In order to obtain meaningful information on peak levels when the concentrations are varying rapidly (e.g. kerbside measurements), real-time continuous measurements are desirable. On the other hand for most measurements of a 'background' type e.g. determination of general pollutant level in a particular region, time resolution of less than one hour is not normally necessary.

Table 4.1 Concentration ranges of pollutant gases in different locations.

Gas	Concentration (ppb v^{-1})				
	Source	Source vicinity	Urban	Rural	Remote
SO_2	2×10^6	10^3	50–500	5 –50	1
NO_x	10^6	10^3	20–500	5 –50	3
H_2S	–	> 50	1– 10	0.1– 1	0.1
NH_3	–	10^3	10–100	5–15	5

The range of concentration of pollutants in the atmosphere is roughly 4 to 5 orders of magnitude. At the present time satisfactory analytical techniques are available for measurement of the common gaseous S and N pollutants at the 0.1 to 1.0 ppm level. Improvement of existing methods and new instrumental techniques now allows reasonably reliable continuous measurement down to 0.01 ppm. The analytical problems involved in the measurement of trace gas concentrations at the ppb level and below are, however, considerable and for most gases have yet to be satisfactorily resolved.

In this chapter analytical methods, both manual and instrumental, for the oxides and hydrides of S and N are discussed. The general methods for the analysis of hydrocarbons given in Chapter 6, are also applicable to organic compounds containing N or S. Some specific methods for organosulphur and nitrogen compounds are discussed in the present chapter. The practical details given in Section 4.3 refer primarily to methods for ambient air measurement but some of the instrumental techniques are suitable for source analysis.

4.2 Basic analytical techniques

The methods for the analysis of gaseous sulphur and nitrogen pollutants may be classified as chemical and physical. The chemical methods, which have been developed from techniques used in the chemical process industries, involve trapping the gas in a suitable medium followed by chemical or electrochemical analysis of the trapped material. Physical methods involve direct measurement of a physical or optical property either of the pollutant itself or following its interaction with another compound. The measurement may be preceded by chromatographic separation.

Since the atmosphere contains a wide variety of trace gases, interference by other gases is a major factor to be considered in the analysis of a given pollutant. Since many chemical methods rely on properties such as the acidity, oxidizing or reducing capabilities of the gases to be analysed, they are subject particularly to interference by other gases. Particulate material in the air may interfere chemically or, especially in the case of continuous instrumental methods, by physical contamination. Interfering substances may be removed either selectively or by discrimination either during sampling or at the analytical stage.

The errors in the analysis of gases in the atmosphere are expressed in terms of the degree of accuracy and precision. Inaccuracy arises from interferences, variable and indeterminate collection efficiency, calibration errors etc., and is usually the main source of analytical error. For most methods the analytical precision is a minor source of error, provided the procedures are carried out carefully. Frequently an unreasonable degree of accuracy and precision in measurement is specified. There is no need to obtain a concentration to the nearest 0.01 unit if differences between effects are not noticeable to the nearest unit. In ambient air, pollutant concentrations vary greatly in time and place and an overall analytical error of 10% will generally be quite satisfactory for the interpretation of field data. In obtaining cause-and-effect relationships in laboratory experiments a greater degree of accuracy may be required, but higher accuracy and precision is easier to achieve in controlled laboratory conditions. It is also unnecessary to obtain exactitude in sampling which is greater than the precision of the analytical procedures and vice versa.

4.2.1 Sampling techniques

The methods used for the sampling of gaseous atmospheric pollutants are surveyed in Chapter 1. For the chemical analysis of gaseous S and N compounds, sample concentration is usually required. This is achieved by passing the air sample through a suitable trapping medium e.g. an absorbing solution or an impregnated filter paper. The latter is convenient for automatic sequential sampling utilizing a filter paper tape. The choice of sampling parameters, e.g. flow rate, absorbent volume, time, etc., will depend on the expected concentration range of the pollutant, the time period over which the average concentration is required, the collection efficiency of the medium and the limitations of the analytical method.

Techniques requiring sample concentration cannot be used for continuous real-time measurement of pollutant concentrations. Physical and electrochemical instrumental methods offer more scope for this type of measurement since the sensitivity of modern instruments is sufficient to dispense with sample concentration, at least for measurements in urban air. Air is simply drawn into the instrument continuously and the amount of pollutant gas entering the sensing device is monitored. Alternatively, physical sensing techniques may be applied to discrete samples, as for example in gas-chromatographic analysis. By making automatic sequential analysis, to give a series of spot measurements, a reasonable approach to continuous monitoring can be made, provided the analysis time is short. Gas chromatographic analysis may also be applied to samples collected at a remote location and transported to the laboratory. However this approach is not recommended for S_2 and N_2 gases in view of possible losses on containment and during transport.

In the design of samplers for gases, it is important to ensure that there is no significant loss or modification of the pollutant during transport from the free atmosphere to the absorption medium or sensing device. Most of the gaseous S and N compounds are chemically reactive and tend to be absorbed on containing materials, particularly in moist environments. Sample probes should be as short, clean and dry as possible and should be of a suitably inert material e.g. borosilicate glass, polytetrafluoroethylene (PTFE) or other fluorocarbon polymer. Most metals and the more common types of plastic tubing, e.g. polythene and PVC, should be avoided. Couplings and valves should, as far as possible, be avoided on sampling lines, but if used should be of high quality stainless steel or inert plastic.

Interfering particulate material is most conveniently removed at the sampling stage by filtration of the gas stream. The filter should be of a suitably inert material to minimize absorption of the gas of interest. Similarly, the filter holder should be constructed so that the gas stream is only exposed to an inert surface such as glass or PTFE. In order to minimize adsorption, it may be necessary to heat the filter particularly when sampling at high relative humidity (see Section 4.3.1.1 for SO_2 analysis).

Selective removal of interfering gaseous substances at the sampling stage has also been widely utilized, particularly in monitoring instruments based on chemical methods of analysis. This is effected by exposure of the sample gas to an absorbent (either a liquid or solid) which removes the interfering component(s) but allows the pollutant of interest to pass on to the collection or sensing unit. Pretreatment may also be used to convert chemically the pollutant to a compound which is more suitable for analysis (e.g. oxidation of NO to NO_2). The efficiency of any sample pretreatment must be carefully assessed to ensure that it is quantitative under the conditions operating. In view of the possibility of enhanced absorption losses or inadvertent modification of the sample, pretreatment of the sample gas should only be applied if significant interference is expected which cannot be removed at a later stage of the analysis.

4.2.2 Analytical methods – chemical

The chemical methods for the analysis of gaseous sulphur and nitrogen pollutants may be broadly classified as acidimetric, colorimetric and coulometric techniques.

4.2.2.1 Acidimetric methods

Acidimetric techniques have been widely used for the routine analysis of SO_2 and involve determination of the free acid (H^+ ion) produced following the absorption of SO_2 in an oxidizing solution (e.g. dilute H_2O_2) where it is converted to sulphuric acid (H_2SO_4). The free acid may be determined by titration or electrically, by conductivity or pH measurements. Automatic measurement of the change in conductivity of a solution exposed to SO_2 provides the basis for continuous acidimetric analysis of the gas. Any other gas which is absorbed rapidly by aqueous solution to yield strong acid will give a response but normally

SO_2 is the most abundant atmospheric constituent which behaves in this way. A more serious interference comes from alkaline substances (e.g. NH_3) which neutralize the acid. Acid and alkaline particulate matter can also interfere and the acidimetric method for the determination of SO_2 has been widely criticized on account of these interferences.

4.2.2.2 Colorimetric methods

A wide variety of colorimetric methods have been applied to N and S gases. The technique involves interaction in solution of the gas or its hydrolysis or oxidation products with a colour-forming reagent, followed by spectrophotometric measurement of the colour. The optical absorbance is proportional to the concentration of the component of interest.

Variations of the basic colorimetric method include turbidimetric measurement of a colloidal suspension (e.g. barium sulphate from the sulphate ion in solution), measurement of optical density of a colour produced on impregnated filter paper (e.g. lead acetate stain method for H_2S) and spectrofluorimetric determination of ions in solution.

Colorimetric methods can be highly specific and sensitive. The colour reagent may be incorporated in the absorbing solution allowing the colour to develop as the sample is taken. Reactions of this type have been utilized in automatic colorimetric analysers in which pumped sample gas and reagent(s) are continuously mixed and the exposed reagent passes to a flow colorimeter where the absorbance is measured (Fig. 4.1). It is necessary to have controlled flows of both reagent and sample gas for reproducible results. Reproducibility and efficiency of absorption of the gas in the reagent are important. A double-beam colorimeter system has advantages for stability, compensating against lamp emission, voltage and reagent optical density. When instruments are designed for air pollution work, reagents for seven or eight days are carried in storage bottles.

An alternative approach to colorimetric analysis involves absorption in a medium which traps the pollutant gas in a stable, non-volatile form (e.g. SO_2 as sulphate ion, NO_2 as nitrite ion). The samples may then be stored for analysis at a later date. Filter paper tape impregnated with a suitable absorbent provides a useful medium for the collection of large numbers of samples for colorimetric analysis of SO_2 and NH_3. The filter samples may be stored dry in sealed bags and extracted into solution when the analysis is to be carried out. For small scale monitoring exercises, simple liquid bubbler absorbers are more convenient.

Of the chemical techniques, colorimetric methods are the most versatile. Manual procedures giving sensitivity to approximately 10 ppb are available for most of the gaseous N and S pollutants. The colour response can be conveniently calibrated using standard solutions of the corresponding ions and any interfering

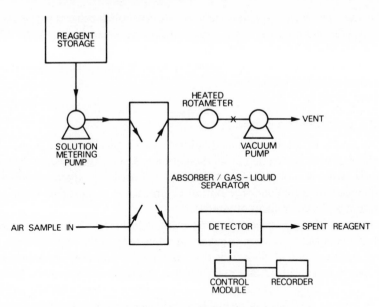

Fig. 4.1 Block diagram of continuous colorimetric analyser.

substances can often be eliminated during sample preparation. The labour involved in the analysis of a large number of individual samples has been reduced by the advent of automatic chemical analysers. These instruments perform automatically the 'wet chemical' operations normally carried out manually in the laboratory. The heart of the instrument is a multichannel proportioning-pump which dispenses sample solution and reagents into a continuous-flow system. Processes such as mixing, heating etc. are performed automatically on the sample stream and finally optical density is measured and the results displayed on a chart recorder. Because of the accurately standardized automatic procedure, precision is frequently better than that obtained in the best manual methods and a very high rate of sample analysis is possible. Technicon, a familiar company in this field, has a large range of analysing systems covering many pollutants.

4.2.2.3 Coulometric methods

The most widely adopted instrumental chemical method for gaseous pollutants is coulometric analysis. This involves measurement of the electrical current produced when strongly oxidizing or reducing pollutant gases react with potassium iodide or bromide solution in an electrochemical cell. Two general types of coulometric analysers have been employed; one involves the principle of coulometric internal electrolysis (galvanic action – Hersch Cell [3]) and the

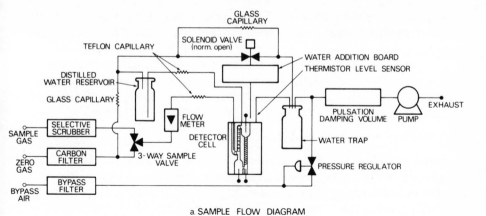

a. SAMPLE FLOW DIAGRAM

With no NO_2:
$I_c = I_a = 0$
With NO_2 present:
$I_a = I_c > 0$

b. ELECTROCHEMICAL REACTIONS OF THE CELL

Fig. 4.2 Coulometric analyser for NO_2 (Beckman).

other is described as an amperometric coulomer. The latter principle was used in early systems for coulometric analysis of O_3 [4] e.g. 'Mast' ozone meter, see Chapter 5 and requires reagent replacement and an applied external voltage. Galvanic cells which utilize a cyclic oxidation–reduction process and require no applied potential, are usually used for the coulometric analysis of N and S gases. In the Beckmann NO_2 analyser (Fig. 4.2), the sample gas is drawn through the detector cell containing buffered KI electrolyte which is circulated past the electrodes. The cell contains a Pt cathode and a C anode with a galvanic potential difference between them. When NO_2 enters the cell it reacts with the iodide ion in the following reaction

$$NO_2 + 2I^- + 2H^+ \longrightarrow I_2 + NO + H_2O$$

The iodine produced is reduced at the cathode in the electrochemical reaction

$$I_2 + 2e^- \longrightarrow 2I^-$$

and a corresponding oxidation reaction occurs at the anode

$$C \text{ (reduced state)} \longrightarrow C \text{ (oxidized state)} + ne^-$$

Thus the iodine liberated allows the passage of a current in the anode–cathode circuit which is proportional to the amount of NO_2 entering the cell. With no oxidizing agent entering the cell the residual current is zero.

A more complex indirect galvanic cell is used for the coulometric analysis of SO_2 and other reducing gases. The sample gas is introduced into the anode side of the detector cell which contains iodide solution having electrogenerated iodine present. The iodine is reduced by SO_2 to iodide in the reaction:

$$SO_2 + I_2 + 2H_2O \longrightarrow SO_4^{2-} + 2I^- + 2H^+$$

resulting in a lowering of the amount of iodine to be reduced at the cathode. A reference cell in a bridge circuit measures the difference, i_d, between the constant electrogenerating current (anode output) and the cathodic output current, i_d being proportional to the amount of SO_2 entering the cell.

If the reaction of the oxidizing or reducing gas in the electrolyte occurs rapidly with 100% efficiency the electrical current is directly related by Faraday's law to the amount of gas entering. Thus for a given controlled sample flow rate, coulometric analysis can give an absolute measurement of concentration without the requirement of calibration against standard mixtures or solutions. The main problem with coulometric analysis of a given component arises from interference by other oxidizing and reducing components which may be present in the sample. For example, in the analysis of NO_2 a positive interference is given by O_3, peroxides, peroxyacylnitrates, Cl_2, etc. and a negative interference by SO_2, H_2S and other sulphides. The success of coulometric instruments for reliable ambient air measurements relies heavily on the design of suitably selective sample scrubbers to remove interfering substances. Several commercial manufacturers claim reasonably selective measurement of NO, NO_2, SO_2, H_2S in the 0.01 to 1 ppm range.

4.2.2.4 Miscellaneous chemical methods

An instrumental technique using specific ion electrodes has been recently introduced (Bran and Lubbe Ltd.) in which gas sample and liquid reagent are pumped into an absorption or reaction chamber and reacted reagent passes at intervals into the measuring chamber where an ion-selective electrode measures the concentration of the selected ion. The main application is in the analysis of hydrogen fluoride (HF) but the system can be used for NH_3, hydrogen cyanide (HCN) and H_2S.

Another new electrochemical technique, which has been applied to the measurement of higher concentrations of pollutants, e.g. in stack gases, is the selective redox cell. The selectivity is based on variation of the electrode potential of the cell. For an oxidation reaction, only gases with oxidation potentials below the electrode potential will be oxidized and, for reduction, only gases with reduction potentials above the electrode potential will be reduced. SO_2 can be detected in the presence of NO_2 by oxidation and NO_2 in the presence of SO_2 by reduction. Analysers based on this system are simple, low in cost, and have sealed replaceable and interchangeable cells (e.g. Dynasciences Sensors, Enviro-metrics Inc., Faristors). Because response times are slow and sensitivity moderate, source measurements are likely to remain their main area of use.

4.2.3 Physical methods

Most of the physical methods currently used for the analysis of gaseous N and S pollutants involve optical measurements of some kind. The optical techniques include chemiluminescence, fluorescence and absorption spectroscopy. Gas chromatographic methods have recently been developed for the S gases but for gaseous N pollutants, application of this powerful technique has not yet been successful.

4.2.3.1 Chemiluminescence

The phenomenon of chemiluminescence occurs when part of the energy of an exothermic chemical reaction is released as light. The factors affecting light emission from chemiluminescent gas reactions are exemplified in the reaction between NO and O_3, which is now widely used in the analysis of nitrogen oxides [5]. The fast reaction between NO and O_3 in the gas phase produces excited NO_2 molecules which lose their energy either by light emission or by quenching collisions with other molecules present:

$$NO + O_3 \longrightarrow NO_2 + O_2$$

$$NO_2 \nearrow \begin{array}{l} NO_2 + h\nu \text{ chemiluminescence} \\ M \searrow NO_2 \text{ (M} = N_2, O_2, H_2O, \text{ etc.).} \end{array}$$

The light emission (I) is given by an equation of the form:

$$I = \text{constant} \times [NO][O_3]/[M]$$

where $[M]$ is proportional to the total pressure. For a fixed pressure, I is dependant only on the NO and O_3 concentrations. If either of these components is held constant (which is effectively achieved by arranging for one gas to be in large excess), the light emission is proportional to the concentration of the other gas. The emission from a given chemiluminescent reaction has a characteristic spectral composition and the spectral region of interest may be selected using

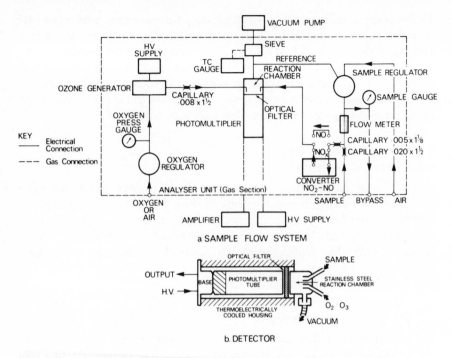

Fig. 4.3 Chemiluminescence NO_x analyser (Thermo-Electron).

optical filters and/or a suitable choice of photomultiplier response characteristics. Thus high selectivity can be achieved for the analysis of a particular gas.

Fig 4.3 shows the basic layout for a chemiluminescence analyser for NO. The sample gas, containing NO and a stream of ionized O_2, is mixed in a reaction chamber which is positioned adjacent to the end-window of a photomultiplier tube. The pressure and flow of gas through the reaction chamber are maintained constant by a vacuum pump in conjunction with critical orifice capillaries and pressure regulators on the inlet lines. Typical operating pressures are 0.01 to 0.05 atm. In operation, the system gives a continuous signal proportional to the amount of NO entering the reaction cell and thence the NO concentration in the sample gas is measured. The system is calibrated using standard NO gas mixtures. The response is linear over a wide range and a sensitivity down to less than 1 ppb can be obtained. When NO_2 is to be measured the sample gas is passed through a converter consisting of a heated stainless steel tube in which NO_2 is decomposed to NO. The total NO + NO_2 is subsequently measured and the NO_2 obtained by difference.

Another type of chemiluminescence detector, which has proved particularly useful in the analysis of gaseous S compounds, is the flame-chemiluminescence or flame photometric detector (FPD) [6]. When S containing compounds are burned in a fuel-rich hydrogen flame an intense chemiluminescence in the 300 to 425 nm spectral region results from the radiative recombination of atomic sulphur:

$$S + S + M \longrightarrow S_2 + M$$

$$S_2 \longrightarrow S_2 + h\nu$$

If the emission is monitored using a narrow band filter (394 ± 5 nm) a specificity ratio of approximately 20 000 : 1 for sulphur compounds compared with other components giving chemiluminescence in the flame (e.g. hydrocarbons) is achieved. The response is roughly proportional to the square of the sulphur concentration in the sample gas but a linear output can be obtained by logarithmic amplification of the photocurrent. The minimum detectable concentration on current instruments is approximately 5 ppb. Analysers, based on FPD, function as total sulphur monitors unless chromatographic separation of the components is carried out. In most air pollution work SO_2 is the predominant sulphur component and total sulphur monitoring is adequate. Around refineries and other sources producing H_2S and mercaptans some discrimination must be made either by GC or selective absorption from the sample gas.

4.2.3.2 Fluorescence

Fluorescent emission occurs when molecules absorb radiant energy at one wavelength and re-emit part of that energy at another wavelength. In gases, fluorescence is a low pressure phenomenon, the emission being normally quenched to undetectable levels at pressures near atmospheric. However given a suitably intense monochromatic light source, the method can in principle provide the basis for selective analysis of trace gases (e.g. SO_2) which have a strong absorption band from which fluorescence emission occurs [7]. Instruments are available for the specific measurement of SO_2 in the 0.5 to 1000 ppm range utilizing fluorescence emission. These employ a pulsed u.v. source which irradiates the sample gas flowing continuously through the optical cell. The fluorescence emission is detected by a photomultiplier viewing at 90° to the excitation beam. Optical filters are used to select narrow bandwidths for the exciting and emitted radiation. At the present time the sensitivity of the method is only sufficient for source analysis but future development may well make it practical for monitoring ambient air.

4.2.3.3 Absorption spectroscopy

Several instruments and techniques for the measurement of gaseous N and S compounds have been based on the absorption of i.r. and u.v. radiation. The absorption spectra are specific fingerprints for compounds absorbing in those regions and the information contained in the absorption spectrum of a sample gas is adequate to give a specific measurement of each of the absorbing compounds in the sample. The problem arises in extracting the information from the spectrum. Simple methods such as non-dispersive i.r. and u.v. analyses are non-specific and have only moderate sensitivity. Non-dispersive i.r. analysis is, however, widely used for source measurement of carbon containing compounds and to a lesser extent for NO. Dispersive i.r. instruments (i.e. spectrometers) are specific but even when long pathlengths are used the sensitivity is barely adequate for monitoring ambient air. Furthermore, i.r. spectrometers are complicated, delicate and expensive instruments and are not readily utilized for continuous operation. Two new methods for reducing spectral data to a simple quantitative output have been recently developed and incorporated into commercially available instruments. These are correlation spectrometry and derivative spectrometry, both of which are primarily used in the u.v. region for the determination of SO_2 and NO_2.

In correlation spectrometry (Fig. 4.4) the incoming light signal (i.e. the light being sampled) is dispersed by a grating spectrometer. Instead of the normal exit slit there is a correlation mask which is a photographic replica of the spectrum of the compound of interest, with slits corresponding to the main absorption peaks. When the spectrum is vibrated across this mask, a beat signal is obtained when the incident light bears the absorption pattern of the compound being monitored. Although absorption of other gases may overlap on some peaks, no other gas will correlate over the whole spectrum and so the signal from the desired component is much enhanced. The Barringer Research Correlation spectrometer [8] has mainly been used as a remote sensing instrument for NO_2 and SO_2 using daylight as a source. The total amount of pollutant in the optical path is measured and expressed in units of ppm meters. Using a high intensity xenon arc-lamp as a source, measurements can be made over a fixed pathlength to give average concentrations in a given location.

The second derivative spectrometer (Spectrometrics of Florida Ltd.) incorporates an oscillating inlet slit in an u.v. grating spectrometer [9]. The varying angle of incidence on the grating produces a signal with amplitude proportional to the second derivative of the absorption spectrum. A substantial increase in the sensitivity and selectivity over direct absorption spectrometry results, giving minimum detectable levels in the ppb range for SO_2, NO and NO_2. Ambient air is continuously aspirated through the absorption cell and each component measured in turn on a 3 to 7 min cycle.

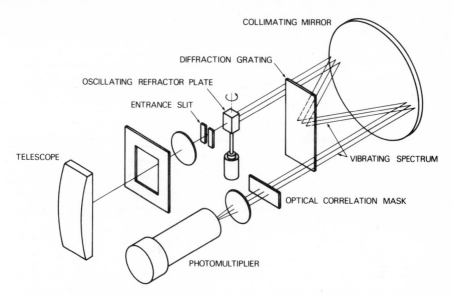

Fig. 4.4   Correlation spectrometry.

At the present time there is much research activity in the field of air pollution monitoring by advanced spectroscopic techniques [10]. In particular the advent of lasers has provided stimulus to the remote optical sensing field. In the future these techniques will almost certainly supersede existing methods for many monitoring applications. Current techniques for remote sensing of pollutants are discussed in Chapter 8.

4.2.3.4 Gas chromatography

Gas chromatography is one of the most versatile and selective methods for gas analysis. The basic principles of the method as applied to the analysis of atmospheric pollutants are described in Chapter 6. The main difficulty in the application of this method to the analysis of S and N pollutants arises from their high reactivity and consequent absorption loss and irreproducible transfer through the chromatographic system. For the measurement of S compounds, however, the pioneering work of Stevens *et al.* [11] has led to considerable advances in technique. By the use of PTFE components and specially developed column materials, together with selective flame photometric detection, quantitative analysis of gaseous S pollutants in ambient air by gas chromatography is now possible.

There have been numerous attempts to separate and analyse NO and NO_2 at ppm levels by gas chromatography [12]. Most workers have experienced serious loss or modification of these gases on the column packings. The best results have been obtained with porous polymer column packings e.g. Porapak. Even if these problems could be overcome the problem of a suitable detector remains. Rare gas ionization detectors or electron capture detectors seem to offer the best prospects.

4.2.3.5 Other physical methods

Although mass spectrometric techniques have been widely used for gas analysis in the laboratory, application to the routine measurement of low concentrations of S and N gases in ambient air has not proved practical. Without prior separation and sample concentration, sensitivity is too low and the mass spectra are impossibly complicated for quantitative analysis.

A technique which is currently being developed for trace gas analysis is photo-ionization mass spectrometry [13]. This involves ionization using high energy u.v. radiation from a rare gas (krypton or argon) resonance lamp followed by mass spectometric detection. As only those molecules which have ionization potentials less than the u.v. excitation energy will be ionized, a measure of selectivity is thereby gained. This may prove a useful method for the determination of NO which has a relatively low ionization potential compared with other low molecular weight gases present in the atmosphere.

4.3 Experimental section

In this section the methods of analysis of the individual gaseous S and N pollutants are surveyed and practical details of several recommended wet chemical methods are given. The practical details include procedures for sampling, analysis and calibration. Instrumental methods are also discussed but it is not possible to evaluate critically the performance of individual instruments from each manufacturer. Instead a brief description of the operation, specification and calibration of selected instrument types is given. Finally, at the end of this section practical details of general methods for the preparation of standard gas mixtures are given.

4.3.1. Analysis of SO_2

Since SO_2 is such a widespread pollutant many methods have been devised for the analysis of this gas at levels found in the atmosphere [14]. Among the principles employed are acidimetry, colorimetry, electrochemistry, flame photometry, emission and absorption spectroscopy. The main factors determining the optimum technique are the type of environment to be sampled, the availability of resources, both financial and manpower, and the operational requirements, i.e. continuous, periodic or 'spot' recordings. Many of the above principles have been adapted for automatic operation, but it should be pointed out that the most severe limitation on all commercial instruments and on many manual methods for SO_2 analysis is the lack of sensitivity. Generally the lowest realistic detection limit of instrumental methods is about 0.01 ppm. Since the SO_2 concentration in most non-urban environments is < 0.02 ppm an improvement in sensitivity of at least a factor of 10 is at present required to satisfy the needs of environmental researchers.

4.3.1.1 Chemical methods

The recommended procedures for manual determination of SO_2 are the West–Gaeke colorimetric technique [15] and the hydrogen peroxide method [16]. There are a number of variations of the latter method inasmuch as the H_2SO_4 which is formed when SO_2 is trapped in aqueous H_2O_2 can be determined either titrimetrically or colorimetrically as SO_4^{2-}. The latter is to be preferred since it is more specific, and using the procedure given below SO_2 concentrations down to 0.001 ppm can be measured. Determination as SO_4^{2-} provides a useful method for the measurement of SO_2 collected on impregnated filter tape [17]. Analysis of SO_2 based on its redox reactions with halogens in solution is widely used in commercial electrochemical instruments. Manual iodometric techniques have also been described [18]. The main problem with the redox system is interference from other oxidizing and reducing substances and, although manufacturers of commercial devices claim to have reduced the interference to acceptable levels, the additional procedures for the removal of interfering substances make this method unattractive for manual analysis.

West–Gaeke (colorimetric) method (ISC method no. 42401-01-69T [82])

Principle

SO_2 is absorbed by aspirating a measured volume of air through a solution of potassium tetrachloromercurate (TCM). The stable non-volatile dichlorosulphito-mercurate ion is formed in this procedure. Addition of solutions of purified, acid-bleached pararosaniline and formaldehyde leads to the formation of intensely coloured pararosaniline methyl sulphonic acid. The pH of the final solution is adjusted to 1.6 ± 0.1 by the addition of a prescribed amount of 3 M phosphoric acid to the pararosaniline reagent since the extinction coefficient of the product depends on pH ($\epsilon = 47.7 \times 10^3$ litres mol^{-1} cm^{-1} at $\lambda_{max} = 548$ nm and pH 1.6).

The method is applicable for SO_2 concentrations of 0.01 to 5 ppm, the lower limit of detection being $0.3 \mu l$ SO_2 per 10 ml TCM which corresponds to 0.01 ppm SO_2 in 30 litres air. Absorption efficiency of TCM falls off at concentrations of SO_2 below this and therefore lower detection limits cannot be achieved by sampling larger volumes unless the absorption efficiency is determined separately using, for example, radioactive sulphur dioxide ($^{35}SO_2$) [23] or a standard SO_2 gas mixture.

The principal interfering compounds are oxides of nitrogen, O_3, and heavy metals. The effects of these are minimized in the experimental procedure by:
(a) the addition of a solution of sulphamic acid which destroys any nitrite ion formed by the absorption of oxides of nitrogen in the TCM solution [19];
(b) allowing any dissolved O_3 to decay by delaying analysis for 20 min after sample collection [20] and;
(c) addition of ethylenediamine-tetra-acetic acid disodium salt (EDTA) to the TCM solution to complex heavy metals that can interfere by oxidation of the SO_2 before it can react with the TCM [20].

Apparatus

The SO_2 may be sampled either in midget or standard fritted bubblers, a midget impinger or standard impinger. The sample probe should be of borosilicate glass, stainless steel or PTFE and, if a prefilter is used, it should be heated when sampling at a relative humidity (r.h.) $> 70\%$ (see Section 4.3.1.1). A pump with a capacity of up to 2.5 litres min^{-1} (for midget samplers) or up to 15 litres min^{-1} for standard samplers is required for aspiration. The gas volumes are measured with a calibrated rotameter or a wet or dry gas meter. Alternatively a high volume pump and a critical orifice flow-meter may be employed.

Reagents

Analytical grade chemicals should be used. The pararosaniline dye should have an assay of greater than 95%.

(a) Absorbing reagent $-0.04 M$ potassium tetrachloromercurate (TCM) K_2HgCl_4: dissolve 10.86 g of mercuric chloride (Poison), 5.96 g of potassium chloride, 0.066 g of EDTA (disodium salt) in water and make up to 1 litre. The pH should not be less than 5.2 or SO_2 absorption efficiency may be impaired. This solution is stable for six months.

(b) Sulphamic acid: dissolve 0.6 g of sulphamic acid in 100 ml of water (stable for a few days if protected from atmospheric oxidation).

(c) Buffer solution (for assay procedure): 100 ml of 0.1 M sodium acetate—acetic acid (pH = 4.69).

(d) Phosphoric acid (H_3PO_4): 3 M H_3PO_4 – dilute 205 ml H_3PO_4 (85%) to 1 litre.

(e) 0.2% Pararosaniline stock solution. The pararosaniline dye needed to prepare this reagent should yield a TCM-reagent blank of not more than 0.17 absorbance units (A.U.) at 22° C, should give a calibration curve with standard sulphite solutions of slope 0.746 ± 0.04 A.U.$\mu g^{-1} ml^{-1}$ for 1 cm cells, and must have an absorbance maximum at 540 nm when assayed in a buffered solution of 0.1 M sodium acetate–acetic acid. To make the stock solution take 0.200 g pararosaniline dye and dissolve in 100 ml of 1 M HCl in 100 ml glass stoppered graduated cylinder. If the pararosaniline does not meet the requirements it may be purified by repeated solvent extraction with 1-butanol. The assay of pararosaniline in the stock solution is carried out as follows: dilute 1 ml of stock solution to 100 ml in a volumetric flask with distilled water. To a 5 ml aliquot in a 50 ml flask, add 5 ml of 1 M sodium acetate–acetic acid buffer and dilute to 50 ml with distilled water. After 1 h determine the absorbance at 540 nm with a spectrophotometer using 1 cm cells. The assay is given by

$$\% \text{ pararosaniline} = \frac{\text{Absorbance} \times 21.3}{\text{grams taken}}$$

(f) Pararosaniline reagent. To 20 ml of stock solution in a 250 ml flask add an additional 0.2 ml of stock solution for each 1% less than 100% assay in the stock, followed by 25 ml 3 M phosphoric acid (H_3PO_4). Dilute to volume with distilled water. The reagent is stable for at least nine months.

(g) Formaldehyde, 0.2%. Dilute 5 ml of 40% formaldehyde to 1 litre with water. Prepare daily.

(h) Standard sulphite solution. Dissolve 0.400 g sodium sulphite (or 0.300 g sodium metabisulphite) in 500 ml boiled and cooled distilled water. Sulphite solutions are unstable and must be freshly standardized before use. This is achieved by adding excess iodine and back titrating with sodium thiosulphate which has been standardized against potassium iodate or dichromate (primary standard).

(i) Dilute sulphite solution. Pipette accurately 2 ml of freshly standardized sulphite solution and make up to 100 ml with 0.04 M TCM. If stored at 5° C this solution is stable for one month.

Procedure

Place a measured quantity of absorbing reagent (10 to 20 ml for a midget impinger; 75 to 100 ml in a standard absorber) and connect up the sampling probe and metering system. Flow rates for midget impingers should be between 0.5 to 2.5 litres min^{-1} or up to 15 litres min^{-1} with large absorbers. Within these ranges the sampling efficiency should be $> 98\%$. Sample for sufficient time to give between 0.5 and 3.0 μg SO$_2$ per ml absorbing solution. Shield the solution from direct sunlight during sampling and storage. Keep cool during storage, and if a precipitate forms, remove it by centrifugation. For analysis of a 10 ml sample transfer it quantitatively to a 25 ml volumetric flask with approximately 5 ml of distilled water for rinsing. For high concentrations of larger volumes, aliquots may be taken at this point. Leave the sample for 20 min to allow any O$_3$ present to decay. Meanwhile prepare a reagent blank using 10 ml of exposed reagent in a 25 ml flask. To each flask add 1 ml of 0.6% sulphamic acid and allow 10 min for the destruction of any nitrite ion from oxides of nitrogen. Accurately pipette 2 ml of formaldehyde (0.2%), then 5 ml of the pararosaniline reagent. Make up to 25 ml and determine the absorbance of both solutions against distilled water after 30 min.

Calibration

Accurately pipette graduated amounts of the dilute sulphite solution (e.g. 0, 1, 2, 3, 4, 5 ml) into 25 ml flasks and make up to approximately 10 ml with 0.04 M TCM. Add the remaining reagents as described in the procedure and measure the absorbances. The total absorbances, plotted as a function of μg SO$_2$ (total), should give a linear plot which intercepts to within 0.02 absorbance units of the blank. The calibration factor B is the reciprocal of the slope of the line. The concentration of SO$_2$ in the air sample is then given by:

$$\text{SO}_2 \text{ (ppm/v)} = \frac{(A - A_0)\,0.382\,B}{V}$$

where A, A_0 are the sample and reagent blank absorbances, 0.382 is the volume in μl of 1 μg SO$_2$ at 760 Torr and 25° C and V is the sample volume in litres (corrected to 760 Torr and 25° C).

The calibration may be alternatively carried out by sampling from a standard source of SO$_2$ in air obtained, for example, using a permeation tube [21, 22] (see Section 4.3.8.2). This procedure has the added advantage that any losses resulting from adsorption in the sampling probe or inefficiency of the absorber can be taken into account.

Hydrogen peroxide (H$_2$O$_2$)/sulphate method (using automatic colorimetric analysis)

Principle

The air sample containing SO$_2$ is sucked through a heater to raise its temperature by 10° C before passing through an absolute prefilter to remove particulate material and into a simple bubbler containing 1 volume H$_2$O$_2$ where the SO$_2$ is rapidly oxidized to involatile H_2SO_4. By heating the air losses of SO_2 by adsorption are reduced to a negligible level even when sampling air of up to 96% r.h. The sulphate is determined by the method of Persson [24] which was developed for use with a Technicon Autoanalyser. The decrease in the light absorbance of a barium–thoranol complex, when barium is removed from it by sulphate ions, is measured at $520 \mu m$. The reactions which are carried out in a weakly acid solution (perchloric acid) of aqueous iso-propanol may be represented as:

$$Ba^{2+} + thoranol \rightleftharpoons Ba–thoranol \text{ complex}$$

$$Ba–thoranol \text{ complex} + SO_4^{2-} \longrightarrow BaSO_4 + thoranol$$

where thoranol is 1-(orthoarsenophenylazo)-2-napthol-3,6-disulphonic acid sodium salt (known also as Thoron, Thorin, naptharson, APANS). The method is very susceptible to foreign ions, either anions which complex barium, or cations, particularly polyvalent cations. Ammonium ions do not interfere, however. Dilute H_2O_2 solutions do not interfere with the method and it is possible to analyse the bubbler samples without pretreatment. The sulphate content of 100 vol. H_2O_2 used to prepare absorbent solutions may be significant. Therefore, standards and autoanalyser wash solutions are prepared from the same batch as used in making up absorbent solutions.

A limit of detection for SO_4^{2-} in solution of $0.1 \mu g\,ml^{-1}$, which in 40 ml of absorbent corresponds to $2.7 \mu g\ SO_2$, is attainable. Analytical precision at this level is of the order of 10% increasing to 0.4% at $3 \mu g\,ml^{-1}$ sulphate ion. Flow rates of up to 30 litres min^{-1} through the absorber can be used and give a limit of detection of the order of 0.001 ppm SO_2 from a 30 min sample.

Apparatus

The sampling unit consists of a heated pyrex inlet tube, a filter holder and a standard impinger assembled as shown in Fig. 4.5. The prefilter can be either Whatman 41 or a 'Microsorban' polystyrene absolute filter (50 mm diameter). The heater consists of a standard $100\,\Omega$ resistor, wire, wound on a hollow ceramic former which fits closely round the sample tube, and is supplied with 30 to 40 V a.c. In order to prevent condensation the filter holder and bubbler entry tube are enclosed in a box of expanded polystyrene during sampling. The system is aspirated with a pump capable of delivering > 30 litres min^{-1}, through a conventional metering system.

A Technicon autoanalyser is used for the analysis of the sulphate ions. The analyser should preferably be equipped with silicone rubber pump tubes since the more usual 'solvaflex' tubes are attacked by the isopropanol solutions and have a normal operation time of only two days before replacement is necessary. A diagram of the analyser flow system is shown in Fig. 4.6.

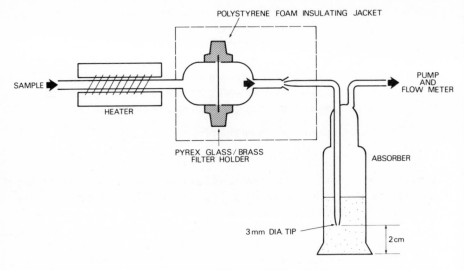

POLYSTYRENE FOAM INSULATING JACKET

SAMPLE ▶

HEATER

PYREX GLASS / BRASS
FILTER HOLDER

PUMP
AND
FLOW METER

ABSORBER

3mm DIA. TIP

2cm

Fig. 4.5 Sampler for the collection of SO_2 in H_2O_2 solution.

Reagents

Analytical grade chemicals and water should be used.

(a) Absorbing solution, 1 vol. H_2O_2. Dilute 10 ml of 100 vol. H_2O_2 (30%) to 1 litre with distilled water. Store in a polythene container in the dark. (Prepare weekly.)

(b) Stock barium perchlorate [$Ba(ClO_4)_2$] 0.1 M perchloric acid. Weigh 0.90 g $Ba(ClO_4)_2$ and dissolve in 1 litre of 0.1 M perchloric acid $HClO_4$, prepared by dilution of 8.6 ml of $HClO_4$ (70% sp. gr. 1.66) with distilled water.

(c) Working barium perchlorate/$HClO_4$. Dilute 10 ml of stock solution to 1 litre with iso-propanol.

(d) Thoranol. Weigh out 0.20 g of Thoranol and dissolve in 1 litre distilled water.

(e) Standard potassium sulphate (K_2SO_4). Weigh out accurately 0.183 g of K_2SO_4 and make up to 1 litre in a volumetric flask using 1 vol. H_2O_2 solution prepared with the same 100 vol. H_2O_2 as above. This solution contains $100 \mu g\, ml^{-1}\, SO_4^{2-}$.

Procedure

Place a 40 ml aliquot of absorbent solution in the impinger and assemble the sampling train with a fresh filter. Switch on the heater and allow it to warm up for a few minutes before sampling. When sampling dry atmospheres (< 60% r.h.)

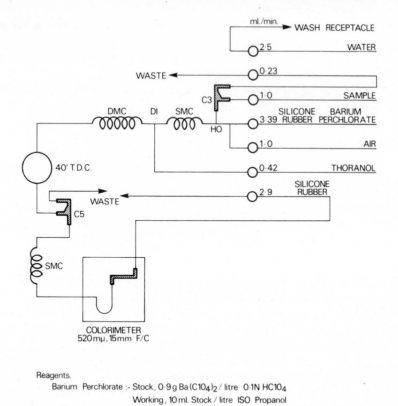

Reagents.

Barium Perchlorate :- Stock, 0·9 g Ba(C10₄)₂ / litre 0·1N HC10₄

Working, 10 ml. Stock / litre ISO Propanol

Thoranol :- 0·2 g/l

Fig. 4.6 Technicon autoanalyser manifold for the determination of SO_4^{2-} 0 to 10 ppm.

heating is not necessary. Sample at the chosen flow rate (1 to 30 litres min^{-1}) for sufficient time to obtain between 1 and 10 μg ml^{-1} SO_4^{2-}. If large volumes are sampled, particularly at low r.h. evaporative losses must be determined by weighing the collection vessel at the beginning and end of sampling. The SO_4^{2-} samples are quite stable and may be stored for several weeks if well stoppered.

For analysis of a series of samples a small amount ($\sim$ 5 ml) of each sample is placed in sample cups spaced in alternate positions on the analyser carousel. The cups in the intermediate positions are filled with unexposed reagent. Standards containing known amounts of SO_4^{2-} in the range 0 to 12 μg ml^{-1} are prepared by the appropriate volumetric dilution of the standard K_2SO_4 solution. These are also placed in alternate positions on the carousel with unexposed absorbing reagent in between. The samples are then run on the analyser and the peak height corresponding to the depletion in absorbance of each solution at $520\,\mu m$ measured from the recorder trace. A calibration plot is prepared from the standards and the sulphate content of the samples (in $\mu g\,ml^{-1}$) read off from the graph. The plot should be linear up to approximately $12\,\mu g\,ml^{-1}$ SO_4^{2-} but is usually curved at higher concentrations. Separate standards should be run with each batch of samples. The SO_2 concentration in the air sample is given by

$$SO_2\ (ppm/v) = \frac{\mu g\,ml^{-1}\ SO_4^{2-} \times 0.255 \times V_a}{V}$$

where V_a is the volume of absorbing solution corrected for evaporation loss, V is the volume of air (litres at 760 Torr and $25°$ C) and 0.255 is the volume of SO_2 (in μl at 760 Torr and $25°$ C) corresponding to $1\,\mu g$ SO_4^{2-}.

Variations

The above analytical procedure using an autoanalyser is to be recommended when high sensitivity and analysis of a large number of samples is required. However, the H_2SO_4 formed when SO_2 is collected in dilute H_2O_2 may be determined by a number of different manual methods, e.g. the barium sulphate turbidimetric method, [25] or the barium chloranilate colorimetric method [26].

The Thoron method using the autoanalyser may also be used for measurement of SO_2 collected on filter paper tapes. The tapes (Whatman 41) are impregnated with 25% potassium carbonate in a 10% glycerol/water mixture and dried under an i.r. lamp. The SO_2 is collected as sulphite which is oxidized to SO_4. A collection efficiency of greater than 95% can be achieved at face velocities of up to $70\,cm\,s^{-1}$. The filters are extracted at a $70°$ C in a 1 vol. H_2O_2 solution which is then cooled and passed through a cation exchange resin (Zeocarb 225) to remove potassium ions which interfere. The sample solutions are then analysed as above.

Instrumental chemical methods for analysis of SO_2

Electrical conductivity analysers

A variety of continuous automatic conductivity analysers for SO_2 are marketed. The absorbing liquid is usually aqueous H_2O_2 which flows through the absorption cell and the change in conductivity resulting from H_2SO_4 formation is measured. The method is non-selective and is therefore not recommended for precise measurements of SO_2 in the atmosphere, but can be useful for measurement of SO_2 in industrial environments. Sensitivity ranges of between 0 to 0.2 ppm and between 0 to 20 ppm SO_2 are available.

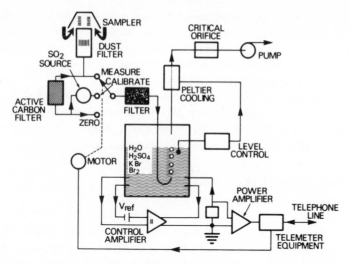

Fig. 4.7   Philips PW 9700 SO_2 monitor.

Coulometric analysers

The reducing action of SO_2 on free halogen/potassium halide solutions (I or Br) is utilized for coulometric SO_2 analysers. The available commercial instruments differ slightly in mode of operation and sensitivity. The Philips PW 9700 (Fig. 4.7) claims the lowest detection limit of 4 ppb SO_2 and is designed for three months unattended operation. Interference from H_2S, O_3 and Cl_2 is reduced to $< 1\%$ on a mol for mol basis by a silver wire scrubber. The main interferents are NO_2 ($< 5\%$) and mercaptans (100%). The instrument incorporates an automatic valve for sample, calibration and zero-check selection. In the zero position the sample air flows through activated charcoal to remove SO_2, giving a 'zero' signal from the coulometric cell. In the 'calibrate' position a known amount of SO_2 is introduced to the air stream from a built-in permeation tube. Coulometric SO_2 monitors are also marketed by Beckmann, Process Analysers Inc ('Tritilog' for total oxidizable sulphur), and T.E.M. Ltd (Model 222 SO_2 analyser). The response time (95% of final value) of coulometric analysers for SO_2 is of the order of 3 to 5 min. Sensitivity ranges of between 0 to 0.2 and between 0 to 200 ppm are available.

4.3.1.2 Physical analysis of SO_2

The recommended procedure for physical analysis of SO_2 in the atmosphere is gas chromatographic separation followed by flame photometric detection (FPD). A 'plug' sample is analysed, but with an automatic gas sampling valve sequential samples may be taken at intervals down to 3 min depending on analysis time. An alternative arrangement, which is used in most commercial FPD analysers, is to dispense with chromatographic separation and continuously pass sample air into the FPD detector, thereby obtaining measurement of 'total S'. A line filter can be incorporated to remove particulate S.

For measurement of high concentrations of SO_2 (i.e. > 1 ppm) the newly developed u.v. fluorescence technique shows exceptional promise for true selective measurement. It is predicted that future development will lead to the extension of the range of this instrument to atmospheric concentrations.

Gas chromatographic method for SO_2 *analysis*

Principle

An air sample is aspirated into the sample loop of a gas sampling valve. The valve is then switched to inject the sample on to a chromatographic column where SO_2 is separated from other gaseous S compounds. The eluent passes into a FPD detector and the detector signal amplified and recorded.

Detector response is roughly proportional to the square of the gaseous S_2 concentration for concentrations up to approximately 1 ppm [27, 28]. The range of electrical response used is approximately four orders of magnitude giving a minimum detectable concentration of 5 to 10 ppb. The actual detection limit will also depend on the retention time of SO_2 on the column and baseline stability after injection.

The major factor affecting reproducibility is the potential adsorption problem. The use of PTFE flow components is imperative and contact of the sample gas with metal should be avoided in all instances. With a suitably designed system, a reproducibility as high as ± 1.5% can be achieved at concentrations of the order of 0.05 ppm. There are no known interferences with this method for the analysis of SO_2 in air. The response may be affected by high concentrations of hydrocarbons in special environments.

Apparatus

Three major components are required.

(i) A sample injection system operated manually or automatically activated with a timer.

(ii) A chromatographic column for the separation of S compounds.

(iii) A flame photometric detector with associated amplifier and recorder.

In addition 1/8 in PTFE tubing and couplings for the sample gas lines and columns and a sample pump are required. Fig. 4.8 shows a flow diagram of the system.

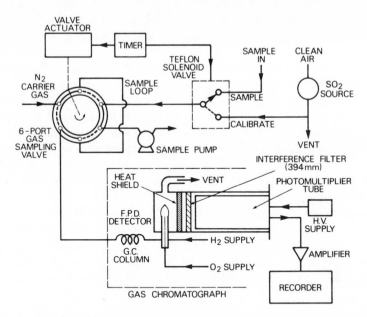

Fig. 4.8 Automatic gas chromatograph − flame photometric detector for SO_2 and other S compounds.

Suitable six-port gas sampling valves (GSV) are the Chromatromix models R6031SV (manual) or R6031SVA (PTFE-bodied, automatic). In addition a PTFE three-way solenoid valve is required if calibration samples are to be taken automatically. An industrial cam timer is programmed to activate the automatic GSV. Automatic GSV's are normally pneumatically operated at 40 to 60 psig. The sample loop should be of PTFE and 10 ml in volume. The sample pump should be capable of drawing 1 litre min^{-1} through the sample loop.

Three different types of column have been successfully employed for the separation and analysis of ppb concentrations of SO_2 in the atmosphere. All columns are packed in 1/8 in PTFE tubing and the following stationary phases can be used:

(a) 36 ft length packed with 40 to 60 mesh PTFE powder flow coated with approximately 10% polyphenyl ether and 0.5% H_3PO_4. Practical details are given by Stevens *et al.* [11]. After conditioning at 140° C for 6 h this column will separate H_2S, SO_2, methyl mercaptan (CH_3SH), ethyl mercaptan (C_2H_5SH) and dimethyl sulphide (CH_3SCH_3). At a flow rate of 100 ml min^{-1} and at a temperature of 50° C, the retention time of SO_2 is approximately 2 min.

(b) 1.4 m length packed with graphitized carbon−black modified by treatment with 0.5% H_3PO_4 and 0.3% Dexsil (Bruner *et al.* [29]). At a flow of

$125\,\text{ml min}^{-1}$ and $60°$ C a separation and analysis of H_2S, SO_2 and CH_3SH can be obtained in 1 min. The column should be conditioned at $100°$ C for 24 h before use.

(c) Deactivated silica gel columns: Hartmann [29] has described a column for the analysis of ambient SO_2 (and H_2S) consisting of a 3 inch length packed with 100/120 mesh commercial deactivated silica gel (Deactigel) which was further deactivated by washing in turn with concentrated HCl, H_2O and acetone (procedure of Thornsberry [30]). At a flow of $80\,\text{ml min}^{-1}$ and at $50°$ C the retention time of SO_2 was approximately 2 min. Deactivated silica gel columns for the analysis of ppb concentrations of sulphur compounds are commercially available e.g. Supelco 'Chromasil 310'. Similarly columns (a) and (b) are available from some chromatographic suppliers.

Any commercial chromatograph equipped with a flame photometric detector can be used to house the column. The column should be connected directly into the FPD burner base where the fuel gas (H_2) and make-up O_2 is added. (N_2 may be used as make-up gas if air is used as carrier gas.) The other end of the column is connected to the GSV by as short a length of PTFE tubing as possible.

Reagents

The only materials required are the column packings, carrier gas (N_2 or air) H_2 and O_2 in cylinders fitted with good quality pressure regulators. For calibration, a gas containing an accurately known concentration of SO_2 in the range 0.01 to 1 ppm is required; this is most conveniently obtained from permeation tubes (see Section 4.3.8.2).

Procedure

After setting up the sampling and chromatographic equipment, set the gas flows to the required rates. The optimum gas flows depend somewhat on burner design; normally about $100\,\text{ml min}^{-1}$. H_2 is used with a total carrier plus make-up of the same order ($100\,\text{ml min}^{-1}$). Ignite the FPD burner and establish a satisfactory base line on a high sensitivity. Before carrying out air sampling, some time should be spent establishing optimum operating conditions and response characteristics using the SO_2 calibration source. In particular, since the response relationship of the FPD is of the form

$$R = [S]^n$$

(R = response, e.g. peak area; $[S]$ = S gas concentration), the value of n, which depends on the type of S gas and the operating conditions, must be determined.

Using a calibration gas of constant SO_2 concentration and repeated injection of samples, optimize conditions of gas flow and temperature to give maximum response consistent with resolution of SO_2 and minimum baseline noise. Inject several samples for each test, as 3 to 4 injections may be required to give a constant signal. The response is calibrated by injecting SO_2 at different concentrations within the operating range and recording the peak area. Plot the SO_2 concentration versus response (in A) on log–log paper and determine n from the slope. The plot should be approximately linear in the range 0.01 to 1.0 ppm and n should lie in the range 1.7 to 2.1. Some FPD detectors are equipped with linearizer amplifiers in which case the response is linearly related to the SO_2 concentration.

The equipment can now be used for ambient air sampling. For automatic operation the timing sequence will depend on the analysis time and the frequency of sampling required. At least 60 s should be allowed to purge the sample loop with the air sample, at a flow rate of a few hundred ml min^{-1}. After injection the loop is purged with carrier gas until the next sample is required, and during this period the automatic GSV control system should be de-energized.

Calibration samples may be run at any desired interval if a continuous SO_2 source is incorporated into the system. Timing of the three-way PTFE solenoid valve should be appropriately synchronized with the GSV. For accurate continuous monitoring it may be desirable to run alternate air-calibration samples so that the column is regularly conditioned with SO_2 even if the ambient level is very low. In this way, possible absorption losses are minimized.

Commercial instruments for SO_2 monitoring

Ambient air – total S

Instruments supplied by Bendix (Model 8300) and Meloy Laboratories (Model SA 160, etc.) monitor total S by continuous injection of an air sample into an FPD. Both systems have similar specifications, i.e. log/linear amplification with six linear switched ranges 10^{-9} to 10^{-4} A FSD corresponding to a working concentration range of 0.01 to 1.0 ppm SO_2. Variable response times of down to 1 s are specified. Automatic zero and calibration modes are provided for continuous unattended operation, but a source of SO_2 must be supplied separately.

Commercial chromatographic sulphur gas analysers, operating basically on the procedures discussed above, are also available (e.g. Tracor Model 270 HA). These instruments can be used for the specific measurement of SO_2 and also H_2S, CH_3SH, carbon oxysulphide (COS), carbon disulphide (CS_2) with a minimum detection limit of 5 to 10 ppb.

Source monitoring for SO_2

Flame photometric detectors can be used for source monitoring (i.e. SO_2 concentrations in the range 10 to 10 000 ppm) if an air dilution system is employed (e.g. Meloy Laboratories Model FSA190 flue gas analyser). However a more promising technique for this application is u.v. fluorescence analysis of SO_2. Two commercially available instruments utilizing this technique are available (Thermo Electron Model 40, Celesco Model 5000). The linear range of both instruments is 0 to 5000 ppm with 1% precision and a response time of 4 s. Complete selectivity for SO_2 is claimed.

4.3.2 Analysis of SO_3

SO_3 in stack gases can be measured by conversion in solution to SO_4^{2-}. Two major problems arise; firstly, the difficulty of quantitative sampling from the hot, moist environment of the stack and, secondly, SO_2 is usually present in the flue gases at higher concentrations than SO_3, and can interfere.

An instrumental method has been developed [31] (EEL 147 SO_3 monitor) in which the gas is sampled continuously through an air-heated probe kept between 200 to 500° C so that the SO_3 does not condense nor the SO_2 oxidize. After filtration the gas is extracted with 4 : 1 isopropanol/H_2O solution which converts the SO_3 to H_2SO_4. Some SO_2 dissolves in solution, the remainder being swept away in a gas/liquid separator. The liquid is transferred by air injection into the reaction bed and dissolved SO_2 is stripped by the air. The solution passes through a bed of barium chloranilate crystals where the H_2SO_4 reacts to form barium sulphate and the soluble, highly coloured acid chloranilate ion. The colour which is proportional to SO_3 concentration is measured spectrophotometrically. The range of detection is 0.1 to 250 ppm SO_3. The main interference comes from SO_2, a small amount of which may be oxidized in solution to give H_2SO_4. Manual methods based on collection in aqueous isopropanol have also been reported [32].

An interesting gas chromatographic method for the determination of SO_3 in the presence of SO_2 has been reported [33]. The gas sample containing SO_3 is passed continuously through a bed of oxalic acid crystals where SO_3 reacts to form CO and CO_2, whilst SO_2 does not react. The CO in the effluent is then analysed by conventional gas chromatography using a katharometer detector. Any CO in the original sample will, however, interfere.

In the ambient atmosphere gaseous SO_3 does not exist in the free state but forms H_2SO_4 aerosol.

4.3.3 Analysis of H_2S

4.3.3.1 Chemical methods

A widely used colorimetric method for the analysis of H_2S in ambient air is based upon absorption in an alkaline cadmium hydroxide suspension, followed by conversion of the precipitated sulphide to methylene blue with N,N-dimethyl-p-phenylenediamine and ferric chloride with spectrophotometric determination [34]. Low recoveries have been encountered when applied to low concentrations, the major cause being photodecomposition of the sulphide during sampling and storage [35]. Significant improvement was obtained by adding 1% STRactan 10 (arabinogalactan) to the absorbent. The sensitivity is sufficient for determination of H_2S down to 1 ppb.

A variety of methods for analysis of H_2S based on collection with filters impregnated with heavy metal salts ($Pb(^{2+})$, $Hg(^{2+})$ and $Ag(^{+})$) have been reported [36]. The formation of sulphides results in a stain which is measured either by its optical density (compared to a blank section of filter) or by reflectance. Lead acetate has been widely used as the substrate in semi-continuous monitors utilizing filter paper tape but serious interference results from bleaching of the stain by O_3, SO_2, NO_2 and light, and the sensitivity is humidity dependent. These shortcomings, which are especially important at the concentrations normally encountered in air pollution work (0 to 40 ppb), led Pare [37] and later Hochheiser and Elfers [38] to develop the mercuric chloride ($HgCl_2$) paper tape method. The tapes (Whatman No 4) are impregnated with a mixture of $HgCl_2$, urea and glycerol. After sample collection the tapes are developed by exposure to NH_3 vapour for up to 12 h and the optical density of the reaction spots measured. The effective concentration range for 2 h samples collected at 5 litres min^{-1} is 0.5 to 15 ppb H_2S with minimal interference from O_3, NO_2 and SO_2.

A manual method based on lead acetate filters has been developed by Okita et al. [39]. Impregnated cellulose membrane filters are dissolved after sampling in a mixed organic solvent and the resultant brown suspension is measured with a spectrophotometer. SO_2 and O_3 interference can be overcome by prefilters and NO_2 does not interfere below 0.2 ppm. The minimum detectable H_2S concentration is 2 ppb; no response was obtained by Okita et al. for lower concentrations, regardless of sampling time.

Natusch et al. [40] have described a method for measuring trace levels of atmospheric H_2S with a lower detection limit of 5×10^{-6} ppm. H_2S is collected on a silver nitrate ($AgNO_3$) impregnated filter paper. The silver sulphide (Ag_2S) is then dissolved in sodium cyanide ($NaCN$) solution and analysed fluorimetrically using very dilute fluorescein mercuric acetate. The reduction of fluorescence intensity, which is proportional to the sulphide concentration, was measured on a Perkin-Elmer Model 203 spectrofluorimeter. Fluorimetric analysis has also been applied to H_2S trapped in aqueous alkaline solution [41]. Whilst the sensitivity of this latter method is adequate for background determinations, the collected sulphide ion is unstable and this necessitates that analysis follow soon after sampling. The main drawback of fluorimetric analysis is the high cost of fluorimeters.

Microcoulometric titration by bromine in an electrochemical cell is the basis of a method for continuously monitoring sulphur compounds including H_2S, primarily in the process industries [42]. Sensitivity for H_2S is in the 5 to 30 ppb range. Selective filters may be used to monitor ambient air for SO_2, H_2S, mercaptans, alkyl sulphides and disulphides.

Manual method for H_2S analysis (methylene blue method) (ISC method 42402-01-70T) [82]

Principle

H_2S is collected by aspirating a measured volume of air through an alkaline suspension of cadmium hydroxide $Cd(OH)_2$. The sulphide is precipitated as cadmium sulphide (CdS) to prevent air oxidation of the sulphide which occurs rapidly in aqueous alkaline solution. STRactan 10 is added to minimize photodecomposition of the precipitated CdS. The collected sulphide is determined by spectrophotometric measurement of the methylene blue produced by reaction with a strongly acid solution of N, N-dimethyl-p-phenylenediamine and ferric chloride.

The methylene blue reaction is highly specific for low concentrations of sulphide. Strong reducing agents (e.g. SO_2) inhibit colour development and NO_2 and O_3 can give a slight negative interference. The minimum detectable amount of sulphide is $0.008\,\mu g\,ml^{-1}$ corresponding to an H_2S concentration of about 1 ppb in a 2 h sample collected at the maximum recommended sampling rate of 1.5 litre min^{-1}. The method is applicable to concentrations of H_2S up to 100 ppb using the procedure below and higher concentrations can be measured with larger collection volumes and shorter sampling times. At low concentrations (< 10 ppb) collection efficiency is variable and is affected by the type of scrubber, the bubble pattern and H_2S concentration.

Apparatus

A midget impinger is used to contain the absorbent and is aspirated with a pump having a minimum capacity of 2 litres min^{-1}. A rotameter or gas meter can be used to measure sample volume.

Reagents

(Use analytical grade chemicals and keep solutions under refrigeration.)

(a) Amine-H_2SO_4 (stock): To 30 ml of distilled water add 50 ml of concentrated H_2SO_4 (r.d 1.84). After cooling add 12 g of N, N-dimethyl-p-phenylenediamine dihydrochloride and mix until complete solution.

(b) Amine-H_2SO_4 (working): Dilute 25 ml of the above stock to 1 litre with 1 : 1 H_2SO_4.

(c) Ferric chloride solution: Dissolve 100 g ferric chloride hexahydrate ($FeCl_3 . 6H_2O$) in H_2O and make up to 100 ml.

(d) Ammonium phosphate solution: Dissolve 400 g of diammonium phosphate $(NH_4)_2HPO_4$ in H_2O and dilute to 1 litre.

(e) Absorbing solution: Dissolve 4.3 g of cadmium sulphate octahydrate ($3CdSO_4 . 8H_2O$) and 0.3 g of NaOH in separate portions of water, mix, add 10 g of STRactan 10 (arabinogalactan) and dilute to 1 litre. The solution should be freshly prepared (it is only stable for 3 to 5 days) and shaken vigorously before each aliquot of absorbing reagent is taken.

(f) Standard sulphide solutions: Aqueous sulphide solutions are unstable, being subject to air oxidation. Stock solutions must be made up with freshly boiled and cooled distilled water and, for accurate work, should be standardized for each calibration routine. During solution preparation and handling, oxidation can be minimized by flushing receptacles with O_2-free N_2. A stock solution containing approximately 400 μg sulphide ion ml^{-1} is made by dissolving in 1 litre 0.1 M NaOH, either gaseous H_2S (300 ml from a gas syringe through a septum) or sodium sulphide monohydrate crystals (approximately 3 g, weighed after washing with distilled water and drying quickly on filter paper). Standardize the solution with standard iodine and thiosulphate solutions. Dilute standard sulphide solutions (4 μg sulphide ion ml^{-1}) are prepared by diluting 10 ml of the freshly standardized stock to 1 litre with boiled distilled water.

Procedure

Aspirate the air sample through 10 ml of the absorbing solution in a midget impinger at 1.5 litre min^{-1} for a selected period of up to 2 h. Excessive foaming may be controlled by the addition of 5 ml ethanol just prior to sampling.

For analysis add 1.5 ml of the amine working solution to the absorbing solution in the impinger. Add 1 drop of $FeCl_3$ solution (if SO_2 is likely to exceed 10 μg ml^{-1} in the sample solution add 2 to 6 drops) and transfer the solution to a 25 ml volumetric flask. Add 1 drop of ammonium phosphate solution (or more if necessary) to discharge the yellow colour of the ferric ion and make up to volume with distilled water. Allow to stand for 30 min (50 min if extra $FeCl_3$ was added) for the colour to develop. The colour is measured at 670 nm against a reagent blank prepared with unexposed absorbing solution.

Calibration

Calibration may be carried out using standard sulphide solutions or by sampling gas mixtures containing known concentrations of H_2S from permeation tubes. The latter method is to be preferred for concentrations < 10 ppb since the variable collection efficiency may then be taken into account in the standardization.

(a) Aqueous sulphide method: Place 10 ml of the absorbing solution in each of a series of 25 ml flasks and add the diluted standard sulphide solution, equivalent to 1, 2, 3, 4 and 5 μg H_2S to the flasks. Add 1.5 ml amine working solution, mix and add 1 drop of $FeCl_3$ solution to each flask. Make up to volume and allow 30 min before determining the absorbance at 670 nm against a sulphide free blank. Prepare a standard plot of absorbance versus μg H_2S ml^{-1}.

(b) Permeation tube method: Use a permeation tube which emits approximately 0.1 μl H_2S min^{-1} at 25° C and 1 atm. Permeation tubes containing H_2S are calibrated under a stream of dry N_2 to prevent deposition of sulphur on the tube walls. A total dilution gas flow of at least 20 litres min^{-1} is required to provide concentrations at the lower end of the working range. Samples are taken from the standard gas mixture using a fixed sample volume and a plot of H_2S concentration against absorbance in the final solution is prepared.

Calculation

H_2S concentrations are calculated from the standardization plots after allowance for volumetric factors. Gas volume should be corrected to 25° C and 760 Torr pressure.

Stability of samples

Although CdS is reasonably stable towards oxidation, the analysis should be completed within 24 h of sampling. The absorbing solution should be shielded from light during both sampling and storage. Black paint or aluminium foil wrapping on the impinger will help prevent photodecomposition of the sulphide.

Instrumental chemical methods

An automated H_2S analyser based on the methylene blue colorimetric method is marketed by Technicon Corporation. The detection limit on a 0 to 100 ppb scale is claimed to be 2 ppb. Instruments for semicontinuous measurements of H_2S in industrial atmospheres based on the lead acetate paper tape method are also available commercially (e.g. Fleming Instruments Type 523, Maihack Mono-colour H_2S analyser). The minimum detectable concentration on these instruments is approximately 50 ppb H_2S.

4.3.3.2 Physical methods

Stevens *et al.* [11] and Bruner *et al.* [28] have reported the application of a gas chromatographic, flame-photometric detector (GC–FPD) system for the automated GC measurement of H_2S (together with SO_2, methyl mercaptan and dimethyl sulphide—see Section 3.1.2). The limit of detection by this method is approximately 2 ppb for H_2S.

Infrared analysis of sulphides is non-discriminating, all sulphides absorbing at similar wavelengths. At present sensitivity is inadequate for ambient monitoring using i.r. Remote sensing of ambient H_2S using laser techniques is likely to be developed in the future [10].

Gas chromatographic methods for ambient H_2S analysis

Practical details of a gas chromatographic sulphur analyser, which is suitable for measurements of H_2S are given in Section 4.3.1. Pecsar and Hartmann [43] obtained better resolution of H_2S on a PTFE column (prepared by the method of Stevens *et al.* [11]) by using air as carrier gas. This enabled a shorter (12 ft) column to be used. Calibration of the GC–FPD system is carried out using standard mixtures of H_2S obtained from either permeation tubes [27] or an exponential dilution flask [28].

4.3.4 Analysis of organic S compounds

Interest in the organo-sulphur compounds is largely concerned with the measurement and control of effluents from the pulp and paper industries, and also with the gas industry where they are used as artificial odorants for natural gas. The compounds of interest are primarily the lower molecular weight mercaptans, dimethyl sulphide and dimethyl disulphide.

4.3.4.1 Chemical methods

As the organic sulphides are strong reducing agents, coulometry offers, in principle, a sensitive method of detection. The main problem is to distinguish between the various organic S compounds and also SO_2 and H_2S which in a coulometric detector behave similarly. Adams *et al.* [42] studied the retention efficiencies of a large number of impregnated membrane filters for five different S gases, H_2S, SO_2, methyl mercaptan (CH_3SH), dimethyl sulphide (DMS) and dimethyl disulphide (DMDS). A bromine coulometric microtitration cell (i.e. galvanic cell with electrogenerated bromine) was used to measure the sulphur gases not trapped by the experimental filters. A series of filters was developed which enabled the above five gases to be determined from measurements on sequential samples. Air was sampled through appropriate filter combinations to provide a continuous, stepwise analysis. Sodium bicarbonate (5%) removed SO_2 but left over 90% of the other compounds unaffected. H_2S was separated from the other compounds by using a zinc chloride—boric acid membrane. Silver membrane filters retained H_2S and CH_3SH but left 95% of the other three unchanged. Mercuric nitrate—tartaric acid retained DMS and CH_3SH. Filters impregnated with $AgNO_3$ retained all compounds except SO_2. Minimum detectability was SO_2, 25 ppb; H_2S, 10 ppb; CH_3SH, 15 ppb; DMS, 25 ppb; and DMDS, 5 ppb. An instrument using preselective filtration was used in a field study of sulphur gas concentrations in the vicinity of a Kraft paper mill [35].

For the selective determination of low molecular weight mercaptans in air the colorimetric method using N, N-dimethyl-p-phenylenediamine reagent is recommended [44]. This provides a sensitive and reproducible manual method for concentrations down to 2 ppb. A colorimetric method has also been developed for the measurement of carbon disulphide in industrial atmospheres [45]. This involves collection in an ethanolic solution of diethylamine-copper acetate contained in a fritted bubbler, the colour developing in the absorbing solution. Maximum sensitivity is in the ppm range.

Colorimetric determination of mercaptans in air
(ISC method 43901-01-70T) [82]

Principle

Mercaptans are collected by aspirating a measured volume of air through aqueous mercuric acetate—acetic acid solution. A red complex is produced when mercaptans react with N, N-dimethyl-p-phenylenediamine (DMPDA) and $FeCl_3$ in strongly acid solution and this is measured spectrophotometrically.

In addition to the low molecular weight mercaptans, H_2S and dimethyl sulphide can also be determined using DMPDA. These compounds commonly co-exist with mercaptans in industrial emissions and are potential interfering substances. Appropriate selection of sampling and analytical procedures minimizes this interference. Thus H_2S may cause a turbidity in the absorbing solution which must be filtered before continuing the analysis. In the analytical procedure $100\,\mu g$ H_2S may give an absorption at 500 nm equivalent to up to $2\,\mu g$ CH_3SH. Interference from dimethyl sulphide is negligible since this compound is not trapped in aqueous mercuric acetate. Tests have shown that SO_2 (300 ppm) and NO_2 (6 ppm) do not interfere.

The minimum detectable amount of CH_3SH is $0.04\,\mu g\,ml^{-1}$ in the final liquid volume of 25 ml. In a 200 litre air sample this corresponds to $5\,\mu g\,m^{-3}$ or 2.5 ppb CH_3SH. Precision is within 2.6% for the C_1 to C_6 mercaptans. The procedure given is suitable for the range 2 to 100 ppb but volumes can be modified to accommodate higher concentrations.

Apparatus

Sample absorption is carried out in a midget impinger fitted with a coarse frit. An air pump with a flow-meter or gas meter capable of aspirating and measuring a flow of $2\,litres\,min^{-1}$ is required.

Reagents

Analytical grade reagents should be used and kept under refrigeration when not in use:

(a) Amine-HCl (stock) solution: Dissolve 5.0 g DMPDA hydrochloride salt in 1 litre concentrated HCl. If kept cool and dark, the reagent is stable for six months.

(b) Reissner solution: Dissolve 67.6 g $FeCl_3.6H_2O$ in distilled water, dilute to 500 ml and mix with 500 ml of an aqueous solution containing 72 ml freshly boiled, concentrated HNO_3 (r.d. 1.42).

(c) Colour developing solution (prepared freshly): Mix 3 vol. of DMPDA with 1 vol. of Reissner solution.

(d) Absorbing solution: Dissolve 50 g mercuric acetate (Poison) in 400 ml distilled water and add 25 ml glacial acetic acid. Dilute to 1 litre. The mercuric acetate must be free of mercurous salts to prevent precipitation of mercurous chloride during colour development.

(e) Standard lead mercaptide (stock): Weigh 156.6 mg of crystalline lead mercaptide and make up to 100 ml with absorbing solution. This solution contains the equivalent of $10\,g\,CH_3SH\,ml^{-1}$.

Procedure

Aspirate the air sample through 15 ml of the absorbing solution in a midget impinger at 1.0 to 1.5 litres min^{-1} for a period of up to 2 h. Note the air volume sampled. Quantitatively transfer the sample to a 25 ml volumetric flask and dilute to approximately 22 ml with washings and distilled H_2O. Add 2 ml of colour reagent, make up to volume, and mix well. Prepare a blank using 15 ml unexposed reagent and 2 ml colour reagent, diluted to 25 ml. After 30 min measure the colour at 500 nm against the blank.

Calibration and calculation

Transfer aliquots of the diluted standard mercaptide solution into a series of 25 ml flasks, dilute each with 15 ml absorbing solution and develop the colour as for the samples. Determine the absorbance of each sample and prepare a plot of absorbance against $\mu g\,ml^{-1}$ CH_3SH. Alternatively, a standard mixture of CH_3SH in air may be prepared using a permeation tube. The permeation rate and dilution flow should be arranged to give concentrations in the range 1 to 25 ppb. By using the standard sampling procedure at different concentrations a plot of absorbance versus ppb CH_3SH can be prepared. This calibration includes any corrections for collection efficiency of the absorbing solution.

The concentration of mercaptans expressed in terms of CH_3SH is given by

$$\text{ppb } CH_3SH = \frac{A \times 0.510 \times B}{V}$$

where A is the absorbance of the solution measured against the reagent blank, 0.510 is the volume (μl) of $1\,\mu g$ CH_3SH at $25°\,C$ and 760 Torr, B is the calibration factor $(\mu g$ per A.U.$)$ and V is the air sample volume in m^3 at $25°\,C$ and 760 Torr.

4.3.4.2 Physical methods

Undoubtedly the most useful and sensitive method for the determination of organo-sulphur compounds is GC analysis. Flame ionization detection can be used if the sulphur compounds are selectively sampled to separate sulphides from hydrocarbons, e.g. using impregnated filters [46]. The optimum system is the flame photometric detector with its high selectivity and sensitivity for S compounds. A variety of columns and techniques have been reported in the literature for the GC analysis of organic S compounds [47] but for ambient air analysis the techniques developed by Stevens *et al.* [11] and Bruner *et al.* [28] for SO_2, H_2S and low molecular weight organo-sulphur compounds are recommended. Practical details of a suitable GC–FPD system for measurement of H_2S, SO_2, CH_3SH and DMS in air are given in Section 4.3.1.2. It should be noted that the silica-gel based columns for H_2S and SO_2 analysis are unsuitable for the analysis of mercaptans and sulphides since the latter are too strongly retained. However CS_2 and carbon oxysulphide (COS) may be analysed using Chromosil 310 (Supelco Ltd) column packing.

A simple, low cost GC–FPD system for measurement of organo-sulphur compounds is marketed by United Analysts Ltd. (LRS Odour Chromatograph). Designed principally as a portable instrument for 'spot sample' measurement of organic S compounds in natural gas, the system can also be applied to air analysis. The minimum detectable concentration is 5 ppb DMS in a 10 ml sample. The measurement of organo-sulphur compounds could be automated by the use of an automatic gas sampling valve.

4.3.5 Analysis of oxides of nitrogen — NO and NO_2

The methods of analysis of NO and NO_2 are closely related in so far as all chemical methods for NO involve its oxidation to NO_2 and the widely adopted chemiluminescence method provides a useful method for NO_2 following its conversion to NO. A recent review by Allen [12] gives a comprehensive survey of methods for the analysis of these gases.

4.3.5.1 Chemical methods

Both colorimetric and coulometric analyses are used for the measurement of atmospheric NO_2. Colorimetric methods involve the determination of nitrite ion formed by hyrolysis of NO_2 in aqueous solutions. The determination of nitrite is based on the Griess—Isolvay reaction in which a pink coloured azo-dye complex is formed between sulphanilic acid, nitrite ion and α napthylamine in acid medium. Various modifications of this method have been proposed but the one that is most widely employed is that introduced by Saltzmann in which the colour is developed in the absorbing solution [48]. The chief disadvantage of the latter method is that samples cannot be stored for more than a day. Although samples collected in alkali absorbent are stable for longer periods, the many methods utilizing this technique (e.g. Jacobs—Hochheiser method [49]) have been shown to give unreliable results due to poor collection efficiency and uncertain stoichiometry for the conversion of NO_2 to nitrite ion. A recently reported technique [50] based on absorption of NO_2 in ethanolamine, has been shown to be suitable for long-period atmospheric sampling. Nash [51] has reported efficient absorption of NO_2 in alkaline solutions after the addition of a small amount of guaicol (o-methoxy-phenol).

The main uncertainty involved in the methods based on the Griess—Isolvay reaction arises from the stoichiometry of conversion of NO_2 to nitirite ion in solution—the nitrite equivalence or 'Saltzman factor', the value of which has been the subject of some controversy (see [12] for a detailed summary). From the various reports it can be concluded that the nitrite equivalence factor appears to be influenced by many parameters including reagent composition, purity, temperature, the NO_2 concentration in the gas sample, the sample flow rate and bubbler design.

Saltzman [48] determined a value of nitrite ion $\equiv 0.72$ NO_2 (gas) for his original procedure and has since obtained further confirmation for this equivalence [52]. Scaringelli *et al.* [53] obtained a value of 0.764 ± 0.005 from several hundred analyses with the original reagent using permeation tubes to prepare gravimetrically standardized NO_2 mixtures. Provided the standard procedure is rigidly followed a value of 0.72 is recommended when using the Saltzman method but if any modification is made the factor should be determined using a standard NO_2 source.

Coulometric analysis of NO_2 utilizes the oxidizing action of NO_2 on potassium iodide solutions. Efficient removal of other oxidizing and reducing pollutants is necessary for meaningful NO_2 measurements.

A number of systems have been devised for the oxidation of NO to NO_2 for analysis of atmospheric NO. The recommended system involves passage of the sample through a tube containing chromic oxide supported on an inert inorganic material. Other oxidizing systems that have been employed are solid manganese dioxide/potassium hydrogen sulphate ($MgO_2/KHSO_4$) [54], acid potassium permanganate bubblers [55], sodium dichromate impregnated filter papers [56] and oxidation in the gas phase by O_3. The former methods can suffer from deterioration of the reagents giving rise to low and variable oxidation efficiency; O_3 oxidation systems require careful design to ensure that excess O_3 does not lead to oxidation of NO_2. Any excess O_3 must be removed in a scrubber to avoid interference with the subsequent analysis for NO_2. The NO concentration may be obtained either by difference from simultaneous measurements of total $NO + NO_2$ in the air sample or directly by removal of the ambient NO_2 prior to NO oxidation.

Saltzman method for analysis of NO_2

Principle

The NO_2 is absorbed in a fritted bubbler containing an azo-dye-forming reagent. A stable pink colour is formed within 15 min which is measured spectrophotometrically [48].

The method is suitable for NO_2 concentrations in the range 0.005 to 5 ppm and with careful work a precision of 1% of the mean can be attained, the limiting factors being measurement of the air volumes and the absorbance of the colour. The efficiency of absorption of NO_2 by the solution is $> 95\%$. Interference from strong oxidizing and reducing agents such as O_3 or SO_2 is only significant if they are present in large excess over NO_2 and colour determination is delayed for more than 1 h. Peroxyacylnitrite (PAN) can give a response of 15 to 35% on a molar basis but the normal levels of PAN are too low to cause significant error.

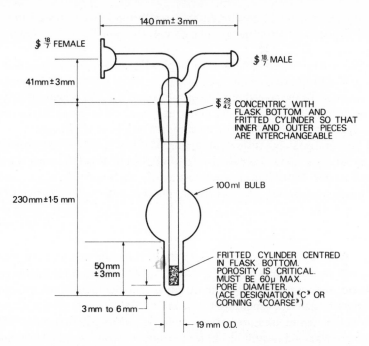

Fig. 4.9   Fritted bubbler for collection of NO_2 in Saltzman reagent.

Apparatus

A special all glass fritted bubbler with a $60\,\mu m$ maximum pore diameter frit is used (see Fig. 4.9). The sample probe should be of glass or PTFE. The air flow of not more than $0.4\,\text{litre min}^{-1}$ is measured on a glass rotameter. A pump capable of drawing the required sample flow for up to 30 min is required. It is desirable to have a tee connection on the pump inlet. The inlet connected to the sampling train should have a trap and stainless steel needle valve. The second inlet should have a valve bleeding in a large excess of clean air to prevent condensation of acetic acid vapour from the absorbing reagent in the pump. Since the rotameter operates at an appreciable vacuum it should be calibrated against another rotameter or wet gas meter placed upstream from the bubbler.

Reagents

All reagents are made from analytical grade chemicals in nitrite-free water, prepared, if necessary, by redistilling distilled water in an all glass still after adding a crystal each of potassium permanganate and barium hydroxide. They are stable for several months if kept under refrigeration in brown bottles. The absorbing reagent should be allowed to warm to room temperature before use.

(a) N-(1-naphthyl)-ethylene-diamine dihydrochlorate, 0.1%. Dissolve 0.1 g of the reagent in 100 ml of H_2O.

(b) Absorbing reagent. Dissolve 5 g sulphanilic acid in almost one litre of a solution of 140 ml glacial acetic acid in water. Gentle heating may be applied to assist solution. To the cooled mixture, add 20 ml of the N-(1-naphthyl)-ethylene-diamine dihydrochloride (1%) and dilute to 1 litre. Avoid unnecessary exposure of the reagent to air since NO_2 will be absorbed giving discolouration.

(c) Standard sodium nitrite solution 0.0203 g litre^{-1}. 1 ml of this working solution produces a colour equivalent to that of 10 μl of NO_2 (10 ppm in 1 litre air at 760 Torr and 25° C. Prepare fresh just before use by dilution from a stronger stock solution containing 2.03 g of reagent grade granular sodium nitrite per litre.

Procedure

Pipette 10 ml of absorbing solution into the bubbler and assemble the sampling train. Draw the air sample through at 0.4 litre min^{-1} or less until sufficient colour has developed (approximately 10 to 30 min). Note the total air volume sampled and correct to standard conditions (25° C at 760 Torr).

The colour can be measured after 15 min on a spectrophotometer at 550 μm against an unexposed reagent blank. If no strong oxidizing or reducing agents are present the colour may be preserved, if well stoppered, with only 3 to 4% loss in absorbance per day.

Calibration

The most convenient method is standardization against nitrite solution in which case the empirical stoichiometric factor, 0.72 mol $NO_2^- \equiv 1$ mol gaseous NO_2 must be employed. It should be noted that if different construction and operation of the fritted bubbler is used, a different empirical factor may result. The alternative calibration method involves sampling from an accurately known NO_2 gas source, e.g. from a permeation tube.

To calibrate with sodium nitrite ($NaNO_2$) solution add graduated amounts of the standard $NaNO_2$ up to 1 ml (carefully measured in a graduated pipette or small burette) to a series of 25 ml volumetric flasks and make up to volume with absorbing reagent. Mix, allow 15 min for complete colour development and read the colours at 550 mμ. Plot the absorbances of the standard colours against μl NO_2 per ml of absorbing reagent (i.e. 0.4 × ml standard nitrite solution with stoichiometric factor included). Beer's law is obeyed. Determine the slope of the line, K, which is μl NO_2 per 10 ml absorbing reagent giving unit absorbance, (for 1 cm cells $K = 0.73$). The NO_2 concentration is then given by

$$NO_2 \text{ (ppm)} = \text{absorbance} \times K/V$$

where V = volume of air sampled, at standard conditions, in litres ml^{-1} of absorbing reagent. For exact work, the sampling efficiency (determined from prior tests using two bubblers in series) and the fading of the colour (3 to 4% per day) may be taken into consideration.

Analysis of NO (ISC method 42601-01-71T) [82]

Principle

After removal of NO_2 from the gas stream, NO is quantitatively oxidized to NO_2 by solid chromic oxide [57]. The NO_2 is then determined by absorption in Saltzman's reagent (see previous section).

This method is useful for the manual determination of atmospheric NO in the range 0.005 to 5 ppm. In addition to the factors affecting the measurement of NO_2, the precision of the method depends on the efficiency of the NO $\rightarrow$ NO_2 converter, which, under controlled conditions, varies between 98 to 100%.

Apparatus and reagents

The NO_2 absorber consists of a 2 cm i.d. × 5 cm long polyethylene tube filled with 10 to 20 mesh inert porous material such as firebrick, alumina, zeolite, etc. coated with triethanolamine. (The granules are soaked in aqueous ethanolamine (20%) and dried for 0.5 to 1 h.) For the efficient operation of the chromic oxide oxidizer the r.h. of the sample should be between 40 and 70%. The r.h. is controlled by a humidity regulator consisting of a 2 cm i.d. × 5 cm long tube filled with crystals of a mixture of anhydrous and hydrated sodium acetate (50/50); 13 ml of H_2O are added dropwise to 40 g anhydrous sodium acetate with stirring).

The oxidizer consists of a 15 mm i.d. glass tube filled to a depth of 7 to 10 cm with pellets of chromic oxide adsorbed on an inert support. Glass, firebrick or alumina, mesh size 15 to 40, is soaked in a solution of 17 g chromium trioxide in 100 ml of H_2O. The pellets are drained, dried at 110° C and exposed to 70% r.h. in a dessicator containing saturated sodium acetate. The reddish colour should change to golden orange when equilibrated. Glass wool plugs should be used to hold the solid reagents in place. The NO_2 produced from NO is collected in a fritted bubbler containing Saltzman's reagent as described in Section 4.3.2.1.

Procedure

Assemble a sampling train comprising, in order, rotameter, NO_2 absorber, humidity regulator, oxidizer, fritted absorber and pump. Pipette 10 ml of Griess–Saltzman reagent into the fritted absorber and draw sample air through the system at 0.4 litre min^{-1} until sufficient colour has developed for measurement. Record the total air volume sampled and correct to 25° C and 760 Torr. Measure the absorbance of the solution at 550 nm after 15 min.

Calibration and calculation

Calibration may be carried out using standard $NaNO_2$ solution as described for NO_2 in Section 4.3.5.1. In this case the Saltzman factor of 1 mol NO ($\equiv NO_2$) = 0.72 mol nitrite ion must be employed. Alternatively a gaseous standardization technique may be employed using standard mixtures of NO in air. These can be obtained by dilution of high concentration NO/N_2 mixtures (100 to 1000 ppm) either dynamically (into a flowing air stream) or in using the bag method (see Section 4.3.8.1). NO concentrations should be chosen to lie within the expected range of the samples. The calculation of the NO concentration is analogous to that for NO_2 described in Section 4.3.5.1.

Determination of average (24 h) ambient NO_2 concentration

Principle

This manual procedure involves the absorption of NO_2 in a triethanolimine solution or a triethanolamine impregnated molecular sieve followed by colorimetric analysis using a modified Griess–Saltzman reagent [50]. The absorber has a trapping efficiency of over 95% at NO_2 concentrations at the pphm level. The factor for the conversion of NO_2 (gas) to nitrite ion was 0.85 for the procedure outlined below, when evaluated against the Saltzman factor of 0.72. The method may be used for short duration sampling and samples may be stored for up to 4 weeks without significant loss of NO_2.

The presence of SO_2 has a slight interference at the 0.1 ppm level with the solid absorber but has no effect on the liquid absorber at the levels normally present in ambient air. NO does not interfere. The effect of O_3 has not been extensively investigated but field trials do not indicate serious interference.

Apparatus

For the liquid absorber use fritted bubblers (70 – 100 μm frits) with 163 × 32 mm polypropylene tubes. The solid absorbent is packed in glass tubes 2 × 3/16 inch) preferably with a standard luer lock fitting at one end. The solid can be held in place with glass wool plugs. A pump capable of maintaining at least 0.5 atm vacuum and a critical orifice flowmeter comprising a 27 gauge hypodermic needle is used to aspirate the samples.

Reagents

(a) Liquid absorber: Dissolve 15.0 g of the triethanolamine in approximately 500 ml distilled water, add 3.0 ml n-butanol and dilute to 1 litre (0.1 N).

(b) Solid absorber: In a 250 ml beaker, place 25 g of triethanolamine, 4.0 g of glycerol, 50 ml of acetone and sufficient distilled water to dissolve. Dilute to 100 ml with distilled water and add about 50 ml of 12 to 30 mesh molecular sieve 13X. Stir and allow to stand for 30 min, decant the liquid, and transfer the molecular sieve to a porcelain pan. Remove the bulk of the water by drying the molecular sieves under a heating lamp and then oven dry at 110° C for 1 h. Store in an air-tight bottle.

(c) H_2O_2: Dilute 0.2 ml 100 vol. H_2O_2 to 250 ml with distilled water.

(d) Sulphanilamide: Dissolve 10 g of sulphanilamide in 400 ml distilled water. Add 25 ml of concentrated H_3PO_4, mix well and dilute to 500 ml.

(e) N-1-naphthylethylene diamine (NEDA): Dissolve 0.5 g NEDA dihydrochloride salt in 500 ml distilled water.

(f) Standard nitrite stock: Dissolve 0.1500 g of analytical grade sodium nitrite in distilled water and dilute to 1 litre (i.e. 100 g ml^{-1}).

Procedure

Attach a sampling probe (PTFE) to either the bubbler or the absorption tube and connect to the critical orifice flow-meter. 50 ml of absorbing solution is required in the liquid absorber. Switch on the pump and rapidly check the flow rate entering. Record the flow rate which should be between 150 to 200 ml min^{-1}. Sample for 24 h then recheck the flow rate.

To the liquid absorber, add distilled water to 50 ml to replace evaporation losses. Transfer a 10 ml aliquot to a 25 ml graduated tube. To another tube add 10 ml of unexposed reagent. To each add 1.0 ml of the H_2O_2 solution, 10 ml of the sulphanilamide solution, 1.4 ml of NEDA solution with thorough mixing after each addition. After 10 min measure the colour at 540 mm against the prepared blank.

Transfer the solid absorber and glass wool plugs to a 50 ml test tube. Wash the glass tube with approximately 10 ml of water and add washings to the sieve. Make up to 50 ml with absorbing solution. Cap and shake vigorously for about 1 min. Allow to stand and shake again after 10 min. Allow the solids to settle and transfer a 10 ml aliquot to a 25 ml graduated tube. Develop colour as for the liquid absorber.

Standardization and calculation

Dilute stock standard 50 : 1 with absorbing solution and add 0, 1, 3, 5 and 7 ml aliquots to a series of tubes. Make up to 10 ml with absorbing solution and develop colour as above. A plot of absorbance versus concentration is prepared so that the amounts of nitrite ion in the samples can be determined. The NO_2 concentration is then given by

$$NO_2 \text{ (ppm)} = \frac{\mu g\ NO_2^- \text{ (in aliquot)} \times 5}{1.88 \times 0.85 \times V}$$

where V is the volume of air sampled (at 760 Torr and 25° C), 1.88 is the factor for converting $\mu g\ NO_2$ to $\mu l\ NO_2$ at 25° C and 0.85 is the nitrite ion (μg) equivalent to NO_2 gas (μg) i.e. stoichiometry factor. The system may also be calibrated using permeation tubes.

Instrumental chemical methods

A number of continuous analysers of oxides of nitrogen based on both electro-chemical principles and colorimetric principles are commercially available.

A widely used instrument for ambient air monitoring is the Beckmann Model 910 NO_2 analyser (and model 909 NO analyser). The detector utilizes galvanic coulometry of the iodine liberated from buffered KI by NO_2. The recirculating cell is designed to give a claimed efficiency of 100% for NO_2 absorption, and oxidation and reagent replacement are not necessary. Selective scrubbers are used to reduce interferences from other oxidizing and reducing substances. In the NO analyser a small ozonizer produces sufficient O_3 to oxidize NO to NO_2 and the excess O_3 is removed by the scrubber. The minimum detectable concentration is 0.004 ppm with a useful range up to 1 ppm (3 switched ranges). Response time is 10 min to 90% of full scale. Concentrations of 1 ppm each of O_3, SO_2, H_2S, mercaptans will give a response equal to 0.1 ppm NO_2; these interferences will be significant in some environments.

Continuous colorimetric instruments for NO/NO_2 analyses are marketed in the US by Pollution Monitors Inc (Model PM 100) WACO Ltd, and Precision Scientific Ltd (AERON). These instruments utilize a modified Griess–Saltzman method for the automatic colorimetric determination of NO_2; total oxides of nitrogen are measured with a separate oxidizer. Minimum detectable concentrations are of the order of 0.005 ppm with response times of the order of 4 min. A version of the Technicon Air-Monitor IV for analysis of NO and total oxides of nitrogen is also available. A minimum range of 0 to 50 ppb offers some advantage in sensitivity over other automated chemical systems.

4.3.5.2 Physical methods

A wide variety of physical methods have been employed for the analysis of NO and NO_2. Among the optical methods, chemiluminescence has emerged as the most sensitive, selective and practical instrumental system and is discussed in more detail below. Apart from some recent sophisticated developments (e.g. laser techniques, correlation spectrometry, second derivative spectrometry) all spectroscopic methods have too low a sensitivity for useful atmospheric monitoring. Furthermore, interference by water vapour and the consequent requirement of efficient drying of the sample gas, limits the usefulness of the more readily available optical techniques, such as non-dispersive i.r. spectroscopy, for the measurement of higher concentrations of NO.

There has been a substantial amount of effort toward the development of a gas chromatographic method for the analysis of oxides of nitrogen. The major problems encountered are, firstly, the difficulty in obtaining a column material which will quantitatively elute low concentrations of NO and NO_2. The high reactivity of these gases makes most chromatographic materials unsuitable, porous polymer packings offering the best prospects. Secondly, although many detectors have been tried, none has been shown to give a consistent performance at the required sensitivity and selectivity.

Of the advanced spectrometric methods, two are available in commercial instruments for measurement of ambient NO and NO_2. The Barringer Research correlation spectrometer is designed for remote sampling of NO_2. The second derivative spectrometer (Spectrometrics of Florida model d^2III) has the rather high minimum detection limit for NO_2 of 0.04 ppm but NO can be measured down to 0.006 ppm. It is likely that the sensitivity of these instruments will be improved in the future.

Chemiluminescence detectors for analysis of NO and NO_2

The chemiluminescent reaction between NO and O_3 (see Section 4.2.3.1) is rapidly becoming accepted as the most reliable and precise method for detecting oxides of nitrogen. The response is linear from the limit of detection (< 0.001 ppm) up to at least 1000 ppm and the high sensitivity and relative freedom from interference by CO, CO_2, hydrocarbons, SO_2, O_3 and water vapour make this an excellent technique for the continuous analysis in both sources and the ambient atmosphere. Many commercial instruments based on this system and which are suitable for exhaust gas analysis [1 to 1000 volumes per million (vpm)] and ambient air monitoring (0.001 to 10 ppm) are now on the market.

Current models of these instruments employ catalytic decomposition of NO_2 to NO before detection, providing monitoring of NO and total oxides of nitrogen. Early models used a high carbon stainless steel converter operating at 650 to 750° C but there is a possibility of oxidation of NH_3 to NO in this system. This problem has been solved by the development of other catalytic materials which effect the conversion of NO_2 to NO at lower temperatures (e.g. Molybdenum at 450° C). It should be noted, however, that any nitrogen compound which is thermally decomposed to NO at elevated temperatures (e.g. organic nitrites, oxyacids of nitrogen) will give a response in the 'total oxides of nitrogen' mode of a chemiluminescence NO analyser. In most environments NO_2 is the predominant nitrogen compound which is detected. A discussion of the operation of catalytic total oxides of nitrogen converters has been given recently by Breitenbach [58].

Instruments for monitoring low concentrations in ambient air are normally equipped with a thermo-electrically cooled photomultiplier to minimize the thermal noise and dark current. For long term monitoring, thermostating of the multiplier is necessary to eliminate thermal drift of the zero. Alternatively zero drift may be compensated by mechanical chopping of the chemiluminescence signal and phase sensitive detection. Both systems have been utilized in commercial instruments. When monitoring stack-gas or automobile exhaust gas, the chemiluminescence output is sufficiently high to dispense with elaborate control of the zero current; the need to prevent water condensation within the instrument, however, complicates the sampling of undiluted combustion products. All common desiccants have been found to remove NO_2 to varying degrees although the drying of gases containing NO is less of a problem.

Other chemiluminescence reactions of NO are suitable for use in methods for detecting oxides of nitrogen. O atoms undergo a fast reaction with NO

$$O + NO \longrightarrow NO_2 \longrightarrow NO_2 + h\nu.$$

Furthermore NO_2 is rapidly converted to NO by O atoms

$$O + NO_2 \longrightarrow NO + O_2$$

and these two reactions provide the basis of a very sensitive detector for NO + NO_2 [5]. The use of flame chemiluminescence has also been investigated in a possible detection method for NO and NO_2 [59]. The reactions, which take place in an oxygen rich H_2-O_2 flame, are:

$$H + NO_2 \longrightarrow NO + OH$$

$$H + NO \longrightarrow HNO \longrightarrow HNO + h\nu.$$

Commercial instruments based on these detection systems for determination of oxides of nitrogen are not, however, available at the present time.

An interesting indirect chemiluminescence method for the determination of NO and NO_2 in the atmosphere has been developed by Guicherit [60]. The equilibrium O_3 concentration produced by continuous u.v. irradiation of NO_2

$$NO_2 + O_2 \underset{}{\overset{\text{u.v. (360 nm)}}{\rightleftharpoons}} NO + O_3$$

is measured by chemiluminescence of O_3 with Rhodamine B (see Chapter 5). The equilibrium O_3 concentration is a function only of wavelength, light intensity and temperature in the photolysis cell and by keeping these parameters constant, NO_2 was determined with a lower limit of detection of 0.002 ppm. NO was measured after oxidation to NO_2.

Operation of chemiluminescence analysers for estimation of oxides of nitrogen in the atmosphere

Any commercial chemiluminescence analyser with a maximum sensitivity of < 0.2 ppm FSD can be used. Suitable instruments are marketed by Thermoelectron Corporation (Models 12A and 14D), REM 'NO$_x$ Analyser', Monitor Labs. Inc. Model 8440 and others. The operational specifications of these instruments are similar with response times of a few seconds or less, multiple ranges, and various types of output circuitry for continuous analogue or stored response (for NO and total oxides of nitrogen). Some instruments operate at atmospheric pressure thereby dispensing with elaborate pumping and pressure controls but with some loss of sensitivity.

A continuous flow of sample gas is pumped into the instrument at a flow rate of the order of 1 litre min^{-1}. Sampling probes for NO and NO_2 should be of PTFE tubing or borosilicate glass. Tube lengths should be kept to a minimum for two reasons. Firstly, NO_2 has a tendency to absorb on even the most inert materials, particularly at high r.h. Secondly, the ratio of NO to NO_2 in ambient air when O_3 is present is a function of light intensity. In the dark NO is rapidly oxidized to NO_2 by ozone and therefore the NO/NO_2 ratio will change if the residence time in the sampling system is too long [61]. The half-life for oxidation of NO is given by

$$t_{1/2} (NO) \simeq 0.03 [O_3]^{-1} \text{ min } ([O_3] \text{ in ppm})$$

and therefore for ambient air monitoring (O_3 concentration approximately

0.05 ppm) the residence time should not exceed a few seconds. For continuous monitoring it is desirable to insert a filter to prevent dust particles from entering the sample lines, where they may interfere with the flow control capillaries, etc. The filter holder should be of an inert material (e.g. PTFE) and a cellulose membrane filter may be used with minimal loss of NO_2.

After start-up, a period of up to 2 h will be required for temperature equilibration of the multiplier cooler and the heater for conversion of oxides of nitrogen into NO. Before use it will be necessary to calibrate the response of the instrument using a source of NO of known concentration. The high linearity and stability of the detector make this a relatively simple procedure since calibration can be made at a single concentration. Initially, however, it may be desirable to check the linearity by controlled dilution of the standard gas. For calibration purposes, standard mixtures of NO in N_2, stored under pressure in corrosion resistant steel cylinders, are commercially available. If a regulator is used it should be of high quality stainless steel. The NO content of the calibration gas may be checked by chemical analysis after oxidation to NO_2 (not highly recommended in view of the uncertainties in the oxidation efficiency and the analysis) or by titration agains a constant O_3 source which can in turn be calibrated using neutral KI. The details of a method for gas phase titration of NO against O_3 are given by Hodgeson et al. [62].

The efficiency of the conversion of NO_2 to NO can also be checked if an O_3 source of constant composition is available. A known flow of NO is injected into an air stream containing an excess of O_3. The concentrations and flow rates should be such that oxidation of NO_2 by excess O_3 does not occur (this will be important only at low flow rates and with a large excess of O_3). If the converter efficiency is 100% the response of the analyser in the total oxides of nitrogen mode should be independent of the relative amounts of NO and NO_2 in the sample.

After calibration and converter efficiency evaluation, the instrument is ready for monitoring operation. Calibration checks need not be more frequent than once per week providing operating conditions are maintained constant.

4.3.6 Analysis of NH_3

4.3.6.1 Chemical methods

Measurements of NH_3 in amounts of the order of the MAK value (50 ppm) are normally carried out acidimetrically, by absorption in dilute H_2SO_4 of known concentration [63]. For measurements in ambient air the presence of other acidic and basic components (principally SO_2) make this method impractical and more specific methods must be used. Two colorimetric methods can be applied for measurement of NH_3 down to the ppb level. The first method [64] uses Nessler's reagent (formula $HgI_2.2KI$) which gives an intense colour when added to solutions containing the ammonium ion. However a disadvantage is the interference of a number of compounds which may be present in ambient air (e.g. H_2S, other sulphides and formaldehyde). For accurate analysis it is necessary to carry out a time consuming distillation step in which the acid sample solution is made slightly alkaline and the NH_3 distilled into a second solution prior to addition of the colour reagent. These disadvantages are eliminated in the second method which is based on the indophenol reaction, i.e. the formation of a blue dye when NH_3 reacts in phenol-sodium hypochlorite solutions. A modified indophenol-blue method in which the reaction is catalysed by sodium nitroprusside is the recommended technique for the analysis of NH_3 in ambient air [65, 66].

Recently the use of a selective electrode for the determination of NH_3 in aqueous solutions has been reported [67]. The electrode (Orion Research Inc.) was capable of detecting $0.03\,\mu g$ NH_3 ml^{-1} in natural waters with a response time of 5 min, and was comparable to the indophenol-blue method in precision. This could provide a useful method for the analysis of NH_3 collected in aqueous solution, from air samples.

When sampling the atmosphere for gaseous NH_3 the most likely interference is from ammonium compounds in particulate matter, which are nearly always present in ambient air. The particulate matter may be removed by filtration during sampling but certain complications may arise, particularly when sampling for long periods. Firstly the collection of acid particulate matter on the filter may serve to remove gaseous NH_3 from the sample stream. Secondly, some particulate ammonium salts (e.g. ammonium nitrate, NH_4NO_3), which will be collected on the filter, have an appreciable vapour pressure at ambient temperatures and may volatilize into the collection solution. Thus, as a result of these sampling problems, the accuracy of the measurement of low concentrations of NH_3 in the atmosphere is subject to some uncertainty.

Analysis of NH_3 by the catalysed indophenol-blue method

Principle

NH_3 is collected by aspiration of air through dilute H_2SO_4 in a standard impinger. The resultant ammonium ion is determined by spectrophotometric measurement of the blue indophenol dye formed in the sodium pentacyanonitrosyloferrate catalysed phenol-hypochlorite reaction with NH_3 in alkaline solution [66]. A filter placed upstream of the impinger prevents collection of particulate matter containing ammonium compounds which would interfere. Other common pollutant gases which may dissolve in dilute H_2SO_4, e.g. SO_2, O_3, NO_2 do not interfere at their normal atmospheric levels. The sensitivity of the method is approximately $0.02\,\mu g$ NH_4^+ in the sample solution of $50\,ml$ which corresponds to a detection limit of approximately $1\,\mu g\,m^{-3}$ NH_3 (1.3 ppb at 760 Torr and $25°\,C$) when sampling at the recommended rate (30 litres min^{-1}) for 30 min.

Apparatus

A standard impinger is used to contain the absorbing solution. A sampling probe with a filter assembly similar to that described for the sampling of SO_2 (Section 4.3.1) should be placed upstream of the impinger. Although NH_3 does not become absorbed on surfaces so readily as SO_2, it is recommended that the sample probe is heated approximately $10°\,C$ above ambient when sampling in humid atmospheres. A pump capable of drawing 30 litres min^{-1} air through the sample train is required. A calibrated rotameter and/or a gas meter (wet or dry) is used to measure the volume of air sampled to within 2%.

Reagents

Analytical grade reagents should be used. The colour developing solutions should be freshly prepared every 1 to 2 days and stored in the dark. Solutions should be prepared from distilled water which has been passed through a cartridge of ion exchange resin to remove the NH_4^+.

(a) Absorbing solution: $0.0025\,M$ H_2SO_4. Prepare by dilution from $1\,M$ H_2SO_4 made up by mixing 28 ml concentrated H_2SO_4 (sp. gr. 1.84) with 500 ml de-ionized H_2O and making up to 1 litre.

(b) Phenol-nitroprusside reagent: dissolve 5 g phenol plus 25 mg of sodium pentacyanonitrosyloferrate (sodium nitroprusside $-$ $Na_4(Fe(CN)_5NO.2H_2O)$) in 500 ml de-ionized H_2O.

(c) Alkaline hypochlorite: dissolve 5 g NaOH in 500 ml de-ionized water. Add 4.2 ml sodium hypochlorite solution (12% available chlorine) and make up to 1 litre with H_2O.

(d) Standard ammonium sulphate $[(NH_4)_2SO_4]$ solution (stock): weigh out $0.370\,g$ $(NH_4)_2SO_4$ and dissolve in 1 litre de-ionized water in a volumetric flask. This solution contains $101\,\mu g$ ammonium ion ml^{-1}.

(e) Dilute $(NH_4)_2SO_4$ solution in a volumetric flask dilute 10 ml of the stock solution to 1 litre with de-ionized water.

Procedure

Place 50 ml of the absorbing solution in the impinger and a fresh filter (Whatman 41 and/or 'Microsorban') in the filter holder. Assemble the sampling train and aspirate the bubbler at 30 litres min^{-1} for 30 min. (Slower flow rates and/or longer times may be employed.) Note the volume of air sampled and correct to 760 Torr and 25° C. Appreciable evaporation of the absorbing solution may occur; this can be corrected by weighing before and after sampling.

For analysis transfer a 5.0 ml aliquot of the absorbing solution into a 25 ml stoppered flask and add from a pipette 10 ml of phenol-nitroprusside reagent. Shake in order to mix well. From a pipette add 10 ml of alkaline hypochlorite solution and mix well. Prepare a reagent blank in the same manner using unexposed absorbing solution. Allow 30 min at room temperature (which should be greater than 20° C) for colour development before measuring absorbance at 625 nm.

Calibration and calculation

Into a series of 25 ml volumetric flasks place 0, 0.5, 1, 2, 3 and 5 ml of the dilute ammonium sulphate solution. Add in turn exactly 10 ml each of the phenol nitroprusside reagent and the alkaline hypochlorite, shaking well after each addition. Make up to volume using absorbing solution, allow to stand for 30 min and measure the absorbance. Construct a calibration plot of μg NH_4^+ ml^{-1} versus absorbance.

The concentration of ammonium ion in the sample solutions can be determined from the calibration. The NH_3 concentrations in the air sample is given by

$$C_{NH_3}(\text{ppm}) = \mu\text{g ml}^{-1} \text{ } NH_4^+ \times 1.36 \times V_a/V_s$$

where V_a is the final volume of the absorbing solution after correction for evaporative losses, V_s is the volume of air (in litres) sampled at 760 Torr and 25° C and 1.36 is the factor for conversion of μg ammonium ion to μl NH_3.

Variations

NH_3 may be sampled on filter paper (discs or tape, Whatman 41) impregnated with 5% potassium bisulphate solution [68]. A flow rate of 10 litres min^{-1} through a 2.5 cm diameter filter gives a good collection efficiency. A prefilter must be used to remove particulate matter. The NH_3, which is trapped as NH_4^+, can then be extracted from the paper with H_2O at 70° C, and determined using the indophenol blue method. Potassium hydrogen sulphate ($KHSO_4$) is also removed from the tape during extraction and can cause attenuation in response. This can be taken into account by maintaining in all samples and standards a level of $KHSO_4$ approximately equivalent to that in the paper extracts (0.1 to 0.2 mg ml^{-1}).

4.3.6.2 Physical methods

At present, there are no widely used direct physical methods for the measurement of NH_3 in the atmosphere. NH_3 may be measured by i.r. or u.v. absorption spectrometry. The minimum detectable concentration of NH_3 by dispersive i.r. absorption in a 10 m path length is approximately 20 ppm using the 10.77 nm absorption maximum. Gaseous NH_3 exhibits several strong absorption bands in the u.v. between 190 and 230 nm. The determination of NH_3 in air by direct u.v. spectrophotometry at 204.3 nm in 10 cm quartz cells has been reported, [69] with a lower detection limit of 7 ppm. The molar extinction coefficient of NH_3 at this wavelength is 2790 litres $mol^{-1} cm^{-1}$ but many other compounds absorb strongly in this region (e.g. SO_2, H_2S, and organic compounds) and may interfere. These spectroscopic methods are only useful, therefore, for measuring NH_3 concentrations at the TLV level (TLV NH_3 = 50 ppm).

An indirect determination of NH_3 using chemiluminescence analysis of NO has recently been reported [58]. When NH_3 is heated in the presence of air and a catalyst, oxidation to NO occurs and the latter is measured by its chemiluminescence with O_3. NO_2 and other compounds of N_2 are also converted to NO under the conditions needed for NH_3 conversion. Thus two catalytic converters are necessary for determination of NH_3, one for conversion of NO_2, etc. to NO and a second converter for NH_3, which is then determined by difference. Several metal–carbon composite catalysts were investigated for the analysis of NH_3 in exhaust gases. The sensitivity for NH_3 depends on the relative amounts of NO, NO_2 and NH_3 present in the sample gas.

4.3.7 Miscellaneous N_2 compounds

The most important organo-nitrogen air pollutants, the peroxyacylnitrates, are discussed in Chapter 5. Other gaseous nitrogen compounds, which are encountered mainly in industrial environments, are amines, hydrazine and HCN. For industrial hygiene purposes specific colorimetric methods have been developed for the quantitative measurement of these compounds in air. Primary aliphatic amines can be determined colorimetrically using ninhydrin (triketohydrindine hydrate) [70]. A method utilizing collection in impingers containing HCl has been developed by Hantzsch and Prescher [71]. Photometric determination of small amounts of hydrazine in air using the colour reaction with dimethyl aminobenzaldehyde has been described [72, 73]. By absorption in a midget impinger a detection limit of approximately 3 ppb in a 100 litre air sample can be achieved. HCN has been determined by a number of techniques. A simple impregnated filter paper method for the determination of cyanide in air has been devised [74] based on the Prussian blue reaction. More sensitive methods have been developed by Hanker and co-workers [75, 76]. They found that the chelate compound formed between 8-hydroxy-7-iodoquinoline-5-sulphonic acid and palladous chloride, when converted to the potassium salt, reacts with ferric ion and cyanide to yield a blue green complex which can be estimated colorimetrically at 650 mμ. A more sensitive variation of this palladium chelate procedure involves the measurement of the fluorescence of a co-ordination complex of 8-hydroxyquinoline-5-sulphonic acid with magnesium ion [77]. The action of cyanide on the non-fluorescent potassium salt of the above chelate compound liberates hydroxyquinoline-sulphonic acid which co-ordinates with magnesium ion to form a fluorescent chelate. Thus the intensity of fluorescence provides a measure of the amount of cyanide present.

Nitrous oxide (N_2O) is of no importance as an air pollutant but is interesting because of its natural occurence in low concentrations (~ 0.3 ppm), both in the atmosphere and in sea water. The analysis of this compound is characterized by its lack of any specific chemical reactions at room temperature and only physical methods are suitable for its determination at low concentrations. Several GC procedures for the measurement of atmospheric N_2O have used Porapak Q columns and Katharometer detection [78]; the N_2O having been previously concentrated on a solid absorbent. Lattue *et al.* [79] have developed a method giving an accuracy of 2% for the determination of N_2O in the atmosphere.

4.3.8 Preparation of standard gas mixtures for calibration

Any method for the analysis of gaseous air pollutants is of doubtful validity unless the response can be reliably calibrated. Wet chemical methods involving colorimetric analysis can usually be accurately calibrated using standard solutions of the solvated species (usually an ion) which is being analysed. The response of coulometric systems can be calculated according to Faraday's law provided the redox reaction of the pollutant proceeds with close to 100% efficiency. These procedures for determining the response of the method do not, however, take into account possible errors resulting from variations in sampling efficiency, non-stoichiometric chemistry and losses by absorption on tubing, etc. These difficulties may be overcome by sampling standard mixtures containing known concentrations of the pollutant of interest at the concentration levels for which the method is to be employed. Furthermore, for many physical and instrumental methods particularly those involving chemiluminescence or GC, standard gas mixtures provide the only means of calibrating the method.

Mixtures containing accurately known concentrations of gases at the ppm level and below can be prepared by either static or dynamic methods.

4.3.8.1 Preparation of standard mixtures by static methods

Essentially this involves dilution of a measured volume of pollutant gas with a known volume of air or other diluent gas at a fixed pressure and temperature. This technique can be used satisfactorily for most of the gaseous N and S pollutants.

The dilution is most conveniently carried out in the laboratory at pressures near atmospheric using either a rigid (constant volume) container or a flexible bag. The container and fittings should be of a suitably inert material such that the pollutant is not absorbed rapidly on the walls, etc. Borosilicate glass, aluminium or plastic can be used for rigid containers and fluorocarbon plastic film (e.g. PTFE) for flexible bags. The type of material will be determined to some extent by the pollutant gas being handled. The volume of the container should be as large as possible, preferably of the order of 200 litres. A large size is particularly useful for rigid containers since the partial pressure (and hence the concentration) of the pollutant decreases as the sample gas is removed. The air sampled can be replaced by clean air to maintain constant pressure, and if the sample volume removed is small compared with the total volume only a small concentration change will result. The volume of a rigid container must be accurately determined. Rigid containers have an advantage in that a circulating fan can be installed to assist rapid mixing. Liquid pollutants can be introduced from a microsyringe and gases from a gas-tight syringe. Accurate amounts of gases and vapours can also be measured out from a vacuum system, if available. A known pressure of gas is measured into a previously evacuated glass cell (fitted with PTFE stopcocks if the gas is corrosive or soluble in vacuum grease).

When diluting in flexible bags the following procedure is adopted. The container is purged with diluent gas several times by filling and deflating through the entry port. The bag is then completely deflated by applying a slight vacuum and the diluent supply quickly connected. The bag is filled with a known volume of diluent gas by metering through an accurate rotameter (for a known time) or a gas-meter. During filling inject a measured volume of pollutant into the gas stream. When the container is approximately 0.75 full, stop the flow and mix the contents by kneading the bag several times. Leave for 15 min to allow complete mixing before sampling. The concentration in ppm is simply the mixing ratio as defined by

$$C \, (\text{ppm/v}) \; = \; V_P / V_D$$

where V_P is the volume of pollutant injected and V_D the total volume of gas in the reservoir, both corrected to 760 Torr and 25° C. If liquid pollutant is injected, the corresponding volume of vapour can be calculated from a knowledge of the density of the liquid at the laboratory temperature and the molecular weight of the vapour.

Calibration mixtures can also be prepared as pressurized gas mixtures in cylinders. Pressurized standard mixtures of a variety of pollutant gases are commercially available. Most of the N and S gases are, however, too reactive for storage under pressure in metal containers for long periods. An exception is NO which may be stored as a pressurized mixture in pure N_2. Standard mixtures of NO in N_2 which are prepared gravimetrically and which provide a useful source for the calibration of chemiluminescence $NO-NO_2$ analysers are commercially available. Pressurized standard mixtures of sulphur hexafluoride in air may be employed for the calibration of flame photometric detectors for sulphur compounds. If pressure regulators are used for dispensing these standard mixtures, they should be of the high quality, corrosion resistant type.

4.3.8.2 Preparation of standard mixtures by dynamic methods

Permeation tubes

The use of permeation tubes to standardize methods for the analysis of trace gases was first documented by O'Keefe and Ortman [80] in 1966. The principle of this device is based on diffusion of a gas or vapour through a plastic membrane at very slow rates. A sealed section of inert PTFE tubing containing a liquefied gas or volatile liquid can, when placed in a metered air stream, maintain a constant low concentration of the gas. By accurate measurement of the weight loss from the permeation tube over a period of time the system can be used as a gravimetric calibration standard. The diffusion rate is a non-linear function of temperature and therefore constant temperature conditions must be maintained during gravimetric standardization and use as a calibration source.

Certain compounds, e.g. SO_2 and H_2S react with O_2 or water vapour at the surface of the plastic tubing. Tubes containing these compounds should be stored and standardized under dry N_2. A relatively high concentration in dry N_2 carrier gas is then diluted stepwise in air to provide calibration mixtures.

Perhaps the most important parameter in the standardization of permeation tubes is the time factor required for the diffusion equilibrium to be reached. Two to five days at constant temperature ($\pm 0.1°C$ for 1% precision) under a flowing dry carrier gas stream is required for accurate standardization. The time is dependent on diffusion rate and accuracy of weighing. At intervals during the standardization, the tube can be quickly removed, weighed and replaced, and the process continued until a constant loss rate is achieved.

Permeation tubes may be purchased or prepared according to the method of O'Keefe and Ortmann [80]. For the preparation of gas mixtures in the concentration range likely to be encountered in the atmosphere a loss rate of 0.1 to $0.2 \, \mu l \, min^{-1}$ is optimum. In addition to temperature, the loss rate depends on

Table 4.2 PTFE permeation-tube characteristics for S gases [27]

Gas	Tube length (cm)	i.d. (in)	Wall (in)	Volume loss min^{-1} at 20.3° C and 1 atm.
CS_2	10.3	0.183	0.03	0.377
SO_2	2.3	0.183	0.03	0.364
H_2S	4.0	0.485	0.2	0.120
CH_2SH	7.2	0.183	0.03	0.087

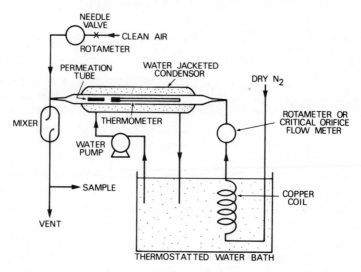

Fig. 4.10 Arrangement for producing standard gas mixtures using a permeation tube.

the pressure difference, wall thickness and surface area (and hence length) of the tube. Some tube dimensions and corresponding loss rates for S compounds [27] are given in Table 4.2.

An apparatus as shown in Fig. 4.10 is required to prepare standard concentrations of a pollutant. Carrier gas (50 to 100 ml min^{-1}) is metered through a needle valve and flow meter or a critical orifice and brought to temperature by passage through a 2 m copper coil immersed in a constant temperature water bath. The gas passes over the calibrated permeation tube which is housed in a water jacketed condenser through which water from the bath is circulated. The temperature should be the same as used for calibration of the tube. The gas stream is then diluted to the desired concentration by varying the flow rate of the diluent air which should be clean and dry. The minimum flow rate will be determined by the requirements of the sampling system and maximum flow rates of 15 litres min^{-1} or more may be used. This flow should be measured with an accuracy of 1 to 2%. The concentration at a given flow is simply given by

$C\,(\text{ppm}) = P/F$

where F is the total flow rate (diluent + carrier) in litre min^{-1} and P is the permeation rate in $\mu l\,min^{-1}$, both corrected to 760 Torr and 25° C. The permeation rate P can be calculated from the rate of weight loss according to

$$P\,(\mu l\,min^{-1} \text{ at } 25° C) = w\,(\mu g\,min^{-1}) \times \frac{22.4}{\text{mol. wt.}} \times \frac{298}{273}\,.$$

When working with low concentrations (i.e. ppb region) it is necessary to condition the system by allowing the trace gas to pass through the system for a period of day(s) before conducting calibrations. An accurate recording instrument is useful for monitoring the outlet gas concentration when attempting to calibrate manual techniques. All tubing into which the gas comes in contact should be of an inert material such as glass or PTFE.

Exponential dilution method

This method produces a gas mixture in which the trace-gas concentration declines with time at a fixed exponential rate. It is useful, therefore, for calibration of instrumental methods which are either continuously recording or capable of giving a number of readings in reasonably rapid succession e.g. GC methods. The basic flow system is shown in Fig. 4.11. An aliquot of pure pollutant (or an accurately prediluted mixture) is injected into the dilution flask which is stirred for complete mixing. The flask is continuously flushed with a metered stream of pure, dry, carrier gas. For these conditions the concentration of the pollutant in the outlet decreases exponentially [81].

$C = C_0 \exp\left(-Qt/V\right)$

where $C_0 =$ initial concentration, $Q =$ volumetric flow rate at the flask pressure (1 atm), and $V =$ the effective volume of the dilution flask, $t =$ time elapsed after injection. The effluent gas is then mixed with a constant diluent flow to obtain a low concentration mixture, the composition of which is varying in a known manner. By a judicious choice of flow rates and volumes, accurate mixtures in the sub-ppm range can be prepared. Thus with a 10 litre exponential dilution flask and carrier and dilution flows of 0.05 and 10 litres min^{-1} respectively, 1 ml of pure pollutant will give a concentration of 0.5 ppm with a half-time

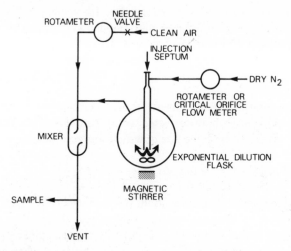

Fig. 4.11 Exponential dilution flask for preparing low concentration gas mixtures.

for decay of 140 min. The application of the exponential dilution technique for the calibration of a chemiluminescence NO detector [5] and the flame-photometric S detector [28] has been described.

4.4 Particulate compounds of S and N

4.4.1 Analysis of SO_4^{2-}

The principle natural source of SO_4^{2-} in the atmosphere is the oxidation of natural H_2S. Additionally, SO_4^{2-} enters the air in marine aerosol and in the erosion of rocks. Sulphates also arise directly from some industrial activities, and from atmospheric oxidation of anthropogenic SO_2. In general, significantly higher levels of SO_4^{2-} are measured in urban areas as compared to rural areas.

Recently, interest in the analysis of airborne SO_4^{2-} has increased substantially. It is believed that oxidation catalysts used for vehicle emission control in North America oxidize the SO_2 formed in combustion of fuel sulphur to SO_3. This condenses with moisture in the vehicle exhaust to generate an aerosol of H_2SO_4. It is anticipated that such an aerosol of acid SO_4^{2-} would present a far greater hazard to health than the neutral SO_4^{2-} normally present in the air. Consequently, as well as stimulating interest in SO_4^{2-} analysis, this has caused a search for methods of differentiating acid SO_4^{2-} from the total.

A recent review article by Tanner and Newman [83] discusses progress and problems in SO_4^{2-} analysis. The SO_4^{2-} must first be collected on a filter paper.

Cellulose paper filters such as Whatman No. 41 have a low background of soluble SO_4^{2-} and are suitable, whilst Mitex or Fluoropore filters, made of Teflon, are recommended for speciation studies [83]. In ambient atmospheres, particulate S is almost entirely SO_4^{2-} and X-ray fluorescence analysis of collected S may be used [83]. Alternatively, if a flame photometric S analyser is available, collected SO_4^{2-} may be volatilized into the analyser by rapid heating and this method has been shown to give results comparable with those obtained by XRF [84, 85]. More commonly, the sulphate is leached from the filter into an aqueous medium and determined by any sensitive specific technique for SO_4^{2-} such as those described in Section 4.3.1.1 for analysis of SO_2 when collected as sulphate in H_2O_2 solution.

Attempts to speciate airborne SO_4^{2-} depend upon selective volatilization or solvent extraction of acid SO_4^{2-}. Such techniques are not expected to be of wide application and the interested reader is referred to the review of Tanner and Newman [83].

4.4.1.1 Experimental procedure for SO_4^{2-} (turbidimetric)

Principle

The method described is an alternative to that given in Section 4.3.1.1. SO_4^{2-} after leaching from filter paper, is mixed with a solution of barium chloride. The resultant precipitate of barium sulphate is measured turbidimetrically [86].

Sampling apparatus

A Whatman No. 41 filter paper in almost any type of filter holder may be used. Membrane and Nuclepore filters have a higher soluble SO_4^{2-} background and are not recommended.

Reagents

(a) HCl 10 N.

(b) Glycerol–alcohol solution. Mix one volume glycerol with 2 volume 95% ethanol.

(c) Standard SO_4^{2-} solution. Dissolve 0.1479 g anhydrous sodium sulphate in de-ionized water and dilute to 1 litre to give a $100 \, \mu g \, ml^{-1}$ stock solution.

(d) Barium chloride. Analar grade, 20–30 mesh.

Procedure

Draw air through the filter at $10 \, \text{litres min}^{-1}$ for 24 h, or longer if required. Extract the exposed filter and a blank by warming to 70° C in de-ionized water (10 ml) for 10 min. Decant the liquid into a 25 ml volumetric flask and repeat with another aliquot of de-ionized water. Adjust to a final volume of 25 ml when cool, filtering if turbid before making the adjustment.

Prepare SO_4^{2-} standards by pipetting 0.5, 1.0, 1.5, 2.0 and 3.0 ml of stock SO_4^{2-} solution into 25 ml volumetric flasks and adjust the volume to 20 ml by addition of de-ionized water by pipette to give standards of 2.5, 5, 7.5, 10 and $15 \, \mu\text{g ml}^{-1}$. Transfer 20 ml of sample and blank solutions to 25 ml volumetric flasks by pipette. To the sample, blank and standard solutions, add 10 N HCl (1 ml) followed by glycerol–alcohol solution (4 ml) and mix thoroughly. Add a constant measure (0.5 spoon spatula) or barium chloride crystals to each. Do not mix, but allow to stand for exactly 40 min. Carefully decant the turbid supernatant into 4 cm cells and determine the absorbance at 500 nm against a reagent blank.

Calculation

Draw a calibration curve of SO_4^{2-} concentration versus absorbance. This normally shows some curvature. Estimate the SO_4^{2-} concentrations of sample ($x \, \mu\text{g ml}^{-1}$) and filter blank ($y \, \mu\text{g ml}^{-1}$) solutions.

Then, atmospheric concentration of SO_4^{2-}

$$= \frac{(x - y) \times 25}{\text{Vol. of air sample (m}^3)} \, \mu\text{g m}^{-3}.$$

4.4.2 Analysis of NO_3^-

Oxidation of the nitrogen oxides in the atmosphere leads ultimately to formation of NO_3^-. This occurs naturally, and also at an accelerated rate in the highly oxidizing atmosphere of a 'photochemical smog'. Analysis for NO_3^- in ambient air is rarely performed, probably since nitrates as such are not of especially high toxicity.

Nitrates may be collected upon any type of filter material, but the use of Whatman No. 41 papers is recommended since this paper has a low background of soluble NO_3^-. A number of colorimetric procedures are available for NO_3^- but it is essential to establish the magnitude of possible interferences before adopting a method.

4.4.2.1 Experimental procedure for nitrate (colorimetric)

Principle

Nitrates are leached from the filter paper and then converted to nitrite by copper-catalysed reduction with hydrazine sulphate in alkaline solution [68]. The nitrite is then determined by a modification of the method of Saltzman [48]. After acidification with *o*-phosphoric acid, the nitrite is reacted with sulphanilamide (4-aminobenzene sulphonic acid) and the resultant diazonium salt is coupled with *N*-(1-naphthyl)ethylenediamine dihydrochloride to form an azo dye which is determined spectrophotometrically at 540 nm. Interference from nitrite is possible. The existence of this species in appreciable concentrations in ambient air is most improbable, but if its presence is suspected it may be determined separately by the above procedure, omitting the hydrazine sulphate reduction, and an appropriate correction made.

Sampling apparatus

A Whatman No. 41 filter paper in any suitable holder.

Reagents

(a) Diazotization reagent. Dissolve sulphanilamide (10 g) and *N*-(1-naphthyl) ethylenediamine dihydrochloride (0.5 g) in 10% v/v *o*-phosphoric acid and make up to volume (1 litre) with the same acid.

(b) Copper sulphate solution. Make the stock solution by dissolving anhydrous copper sulphate (2.5 g) and making up to volume (1 litre) with water. Prepare fresh working solution daily by dilution of stock solution (12.5 ml) to 4 litres.

(c) Hydrazine sulphate. Prepare stock solution by dissolving hydrazine sulphate (27 g) and making up to volume (1 litre) with water. Prepare fresh working solution daily by dilution of stock (25 ml) to 1 litre.

(d) NaOH. Dissolve Analar NaOH (12 g) and make up to volume (1 litre) with water.

(e) NO_3^- standards. Dissolve Analar potassium nitrate (0.163 g) in water and make up to volume (1 litre to prepare a stock solution of $100\,\mu g\,ml^{-1}$ NO_3^-. Prepare a working standard of $10\,\mu g\,ml^{-1}$ by diluting stock solution (100 ml) to 1 litre.

Procedure

Draw air through the filter at $10\,litres\,min^{-1}$ for 24 h, or longer if required. Extract the exposed filter and blank by warming to 70° C in de-ionized water (5 ml) for 10 min. Decant into a 10 ml volumetric flask, repeat the extraction with a further portion of water (4 ml), add the washings to the volumetric flask and when cool dilute to volume.

Dilute the stock standard solution with water to give standards 0, 1, 2, 3, 4 and $5\,\mu g\,ml^{-1}$ NO_3^-. Pipette 2 ml aliquots of sample, blank and standard solutions into small flasks and add copper sulphate solution (6 ml), NaOH solution (6 ml) and hydrazine sulphate solution (4 ml). Shake gently to mix and then heat to $38°$ C using a water bath. Cool and add the diazotization reagent (6 ml). Using a spectrophotometer determine the absorbance of the solutions at 540 nm against a de-ionized water blank.

Calculation

Draw a calibration curve of absorbance versus NO_3^- concentration and estimate the concentration of the sample ($x\ \mu g\,ml^{-1}$) and filter blank ($y\ \mu g\,ml^{-1}$) solutions. Then, concentration of NO_3^- in air

$$= \frac{(x-y) \times 10}{\text{Vol. of air sample (m}^3)}\ \mu g\,m^{-3}.$$

4.4.3 Analysis of NH_4^+ salts

NH_4^+ salts arise primarily from neutralization of NH_3 by acidic substances in the atmosphere. $(NH_4)_2SO_4$ is believed to be a major constituent of urban aerosols.

NH_4^+ salts may be collected from the atmosphere by filtration, and after leaching from the filter paper are determined by the same methods as are applicable to gaseous NH_3 after collection (see Section 4.3.6.1). One problem for which there appears to be no ready solution arises from the significant vapour pressures of salts such as NH_4NO_3 and chloride at ambient temperatures. A very small concentration of these salts may escape collection by filtration, and this may be of significance when determining the low levels present in unpolluted air.

4.4.3.1 Experimental procedure for NH_4^+ (colorimetric)

Principle

After collection upon a filter paper the NH_4^+ salts are leached into water, where they are determined by the catalysed indophenol-blue method.

Sampling apparatus

A Whatman No. 41 filter is recommended, held in any convenient filter holder. This type of filter has a low background of soluble NH_4^+ ions.

Procedure

Draw air through the filter at 10 litres min^{-1} for 24 h, or longer if required. Extract the exposed filter and blank by warming to 70° C in de-ionized water (10 ml) for 10 min. Decant the liquid into a 25 ml volumetric flask and repeat with another aliquot (10 ml) of de-ionized water. Adjust to volume when cool. Transfer an aliquot (5 ml) into a 25 ml stoppered flask and add phenol-nitroprusside reagent (10 ml) as described in Section 4.3.6.1 for analysis of gaseous NH_3, and complete the analyses as described. Calibration is performed in the manner described in Section 4.3.6.1, extending the range of standards to give concentrations of $0-5 \mu g\,ml^{-1}$ by increasing the concentration of the standard NH_4 solution five-fold.

Calculation

Draw a calibration curve of NH_4^+ ion concentration versus absorbance and estimate the concentration of the sample ($x \mu g\,ml^{-1}$) and filter blank ($y \mu g\,ml^{-1}$) solutions. Then, concentration of NH_4^+ in air.

$$= \frac{(x-y) \times 25}{\text{Vol. of air sampled } (m^3)} \mu g\,m^{-3}.$$

References

[1] Environmental Protection Agency Publ. No. AP-50 (1969). *Air Quality Criteria for Sulphur Oxides.*
[2] Environmental Protection Agency Publ. No. AP-84 (1971). *Air Quality Criteria of Nitrogen Oxides.*
[3] Hersch, P. and Deuringer, R. (1963). *Analyt. Chem.* 35, 897.
[4] Brewer, A.W. and Milford, J.R. (1960). *Proc. Roy. Soc.*, A256, 470.
[5] Fontjin, A., Sabadell, A.J. and Ronco, R.J. (1970). *Analyt. Chem.* 42, 575.
[6] Brody, S.S. and Chaney, J.E. (1966). *J. Gas Chromat.* 4, 42.
[7] Okabe, H. (1971). *J. Am. Chem. Soc.* 93, 7095.
[8] Moffat, A.J., Robbins, J.R. and Barringer, A.R. (1971). *Atmos. Environ.* 5, 511.
[9] Williams, D.T. and Huger, R.N. (1970). *Appl. Optics* 9, 1597.
[10] Hodgeson, J.A., McClenny, W.A. and Hanst, P.L. (1973). *Science* 182, 248.
[11] Stevens, R.K., Mulik, J.D., O'Keefe, A.E. and Krost, K.J. (1966). *Analyt. Chem.* 38, 760.
[12] Allen, J.D. (1973). *J. Inst. Fuel* 46, 123.
[13] Driscoll, J.N. and Warneck, P. (1973). *J. Air. Pollut. Control Ass.* 23, 858.
[14] Forrest, J. and Newman, L. (1973). *J. Air. Pollut. Control Ass.* 23, 761.
[15] West, P.W. and Gaeke, G.C. (1956). *Analyt. Chem.* 28, 1816.
[16] Jacobs, M.B. and Greenburg, L. (1956). *Ind. Engng. Chem.* 48, 1517.
[17] Huygen, C. (1962). *Analyt. Chim. Acta* 28, 349.
[18] Katz, M. (1950). *Analyt. Chem.* 22, 1040.

[19] Pate, J.P., Ammons, B.E., Swanson, G.A. and Lodge, J.R. (1965). *Analyt. Chem.* **37**, 942.

[20] Scaringelli, F.P., Saltzmann, B.E. and Frey, S.A. (1967). *Analyt. Chem.* **39**, 1709.

[21] Scaringelli, F.P., Frey, S.A. and Saltzmann, B.E. (1967). *Am. Ind. Hyg. Ass. J.* **28**, 260.

[22] Thomas, M.O. and Amtower, R.E. (1966). *J. Air Pollut. Control Ass.* **16**, 618.

[23] Bostrun, C.E. (1965). *Int. J. Air Pollut.* **9**, 333.

[24] Persson, G.A. (1966). *Int. J. Air Pollut.* **10**, 845.

[25] Volmer, W. and Frohlich, Z.H. (1944). *Analyt. Chem.* **126**(16) 414.

[26] Kanno, S. (1959). *Int. J. Air Pollut.* **1**, 231.

[27] Stevens, R.K., O'Keefe, A.E. and Ortmann, G.C. (1969). *Environ. Sci. Technol.* **3**, 652.

[28] Hartmann, C.H. (1971). Proc. Joint Conference on Sensing of Environmental Pollutants, AIAA Paper No. 71-1046. Palo Alto, California.

[29] Bruner, F., Liberti, A., Possanzini, M. and Allegrini, I. (1972). *Analyt. Chem.* **44**, 2070.

[30] Thornsberry, W.L. (1971). *Analyt. Chem.* **43**, 452.

[31] Jackson, P.J., Langdon, W.E. and Reynolds, P.J. (1970). *J. Inst. Fuel* **43**, 10.

[32] Fielder, R.S. and Morgan, C.H. (1960). *Analyt. Chim. Acta* **23**, 538.

[33] Bond, R.L., Mullin, W.J. and Pinchin, F.J. (1963). *Chem. Ind (London)* **48**, 1903.

[34] Jacobs, M.B., Braverman, M.M. and Hockheiser, S. (1957). *Analyt. Chem.* **29**, 1349.

[35] Bamesburger, W.L. and Adams, D.F. (1969). *TAPPI* **52**, 1302.

[36] Bethea, R.M. (1973). *J. Air Pollut. Control Ass.* **23**, 710.

[37] Pare, J.P. (1966). *J. Air Pollut. Control Ass.* **16**, 325.

[38] Hockheiser, S. and Elfers, L.A. (1970). *Environ. Sci. Technol.* **4**, 672.

[39] Okita, T., Lodge, J.P. and Axelrod, H.D. (1971). *Environ. Sci. Technol.* **5**, 532.

[40] Natusch, D.F.S., Kloris, H.B., Axelrod, H.D., Teck, R.J. and Lodge, J.P. (1972). *Analyt. Chem.* **44**, 2067.

[41] Axelrod, H.D., Cary, J.H., Bonelli, J.E. and Lodge, J.P. (1969). *Analyt. Chem.* **41**, 1865.

[42] Adams, D.F., Bamesburger, W.L. and Robertson, T.J. (1968). *J. Air Pollut. Control Ass.* **18**, 145.

[43] Pecsar, R.E. and Hartmann, C.H. (1971). *Analyt. Instrum.* **9**, H-2, 1.

[44] Moore, H., Helwig, H.L. and Graul, R.J. (1960). *Ind. Hyg. J.* **21**, 466.

[45] Leither, W. (1970). *The Analysis of Air Pollutants*, p. 228. Ann Arbor-Humphrey: Ann Arbor, London.

[46] Okita, T. (1970). *Atmos. Environ.* **4**, 93.

[47] Ronkainen, P., Denslow, J. and Leppanen, O. (1973). *J. Chromat. Sci.* **11**, 384.

[48] Saltzman, B.E. (1954). *Analyt. Chem.* **26**, 1949.

[49] Jacobs, M.B. and Hockheiser, S. (1958). *Analyt. Chem.* **30**, 426.

[50] Levaggi, D.A., Siu, W. and Feldstein, M. (1973). *J. Air Pollut Control Ass.* **23**, 30.

[51] Nash, T. (1970). *Atmos. Environ.* **4**, 661.

[52] Saltzmann, B.E. and Wartburg, A.F. (1965). *Analyt. Chem.* **37**, 1261.

[53] Scaringelli, F.P., Rosenburg, E. and Rehme, K.A. (1970). *Environ. Sci. Technol.* **4**, 924.
[54] Hartkamp, H. (1970). *Schr. Reihe Landesanst Imn-u Bodenutzungshutz Landes N. Rhein/Westfalen,* **18**, 55.
[55] Ripley, D.L., Chingenpeel, J.M. and Hurn, R.W. (1964). *Int. J. Air Pollut.* **8**, 455.
[56] Thomas, M.D. (1956). *Analyt. Chem.* **28**, 1810.
[57] Levaggi, D.A., Kothny, E.L., Belsky, T., DeVera, E.R. and Mueller, R.K. (1974). *Environ. Sci. Technol.* **8**, 348.
[58] Breitenbach, L.P. and Shelef, M. (1973). *J. Air Pollut. Control Ass.* **23**, 128.
[59] Krost, K.J., Hodgeson, J.A. and Stevens, R.K. (1973). *Analyt. Chem.* **45**, 1800.
[60] Guicherit, R. (1972). *Atmos. Environ.* **6**, 807.
[61] Butcher, S. and Ruff, R.E. (1971). *Analyt. Chem.* **43**, 1890.
[62] Hodgeson, J.A., Baumgardner, R.E., Martin, B.E. and Rehme, K.A. (1971). *Analyt. Chem.* **43**, 1128.
[63] Leither, W. (1970). *The Analysis of Air Pollutants,* p. 170. Ann-Arbor-Humphrey, Ann Arbor, London.
[64] Buck, M. and Strathmann, H. (1965). *Z. Analyt. Chem.* **213**, 241.
[65] Chaney, A.L. and Marbach, E.P. (1962). *Chim. Chem.* **8**, 130.
[66] Weatherburn, M.W. (1967). *Analyt. Chem.* **39**, 971.
[67] Thomas, R.F. and Booth, R.L. (1973). *Environ. Sci. Technol.* **7**, 523.
[68] Eggleton, A.E. and Atkins, D.F. (1972). *Results of the Tees-side Investigation,* AERE R-6983, H.M.S.O., London.
[69] Gunther, F.A., Barkley, J.H., Kolbezen, M.J., Blian, R.C. and Staggs, E.A. (1956). *Analyt. Chem.* **28**, 1985.
[70] Williams, D.D. and Miller, R.R. (1962). *Analyt. Chem.* **34**, 225.
[71] Hantzsch, S. and Prescher, K.E. (1966). *Staub* **26**, 332.
[72] Price, J.G., Fenimore, D.C., Simmonds, G.P. and Patkiss, A.Z. (1968). *Analyt. Chem.* **40**, 541.
[73] Porter, K. and Volman, D.H. (1962). *Analyt. Chem.* **34**, 748.
[74] Gettler, A.O. and Goldbaum, L. (1947). *Analyt. Chem.* **19**, 270.
[75] Hanker, J.S., Goldberg, A. and Witten B. (1958). *Analyt. Chem.* **30**, 93.
[76] Hanker, J.S., Gamson, R.M. and Klapper, H. (1957). *Analyt. Chem.* **29**, 879.
[77] Leithe, V.W. and Hofer, A. (1968). *Allg. Prakt. Chem.* **19**, 78.
[78] Bock, R. and Schutz, K. (1968). *Z. Analyt. Chem.* **237**, 321.
[79] Lattue, M.D., Axelrod, H.D. and Lodge, J.P. (1971). *Analyt. Chem.* **43**, 1113.
[80] O'Keefe, A.E. and Ortman, G.C. (1969). *Analyt. Chem.* **38**, 761.
[81] Lovelock, J.E. (1960). *Gas Chromatography, 1960* (ed. Scott, R.P.W.) p. 26. Butterworths, London.
[82] Intersociety Committee for Manual Methods of Air Sampling and Analysis. (1972). *Methods of Air Sampling and Analysis,* American Public Health Association, Washington D.C.
[83] Tanner, R.L. and Newman, L. (1976). *J. Air Pollut. Control. Ass.* **26**, 737.
[84] Roberts P.T. and Friedlander, S.K. (1976). *Atmos. Environ.* **10**, 403.
[85] Husar, J.D., Husar, R.B. and Stubits, P.K. (1975). *Analyt. Chem.* **47**, 2062.
[86] Jacobs, M.B. (1960). *The Chemical Analysis of Air Pollutants,* Interscience Publishers, New York.

CHAPTER FIVE

Secondary pollutants

5.1 Introduction

The term secondary pollutants is applied to pollutants which are formed as a result of chemical reactions of primary gaseous pollutants within the atmosphere. Secondary pollutants may be either gaseous or particulate aerosols. The gaseous pollutants are usually formed in homogeneous gas-phase reactions which in many cases are photochemically initiated. Aerosol pollutants may be formed within the atmosphere as a result of gas phase reactions followed by condensation of the products, or by reactions taking place in the existing atmospheric aerosol phase.

Gaseous pollutants formed within the atmosphere include O_3, oxides and oxyacids of nitrogen (NO_2, N_2O_5, HNO_3, HNO_2), SO_2 (from oxidation of sulphides etc.), H_2O_2 and organic peroxides, aldehydes, organic acids and other partially oxidized organic compounds including nitrogen-containing esters such as the peroxyacylnitrates (PANs). The aerosol components which can be produced in atmospheric reactions are H_2SO_4 and sulphates, inorganic nitrates, chlorides, and their ammonium salts and also organic condensation aerosols. Some of the 'secondary' pollutants formed in the atmosphere are also emitted as primary pollutants e.g. NO_2, SO_2, and sulphates; the classification is not, therefore, rigid. In this section only the gaseous secondary pollutants which have not been dealt with in other chapters will be considered. The aerosols produced by chemical transformations in the atmosphere contribute to the overall burden of particulate matter in the air. The analytical techniques for sampling, characterization and measurement of atmospheric particulate are described in Chapters 1 and 2.

Of the gaseous secondary pollutants, O_3 has undoubtedly received the most attention. O_3 has been known for a long time to be a natural consitituent of the earth's atmosphere. Relatively high concentrations of O_3 exist in the stratosphere where it is produced by photolysis of molecular oxygen, O_2. O_3 from the high altitudes is transported to the lower atmosphere where it is removed by chemical reaction or at the earth's surface. Recognition of O_3 as a pollutant in urban air resulted from the discovery in the late 1940s by Haagen-Smit and his colleagues that abnormally high O_3 concentrations were present during air pollution episodes in the Los Angeles basin. It was subsequently found that the formation of O_3 and other 'oxidants' resulted from sunlight inititated photochemical reactions of nitrogen oxides and unburned hydrocarbons, emitted principally from motor vehicle exhausts. Other undesirable pollutants such as aldehydes, PANs and organic and inorganic aerosols are also produced in these reactions. This particular type of atmospheric pollution was termed 'photochemical smog', and although the phenomenon was originally thought to be confined to the west-coast cities of the United States, it has become apparent in recent years that the formation of photochemical oxidants in urban air is rather widespread. Severe photochemical smog episodes are confined, however, to certain large cities which are subject to special climatic conditions i.e. strong temperature inversions and prolonged sunshine.

The chemistry of photochemical smog is very complex and the reader is referred to other sources for a detailed description [1, 2]. The phenomenon results from oxidation of organic pollutants by a chain reaction involving free radicals, which are generated photochemically following light absorption by certain primary and secondary pollutants. The presence of nitrogen oxides is required to sustain the free radical chain reactions, and the primary pollutant, NO, is oxidized to NO_2 during the process. The hydrocarbons are oxidized firstly to aldehydes and acids which are converted to CO, and eventually to CO_2. This accounts for the large and complex array of partially oxidized organic compounds present in the smog. O_3 is formed as a result of photodissociation of NO_2 to give oxygen atoms, which combine with molecular oxygen:

$$NO_2 + \text{u.v. light} = O + NO$$

$$O + O_2 = O_3 .$$

Since NO is converted back to NO_2 in the free radical reactions, NO_2 is continuously replenished and the O_3 concentration builds up.

The sequence of chemical reactions occurring during smog formation gives rise to a characteristic pattern in the diurnal variation of the concentrations of primary and secondary pollutants in urban air [3]. In the early morning, the concentration of the primary pollutants, NO and hydrocarbons, rises due to the increased emissions from traffic. At this stage the O_3 concentration is usually very low due to overnight destruction at the ground and by reaction with other pollutants. During the morning, as sunlight intensity increases, photochemical oxidation of the primary pollutants causes the concentrations of NO_2 and other oxidation products to rise. The O_3 concentration also rises until a maximum is reached sometime after midday, and falls thereafter due to reactions with other pollutants. The concentrations of other secondary pollutants, both gaseous and particulate, also reach their maximum during the afternoon or early evening. This model diurnal pattern is determined by the chemistry and emission pattern of the pollutants. The diurnal variation in the concentration of the pollutants is also sensitive to changes in meteorological features and the model diurnal pattern may not always be observed in practice. Furthermore, even in unpolluted air, an afternoon maximum in the ground-level O_3 concentration can often be observed due to diurnal changes in the vertical mixing in the atmosphere, giving increased downward mixing of natural O_3 during daytime.

The rate of chemical transformation of primary to secondary pollutants normally depends on the air-concentration of primary pollutants. Thus the concentration of secondary pollutants will be influenced by the same gross atmospheric features, such as air-mass origin and degree of dispersion, as the primary pollutants. However the detailed variation in time and space of the concentration of the secondary pollutants will differ from their precursors. The transformation rates are also influenced by other parameters such as temperature, absence or presence of sunlight and its intensity, r.h. etc. The time scale of the overall chemical processes is normally of the order of a few hours and as a result the very high concentrations and rapid fluctuations associated with primary sources are not encountered with the secondary pollutants. Furthermore, the maximum concentration of the secondary pollutants may be observed some distance from the source of the precursors and a larger geographical region may be subjected to the effects of the pollutants.

Many of the secondary pollutants under discussion have known deleterious effects [3]. The 'cracking' of rubber by O_3 is well known and was in fact used as a diagnostic for O_3 formation in early work on photochemical smog. The most serious effect of elevated O_3 levels, however, is its phytotoxic effect on vegetation, in particular on agricultural crops such as tobacco, market-garden produce and citrus fruit trees. O_3 is also toxic to humans and the TLV of 150 ppb is often exceeded during photochemical episodes both in the U.S.A. and Europe. The peroxyacyl nitrates also show phytotoxic effects on some plants [4]. HNO_3 and H_2SO_4 produced by oxidation of nitrogen and sulphur oxides are both highly corrosive and can cause damage to many materials [5], particularly metals. H_2SO_4 aerosol may present a serious health hazard since it can be readily injested into the lung by virtue of the small particle size [6]. Certain aldehydes and PANs are responsible for the severe eye irritation associated with photochemical smog [7].

The range of air-concentrations likely to be encountered in an air sampling

Table 5.1 Typical air concentrations in parts per billion of gaseous secondary pollutants

Pollutant	Urban smog	Rural	Background
O_3	150 to 500	50 to 100	20 to 50
Peroxides	50 to 150	–	–
Total aldehydes	~ 50	2 to 10	< 1
peroxyacylnitrate	3 to 30	< 1	< 0.1
Nitric acid and nitrates	10 to 100	–	–

programme for secondary pollutants is indicated in Table 5.1, which shows typical levels of gaseous pollutants which have been observed in different locations. As noted above, the concentrations of the secondary pollutants do not fluctuate rapidly with time and, therefore, sampling time resolution of 1 to 2 h is adequate for obtaining a reasonable measure of the pollutant burden. In view of the diurnal variations in the pollutant levels, however, 24 h average measurements only give limited information and, in the case of O_3, may give an entirely false impression of the degree of pollution.

Analytical methods of sufficient sensitivity are now available for measurement of most of the above secondary pollutants in urban and rural air. In discussing the analytical techniques it is convenient to classify the pollutants into three groups i.e. oxidants, organic secondary pollutants and inorganic oxyacids.

Oxidants

These are broadly defined as those compounds present in the air which will oxidize a reference reagent that is not oxidized by atmospheric oxygen [3]. The widely adopted reference compound for the measurement of atmospheric-oxidants is KI, the oxidant concentration being equivalent to the amount of iodine released following exposure to a known volume of air. Using this technique a measurement of 'total oxidants' is obtained, without distinction as to the individual nature of the pollutants, which may include both inorganic and organic primary and secondary pollutants. Furthermore, reducing components such as H_2S, SO_2 and aldehydes, will decrease the degree of oxidation observed leading to a net oxidant concentration. O_3 is the main component giving a positive response, with minor contributions from peroxides, PAN and NO_2.

The 'total oxidants' measurement can be considered to represent the 'condition' of the atmosphere. Low (or negative) oxidants (< 10 ppb) might indicate, for example, high levels of SO_2; oxidant readings of 20 to 50 ppb would be expected in clean air due to natural O_3; oxidant readings in excess of 100 ppb might indicate photochemical pollution. These measurements are, however, of limited use in air pollution studies and this has stimulated investigation into more specific methods for the measurement of individual oxidizing pollutants.

The organic secondary pollutants

These broadly comprise all partially oxidized hydrocarbon compounds i.e. alcohols, aldehydes, organic acids and peroxides and organo-nitro compounds. The overall atmospheric burden of these compounds will contain, in addition to atmospheric reaction products, partially oxidized organic compounds from combustion and other sources. A distinction between the 'primary' and 'secondary' component is difficult. However some compounds, e.g. PANs have an exclusively atmospheric source and, therefore, serve as useful indicators of the overall photochemical reactivity in the atmosphere.

The inorganic oxyacids

These include HNO_2, HNO_3 and H_2SO_4. The precursor oxides NO_2 and other oxides of nitrogen (N_2O_3, N_2O_4, NO_3 and N_2O_5) and sulphur trioxides, which may also be considered as secondary pollutants, have been discussed in Chapter 4. H_2SO_4 only exists in the atmosphere as an aerosol and the analysis of particulate acids is treated in detail in Chapter 4. HNO_3 may exist either in the gaseous phase or as an aerosol but little is known about its partition between the phases under atmospheric conditions at the present time. Although there is evidence for the presence in air of both HNO_3 and HNO_2 vapours, analytical methods for these compounds are still at a very rudimentary stage.

5.2 Basic analytical techniques for the analysis of gaseous secondary pollutants

The basic techniques which have been used to measure the air concentrations of gaseous secondary pollutants are similar to those described in the previous chapter relating to gaseous nitrogen and sulphur compounds. Much of the earlier work on identification and measurement of oxidants, oxidized organic compounds, etc. was made using simple chemical methods. Infrared spectroscopic analysis has also played an important role in elucidating some of the unusual products and chemical reactions in photochemical smog, mainly in laboratory studies. Recently the application of physical methods, particularly chromatographic and chemiluminescence techniques, has become more widespread. The main advantages offered by the latter are improved specificity in the measurement of individual components in the complex array of substances present in the urban atmosphere. The general remarks regarding interferences, accuracy and precision, made in Chapter 4, also apply to the analysis of gaseous secondary pollutants.

5.2.1 Sampling methods

The basic techniques used for sampling trace gases from the atmosphere are surveyed in Chapter 1 and some additional comments relating to the design of samplers to minimize absorption losses and interferences, are given in Chapter 4. For the gaseous secondary pollutants, particular attention should be paid to the minimization of losses by absorption and decomposition during sampling, since these gases are generally themselves chemically reactive. Thus suitable inert materials should be used for sample probes, tubing, etc. Teflon (PTFE) equipment is recommended for the sampling of O_3 and other oxidants, but clean glass tubing may be used provided long sample probes are not used. Metals, rubber and PVC should be avoided. Organic vapours may be sampled with stainless steel, PTFE or glass equipment. However since many organic secondary pollutants are water soluble it is essential that the sampling tubes be kept dry, to minimize absorption losses. In view of the reactive nature of most secondary pollutants, sample concentration by cryotrapping or adsorption on a porous support is not recommended for quantitative work.

5.2.2 Analytical techniques

5.2.2.1 Chemical methods

Classical chemical techniques involving colorimetric, coulometric or acidimetric analysis of air samples are still widely used for the measurement of gaseous secondary pollutants. In fact these are likely to remain the favoured methods for 'total oxidants' which, according to the definition given in Section 5.1 can only be measured by chemical methods.

Colorimetric analysis involves trapping of the gaseous air pollutant in a suitable medium until sufficient of the material of interest has been collected to give a measurable colour with a specific colour-forming reagent. In view of the unstable nature of many gaseous secondary pollutants, a liquid absorbing medium containing a reagent which reacts immediately with the gas(es) of interest to give an involatile stable product, is normally used. The colour-forming reagent may either be present in the absorbing solution or can be added to the sample after collection. Colorimetric analysis is most conveniently applied to batch samples but automated colorimetric instruments are available (e.g. Technicon Mk IV) which have been developed to give a semi-continuous measurement of a number of pollutants including oxidants and aldehydes. Autoanalysers also allow rapid and precise measurement of large numbers of batch samples. This aid is invaluable for any routine survey work.

The main problem with colorimetric analysis and with chemical methods generally is the lack of specificity of the colour-forming reactions. For example, the liberation of iodine from KI solution, which has been widely used for the colorimetric determination of O_3 and total oxidants, is subject to a strong negative interference from reducing gases such as SO_2 and H_2S, which are commonly present in urban air. Chemical methods for the analysis of organic secondary pollutants generally utilize a specific colour reaction of a characteristic funtional group in the organic compound. Thus selective determination of individual compounds of a certain class is not normally possible. Notable exceptions are the aldehydes, formaldehyde and acrolien and also H_2O_2, for which specific colorimetric methods have been developed.

Coulometry using the I_2/I^- redox system is widely used for the measurement of total oxidants and O_3. The air sample containing oxidant is aspirated continuously through an electrochemical cell containing KI solution. The liberated I_2 causes current to flow between the anode and the cathode. Two methods have been used to monitor the electrochemical reaction. The Galvanic coulometric O_3 sensor, first described by Hersch and Deuringer [8], utilizes a cyclic oxidation-reduction process and requires no reagent replacement or applied external voltage. The operating principle of this cell is described in detail in Chapter 4. The

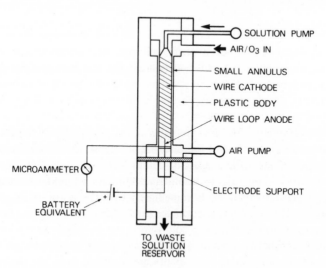

Fig. 5.1   Detector cell for Mast oxidant meter.

second type has been described as an amperometric coulometer and was originally developed as a balloon-borne instrument for measurement of O_3 in the upper atmosphere [9]. A commercially available form of amperometric coulometric detector for atmosphere oxidants (Mast Development Corporation, model 724-2) has been widely used for measurements in polluted air [10–12]. In this detector, which is shown diagrammatically in Fig. 5.1, a platinum wire helical cathode and a wire loop anode are wetted in turn by a solution containing 2% KI and 5% KBr, which is continuously pumped through the cell. Air is drawn through the cell at a constant flow rate of $140 \, cm^3 \, min^{-1}$ and, if O_3 is present in the sampled air, I_2 is liberated in the electrolyte. The liberated I_2 is continuously reduced by an applied cathodic potential of 0.25 to 0.3 V and the current flowing between the cathode and the wire loop anode is proportional to the amount of I_2 liberated. The current is assumed, therefore, to be a linear function of the O_3 concentration and the flow rate. In common with all methods based on oxidation of I_2 solutions, coulometric methods for measurement of O_3 suffer interferences from other oxidizing and reducing substances present in the sample.

5.2.2.2 Physical methods

Chemiluminescence methods

The basic principles of chemiluminescence analysis of atmospheric pollutants have been described in Chapter 4. The only secondary pollutant for which chemiluminescence methods have been developed to the instrumental stage is O_3. However, the development of these methods, which provide a relatively simple, sensitive and highly specific measurement of O_3, has been timely in view of the difficulties inherent in all chemical methods for this pollutant. A detailed survey of the techniques which have been used for chemiluminescence analysis of O_3 is given in Section 5.3.2.2.

Gas chromatography

Gas chromatography is by far the most widely used technique for the analysis of organic atmospheric pollutants. A useful review of the application of gas chromatography to atmospheric analysis has been given by Altshuller [13], and Fishbein [14] has surveyed selected chromatographic procedures for a variety of environmental pollutants including atmospheric gases. The use of gas–liquid or gas–solid chromatography with flame ionization detection (FID) for the analysis of hydrocarbon compounds is discussed in detail in Chapter 6. Measurement of secondary organic pollutants, i.e. oxygenated and nitrogen-containing organic compounds, by the application of normal GC–FID techniques presents two major difficulties. Firstly, the high reactivity and polar nature of these compounds leads to problems of material losses or modification during sample handling and chromatographic separation. Secondly the response of the FID to low molecular weight O and N containing compounds is considerably less than for the corresponding hydrocarbons. The lower sensitivity for O and N containing organics makes sample concentration necessary for the detection of these compounds at the levels present in ambient air using FID. Furthermore, the chromatographic separation of aldehydes, etc., from the vast range of hydrocarbons of different volatility which are present in urban air presents a formidable problem.

For the GC analysis of aldehydes Levaggi and Feldstein [15] have developed a novel method of obtaining the necessary selectivity and sample stability. This was achieved by trapping the aldehydes in aqueous sodium bisulphite solution when a stable aldehyde–bisulphite complex is formed. Hydrocarbons, etc. are not trapped in this medium. The aldehydes are then quantitatively recovered by thermal decomposition of the complex in a specially designed heated injection part on the gas chromatograph.

In recent years, a variety of new types of column material have been developed which possess greatly improved elution characteristics and separating ability for trace amounts of reactive substances. These include modified forms of porous polymeric stationary phases (e.g. Poropak, Durapak, Porasil) and supports with chemically bonded stationary phases. The use of silanized support materials (e.g. HDMS treated Chromosorb) which provide a more inert chromatographic medium, has enabled chromatographic separation and quantitative elution of unstable compounds such as peroxyacetylnitrate (PAN). Unfortunately there have been rather few reports in the literature on the performance of these new columns, as applied to the analysis of organic compounds in ambient air samples.

For the analysis of organo-nitrogen compounds the inadequate sensitivity of the FID has been largely overcome by the application of the electron capture detector (ECD) [16]. The extreme sensitivity of this detector for strongly electron absorbing compounds such as PAN and alkyl nitrates allows direct measurement of their concentration at the ppb level in 1 to 10 ml air sample. Recent review articles on the ECD have been given by Wentworth [17] and Aue and Kapila [18] but a brief description will be given here.

Electron capture detection utilizes the drastic reduction in the electrical conductivity of a gaseous mixture in an ionization chamber when electrophilic compounds are present. The detector consists of an ionization chamber containing a radioactive β emitting source with a stream of inert gas flowing through it. The β activity causes ionization of the gas liberating free electrons. By the application of a low voltage potential, the electrons are caused to migrate to the anode and a constant 'standing' current through the detector results. On the introduction of a trace gas with a high electron-capture cross-section, e.g. compounds containing halogen atoms, nitro or amino groups and some sulphur and oxygen compounds, the electrons undergo reaction to form low mobility ions and the detector current drops. The fall in current is directly proportional to the amount of trace contaminant present, provided the fall does not exceed 20 to 30% of the standing current. When approaching saturation, i.e. the electron concentration is severely depleted by reaction with the electron absorbing component, the response becomes very non-linear. The d.c. potential is usually applied in a pulsed mode to prevent polarization in the detector leading to anomalous response particularly under conditions of detector contamination. In the pulsed mode the voltage is applied in short pulses (0.5 to 1 μs) at intervals of 50 to 500 μs i.e. the voltage is off for most of the time. In this way not only are polarization effects minimized but the electron density in the detector will be greater. Thus the likelihood of electron capturing events and hence the sensitivity is increased. The average electron concentration increases with the interval between pulses up to 500 μs but above this, natural ion recombination and the presence of trace contaminants limits further increases in sensitivity. The linear response range of the ECD can be increased by operating in the pulse frequency feedback mode [19]. In this mode the electron concentration is maintained essentially constant as the electron absorbing component passes through the detector, by changing the pulse frequency in response to the change in standing current. The change in pulse frequency required to maintain the current is directly related to the amount of electron absorbing component present.

The molar response of the detector to a given electron absorbing component, in the linear region, depends on the rate constant for electron attachment to the component. This parameter varies greatly among the various electron absorbing species. Lists of relative response factors for halogenated compounds and methods of predicting them have been given [20, 21]. Calibration of the detector by normal methods of injecting measured quantities of the component is only practical for the more weakly absorbing compounds. The response is so high for strong absorbers that difficulty is encountered in injecting quantitatively the extremely small amounts of material required to calibrate the detector in the linear range. A novel approach to this problem is the application of gas phase coulometry [22]. If the detector is constructed in the form of a long tube then, for strong absorbers, the proportion ionized in the detector approaches 100% provided an appropriate flow rate is employed. The integrated response in A s (i.e. peak area, A) is related to the mass of substance entering the detector (m) in grams as follows:

$$m = \frac{MA}{9.65 \times 10^4} \quad (M = \text{mol. wt}).$$

This gives an absolute calibration of the detector independent of ambient variables of temperature and pressure. This technique has not been applied so far to the organo-nitro compounds of interest in the present context.

The main practical problem in the operation of electron capture detectors arises from unwanted contaminants entering the detector. Impurities, present in the carrier gas or on the detector surfaces, serve to reduce the standing current, thereby decreasing sensitivity and the linear range and also increasing the noise in the detector. For optimum performance with commercial ECDs great care should be taken to follow the recommended operating procedure, with particular attention to the standing current in the detector.

Spectroscopic methods

Spectroscopic techniques whether in emission or absorption offer a means for the direct and continuous detection of trace gases in the gas phase. They have been used for many years to identify and measure pollutant gases both in the laboratory and in the atmosphere. A brief introduction to the use of spectroscopic techniques as applied to gaseous pollutants is given in Chapter 4. Historically absorption spectroscopy has been of considerable significance in the identification and measurement of secondary gaseous pollutants. The presence of elevated O_3 levels in Los Angeles smog was convincingly demonstrated in the mid-1950s using both u.v. [23] and long-path i.r. absorption spectroscopy [24]. The organic peroxyacylnitrates were first discovered at about the same time by use of i.r. absorption spectroscopy for studying the photochemical reactions of hydrocarbons and nitrogen oxides in laboratory smog chambers [25]. The presence of peroxyacetylnitrate in the Los Angeles atmosphere was subsequently verified by means of long-path i.r. absorption spectroscopy [24].

Since these early experiments, considerable advances have been made in the development of spectroscopic techniques to provide the sensitivity, specificity and practicability necessary for measuring air pollutants. Two basic innovations have enabled those advances to be made. Firstly the advent of lasers has removed some of the limitations in terms of energy, spectral purity and coherency of classical radiation sources. A laser can serve as a light source in a non-dispersive analyser if one of the laser emission lines falls on one of the absorption lines of the pollutant to be measured. Measurements can be made either in absorption or emission. Current research is focussed on improving sensitivity and selectivity of laser systems for the measurement of pollutants both in discrete gas samples and by remote sensing. The second innovation is the application of computer techniques for processing complex spectral data to extract the required information. This has enabled the development of a particularly promising technique for the detection of pollutants in the i.r., i.e. Fourier transform spectroscopy with a Michelson interferometer. This technique has been applied both to remote sensing of trace gases in the upper atmosphere [26] and also to measurement of pollutant concentrations in the surface atmosphere [27, 28]. Progress in the field of advanced spectroscopic techniques has been summarized in several recent review articles [29–31].

At the present time, these advanced spectroscopic techniques can only be regarded as research instruments; the cost and expertise required for their operation precludes their use for routine monitoring. Furthermore, development has been largely focussed toward measurement of the major primary pollutants i.e. NO, NO_2, SO_2, CO, hydrocarbons, etc. and it is for these gases that routine application of new optical techniques is envisaged in the near future. O_3 is the only secondary gaseous pollutant for which measurements on a routine basis using spectroscopic techniques are now feasible. However, detection of the less familiar secondary pollutants such as formaldehyde, H_2O_2, formic acid and HNO_3 using new spectroscopic techniques has already been reported [28] and their use in atmospheric research is likely to become more widespread.

5.3 Experimental section

5.3.1 Analysis of 'total oxidants'

Since oxidants are broadly defined as those compounds present in the air which will oxidize a chemical reagent which is not oxidized by molecular oxygen, 'total oxidants' can only be determined by chemical methods. The degree of oxidation of a chosen reference compound is measured, either by colorimetric or coulometric analysis. The results are usually expressed in terms of O_3 which is normally the most abundant atmospheric oxidant.

5.3.1.1 Discussion of analytical methods

Methods based on I_2

The most widely used reference compound for total oxidant determination is KI. The I_2 liberated, in either a neutral or alkaline solution of KI, can be measured colorimetrically from the optical extinction of the I_3^- complex at 352 nm [32, 33]. The alkaline KI method has the advantage that delay is permissible between sampling and analysis. However, the neutral KI procedure has greater simplicity, accuracy and precision and is, therefore, the preferred method. The empirically determined stoichiometry in terms of moles of I_2 produced per mol of O_3 is 1.0 and 0.65 for the neutral and alkaline methods respectively.

Possible oxidizing pollutants which comprise 'total oxidant' are O_3, H_2O_2, organic hydroperoxides and peroxides, peracids, peroxyacylnitrates, NO_2 and Cl_2. Cohen *et al.* [34] have investigated the response of various reagents, including KI, to some of these compounds. Neutral KI shows immediate response to O_3 and peracids but a slow response to other peroxides. The response of NO_2 is rapid but it is only approximately 10% of O_3 on a molar basis (1 ppm $NO_2 \equiv$ 0.10 ppm O_3 in terms of I_2 liberated). An immediate response to H_2O_2 together with O_3 and peracids is obtained when ammonium molybdate is added as catalyst to the neutral KI reagent. (1.0 ml of a 10^{-3} M ammonium molybdate solution is added to KI absorbing reagent after sampling.)

Coulometric measurement of the I_2 released from KI or mixed KI–KBr neutral solutions has also been widely used for measurement of total oxidants e.g. Mast oxidant meter. These instruments only respond to the oxidants which give rapid liberation of I_2. In the Mast instrument, the current required to reduce the liberated I_2 is assumed to be a linear function of the oxidant concentration and the sample flow rate. Several groups [11, 34, 35] have made careful evaluations of the response of Mast instruments to O_3 under laboratory conditions and in the field. In all cases, the response was less than that obtained using the colorimetric neutral KI method as a standard, and in some cases deviations of up to 50% were observed. Clearly results obtained with this type of instrument can only be semi-quantitative and caution should be exercised in comparing total oxidant or O_3 concentrations by this and other methods.

Reducing gases such as SO_2 and H_2S all give serious negative interference with oxidant measurements using KI (probably on a mol–mol basis). The procedures are also sensitive to reducing dusts which may be present in the air or on glassware. Elimination of the interference from SO_2 has been accomplished with a sample pre-filter containing glass fibre impregnated with chromium trioxide [36].

Other methods

An alternative method for manual total oxidant determination, developed by Cohen *et al.* [34] is based on the oxidation of ferrous ammonium sulphate followed by colorimetric measurement of the ferric ion by the addition of ammonium thiocyanate to form the highly coloured $Fe(CN)_6^{3-}$ ion. This method gives instant response for O_3 and most peroxy compounds and is more sensitive than the methods based on I_2 liberation. The high sensitivity and low selectivity combined with good reagent stability make this an ideal method for the determination of total oxidants. However, since most of the measurements of oxidants have been made using KI as the reference compound, the latter should be used if comparison with previous data is contemplated.

The most sensitive method for determining atmospheric oxidants is the NO_2

equivalent method, devised by Saltzman and Gilbert [33]. This method responds primarily to O_3 and involves the addition of a dilute mixture of NO in N_2 to the sample air flow. Any O_3 present reacts rapidly in the gas-phase with NO, converting it to NO_2, which is subsequently trapped and measured colorimetrically in a modified Griess reagent [37]. A second sample is taken to determine concentration of NO_2 initially present in the air (i.e. without NO addition) and the O_3 determined by difference. Difficulties may be encountered with the NO_2 equivalent method in environments where the NO_2 concentrations are greater than that of O_3; small changes in NO_2 levels could be misinterpreted as O_3 responses. Also peroxyacylnitrates give a positive interference to the colorimetry.

5.3.1.2 Neutral KI method for manual analyses of 'total oxidants'

Principle

This method [38] is intended for the manual determination of O_3 and other oxidants in the range 0.01 to 10 ppm. O_3, Cl_2, H_2O_2 and organic peroxides, when absorbed in a neutral buffered (pH = 6.8 ± 0.2) solution of KI, liberate I_2, which is measured spectrophotometrically by determination of the absorption of the tri-iodide ion at 352 nm.

SO_2 and sulphides produce a 100% negative interference on a molar basis. Up to 100-fold ratio of SO_2 to oxidant may be eliminated without loss of oxidant by incorporating a chromic acid paper absorber in the sampling stream. The absorber also oxidizes NO to NO_2, however, and NO_2 gives a positive interference equivalent to 10% of O_3 with neutral KI. Therefore, when SO_2 is less than 10% of the NO concentration the use of the chromic acid absorber is not recommended. NO_2 interference can be corrected for by concurrent analysis of the NO_2 concentration and subtracting one-tenth of this concentration from the total oxidant value. Peroxyacyl nitrates give a response equivalent to 50% of an equimolar concentration of O_3.

The precision of the method is approximately $\pm 5\%$ deviation from the mean. The major error is loss of I_2 by volatalization during longer sampling periods; this error can be reduced by the use of a second impinger. The calibration is based on the assumed stoichiometry for the reaction:

$$O_3 + 3KI + H_2O = KI_3 + 2KOH + O_2. \tag{1}$$

Apparatus

Absorber

All glass midget impingers with a graduation mark at 10 ml are used. Other bubblers with nozzle or open-ended inlet tubes may be employed. Fritted bubblers are not recommended since they produce less I_2. Impingers must be kept clean and dust free. Cleaning should be done with laboratory detergent followed by liberal rinsing with tap and distilled water.

Air metering device

A glass rotameter capable of measuring a flow of 1 to 2 litres min^{-1} with an accuracy of ± 2% is required.

Air pump

Any suction pump capable of drawing the sample flow for up to 30 min through a needle valve or critical orifice is suitable. A trap placed downstream from the absorber is recommended to protect the pump etc. from accidental flooding by the reagent.

Reagents

All reagents are made from analytical grade chemicals and double distilled water. The latter is obtained by distillation in an all glass still with a crystal each of potassium permanganate and barium hydroxide added.

Absorbing reagent

Dissolve 13.61 g of potassium dihydrogen phosphate, 14.20 g of anhydrous disodium hydrogen phosphate (or 35.8 g or the dodecahydrate salt) and 10.00 g of KI successively and dilute to exactly 1 litre with double distilled water. Keep at room temperature for a least 1 day before use. This solution may be stored for several weeks in a glass stoppered brown bottle in a refrigerator. Do not expose to sunlight.

Standard I_2 solution, 0.05 M

Dissolve successively 16.0 g of KI and 3.173 g I_2 in doubly distilled water and make up to 500 ml. Keep for at least 1 day at room temperature before using. Standardization is unnecessary if the weighing is carefully done although, if desired, the solution may be standardized by titration with sodium thiosulphate solution, using starch indicator.

SO_2 absorber

Flash-fired glass fibre filter paper is impregnated with chromium trioxide as follows: drop 15 ml of aqueous solution containing 2.5 chromium trioxide and 0.7 ml conc. H_2SO_4 uniformly over 400 cm^3 of paper and dry in an oven at 80 to 90° C for 1 h; store in a tightly capped jar. Half of this paper serves to pack one absorber. Cut the paper into 6 × 12 mm strips each folded into a V shape. Pack into an 8.5 ml U tube and condition by drawing dry air through overnight. The absorbent has a long life (at least 1 month). If it becomes visibly wet from sampling humid air, it must be dried with dry air before further use.

Procedure

Assemble the sampling train consisting, in order, of the SO_2 absorber (optional), impinger, rotameter and air pump. The sample probe should preferably be of PTFE but glass or stainless steel may be used for short probes. PVC should be avoided except for butt jointing of glass tubing.

Pipette exactly 10 ml of absorbing reagent into the impinger and sample at a flow rate of 1 to 2 litres min^{-1} for up to 30 min. Sufficient air should be sampled to collect the equivalent of 0.5 to 10 μl O_3 in the absorber. Measure the volume of air sampled and correct to 760 Torr and 25° C. Do not expose the reagent to direct sunlight.

For analysis, add distilled water to the impinger to make up to the 10 ml graduation mark (i.e. if evaporation losses have occurred). Within 30 to 60 min of sampling, transfer a portion of the exposed reagent directly to a curvette and measure the absorbance at 352 nm against a reference of double distilled water. Samples having a colour too dark to read may be quantitatively diluted with absorbing reagent. Also measure the absorbance of the unexposed reagent against the reference and subtract the blank from the sample absorbance.

Calibration and calculation

Firstly prepare 0.001 M I_2 standard by pipetting exactly 4 ml of the 0.025 M standard I_2 solution into a 100 ml volumetric flask and diluting to the mark with absorbing reagent. Discard after use. For calibration purposes, exactly 4.09 ml of the 0.001 M I_2 solution is diluted with absorbing reagent just before use to 100 ml i.e. making the final I_2 concentration equivalent to 1 μl O_3 ml^{-1} according to the stoichiometry of Equation 1.

$$1 \, \mu l \, O_3 \, ml^{-1} \equiv \frac{100}{24.47} \, \mu mol \, I_2 \text{ in } 100 \, ml \, (= 4.09 \times 10^{-6} \, mol).$$

In order to obtain a range of concentration values, add graduated amounts of the above calibrating solution up to 10 ml to a series of 10 ml volumetric flasks and dilute to volume with absorbing reagent. Read the absorbances and plot them against the equivalent concentration of O_3 in $\mu l\ O_3\ (10\,ml)^{-1}$ absorbing reagent. The plot follows Beers law. Draw the straight line through the origin giving the best fit and read off the total $\mu l\ O_3\ (10\,ml)^{-1}$ for the samples from the calibration graph. The concentration of O_3 in the gas phase in $\mu l\,litre^{-1}$ or ppm is given by

$$O_3\,(ppm) = \frac{\text{total }\mu l\ O_3\text{ per }10\,ml}{\text{Total volume of air sampled in litres}}$$

Effects of storage

O_3 liberates 90% of the iodine from the buffered reagent immediately and the remaining 10% through a slow set of reactions. Some of the other oxidants cause a slow formation of iodine. Some indication of the presence of such oxidants and of gradual fading due to reducing agents can be made by making several measurements over a period of time.

5.3.1.3 Instruments for measurements of total oxidants

A number of commercial instruments for the measurement of total oxidants, using both colorimetric and coulometric principles, are available. All use either aqueous iodide or mixtures of iodide and bromide as absorbing medium for the oxidizing substances present in the air sampled.

Automatic colorimetric instruments are based on the continuous spectrophotometric measurement at 352 nm of the I_2 liberated in a flowing reagent stream. The air sample stream is exposed to the reagent in a wetted-wall absorber. Fig. 5.2 shows a schematic diagram of a typical system. The reagent reservoir contains approximately 4 litres reagent (neutral buffered 10% KI) which is forced by a constant delivery pump at approximately $4\,ml\,min^{-1}$ through a charcoal column (to remove I_2 from the solution) and a reference colorimetric cell, to the top of the contact column. Sample air is drawn countercurrently through the contact column at approximately $4\,litres\,min^{-1}$. The exposed reagent then flows by gravity through the optical cell where the absorbance is continuously measured. The flow rates of both the air sample and the liquid reagent must be carefully controlled for accurate recording of oxidant concentration. Calibration of the instrument response is carried out by means of a constant known O_3 source (see Section 5.3.2). The method is subject to the same interferences as discussed for the manual colorimetric determination of oxidants (Section 5.3.1), although on some instruments pretreatment of the sample is used to eliminate some of the interfering substances.

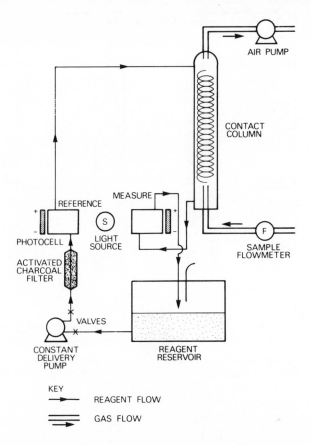

AIR PUMP

CONTACT COLUMN

MEASURE

REFERENCE

S

+
−
PHOTOCELL

LIGHT SOURCE

SAMPLE FLOWMETER

F

ACTIVATED CHARCOAL FILTER

VALVES

CONSTANT DELIVERY PUMP

REAGENT RESERVOIR

KEY

→ REAGENT FLOW

⇒ GAS FLOW

Fig. 5.2   Schematic diagram of continuous colorimetric oxidant analyser.

The most widely known coulometric instrument for the measurement of oxidants is the Mast 'O_3 meter'. Other instruments operating either on the amperometric or galvanic principle (see Section 5.2) are available. Many of these incorporate sample pretreatment to remove reducing substances and oxidants other than O_3 and are marketed as 'O_3' meters.

5.3.2 Analysis of O_3

O_3 is the most important and abundant 'oxidant' present in the atmosphere and a considerable amount of effort has been devoted to the development of specific analytical methods, both chemical and physical, for the measurement of this gas. Although O_3 is a highly reactive gas and will interact with a variety of chemical

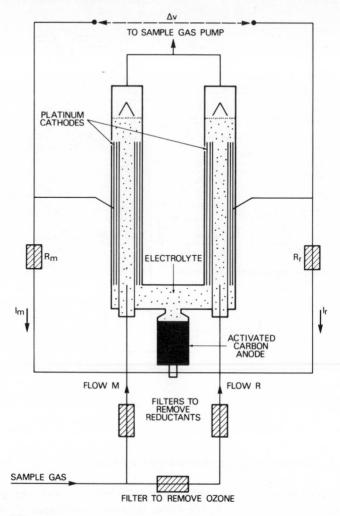

Fig. 5.3 Differential galvanic detection of O_3 in air.

reagents, no truly specific chemical analytical method has yet been developed for the measurement of atmospheric O_3.

5.3.2.1 Chemical methods

Iodide and related methods

The most widely used chemical methods have utilized the liberation by O_3 of I_2 from solutions containing KI. The KI method, however, is not specific for O_3. All oxidizing and reducing agents can potentially interfere, as discussed in Section 5.3.1, and procedures for measurement of O_3 must involve elimination of these possible interferences. A number of sample pretreatment systems have been devised to minimize the effects of the major interferences likely to be encountered in atmospheric measurements, i.e. NO_2, SO_2, and H_2S. Thus the chromic acid scrubber developed by Saltzmann *et al.* [36] (described in Section 5.3.1.2) will eliminate interference of up to a 100-fold ratio of SO_2 to O_3 without loss of O_3.

The liberated I_2 may be measured by titration, by colorimetry or electrochemically using an amperometric cell [9] or a galvanic cell [8]. An interesting electrochemical sensor based on differential galvanic measurement, for atmospheric O_3 measurement which is virtually free of interference, has been described by Lindqvist [39]. Halogen is liberated from an aqueous electrolyte containing NaBr, NaI, Na_2HPO_4 and NaH_2PO_4 at concentrations of 3.0, 0.001, 0.1 and 0.1 M respectively, contained in a recirculating galvanic cell containing two cathodes and one counter electrode (Fig. 5.3). The air sample containing O_3 is divided into two equal streams. By selectively removing reducing agents in both sections and O_3 in only one section, the differential galvanic current is directly related to the concentration of O_3. This current is independent of the presence of other oxidants provided they are not removed by the O_3 scrubber and the O_3 concentration can be calculated directly by Faraday's law, precluding the necessity of calibrating the sensor. No interference of NO_2, Cl_2, SO_2, H_2S, NH_3, HCl, C_2H_4 and 1-C_4H_8 could be observed at concentrations up to 1 ppm. A 12% interference by peroxyacetylnitrate, in mol equivalents of O_3, was observed in the laboratory but this would have little significance in practice. Under laboratory conditions O_3 measurements agreed well with those using the neutral KI method. In operation the instrument showed good stability and a reasonable time constant of about 40 s.

The manual neutral KI method, although of doubtful value for the measurement of O_3 in ambient air, is useful for standardization of O_3 sources for laboratory calibration of instruments. The procedure described in Section 5.3.1.2 gives a relative standard deviation of about ± 5% of the mean. Although there has been some disagreement over the reaction stoichiometry it now seems established that, at a pH of 6.85, the reaction of O_3 with KI proceeds with a stoichiometry of 1 : 1 [40–42].

Other methods

Several colorimetric procedures, which are claimed to be specific for the determination of O_3, have been reported. The measurements have been based either on the bleaching action (i.e. decrease in absorbance) following the reaction of O_3 with a colour reagent or on the formation of ozonolysis products which undergo sensitive colour reactions.

Use of the bleaching action of O_3 on a buffered solution of indigo sulphuric acid in water was first described by Dorta-Shaeppi and Treadwell [43]. Guicherit *et al.* [35] recently evaluated a slightly modified indigo H_2SO_4 method and found a good correlation between O_3 concentrations determined by this method and those determined using the differential galvanic detector of Lindqvist [39]. At pH 6.85 the reaction proceeds stoichiometrically and the decrease in extinction follows Beers law over a concentration 0.05 to 1.5 μg O_3 ml^{-1} solution. The method is simple, rapid, inexpensive and sensitive (5 ppb O_3 can be measured in a 60 litre air sample). Of the components which are most frequently present in ambient air, only NO_2 interferes. This interference, however, is only 4% in O_3 equivalents and will not normally give rise to serious error.

A basically similar method for O_3 determination based on bleaching of Diacetyl-dihydro-lutidine (DDL) has been described by Nash [44]. This method apparently has minimal interference from O_2 and oxidants other than O_3, SO_2 and oxides of nitrogen. However the DDL reagent is less stable and the collection efficiency lower than the indigo H_2SO_4 reagent.

Specific methods for O_3 determination based on spectrophotometric determination of ozonolysis products have been described by Bravo and Lodge [45] and Hauser and Bradley [46]. Both methods are slow, complicated and use corrosive non-aqueous collection media. The latter method, which involves spectrophotometric determination (with 3 methyl-2-benzothiazolone hydrazonhydrochloride) of the pyridine-4 aldehyde formed when O_3 reacts with 1, 2 di-(4-pyridyl) ethylene (DPE), has non-unit stoichiometry (reported values vary from 1.24 to 1.33). A field comparison study in which the DPE method was tested against an O_3-selective galvanic detector, showed deviations of more than a factor of 2 [35].

5.3.2.2 Physical methods

Chemiluminescence methods

In recent years, chemiluminescence detection has been widely adopted as a technique for the measurement of atmospheric O_3. The first chemiluminescence O_3 detectors were developed by Regener and co-workers for use on balloon-borne O_3 sondes [47]. These instruments utilized the chemiluminescent reaction of O_3 with an organic dye, rhodamine B. The solid dye was adsorbed on a silica-gel impregnated surface which was positioned close to the window of a photomultiplier tube. Sample gas is aspirated over the active surface and any O_3 present in the sample reacts with the dye with the emission of light, which is detected by the photomultiplier. This technique is extremely sensitive and highly selective for O_3 (no other components of the atmosphere have been observed to give chemiluminescence with the active surface). The main problems arise from instability and irreproducibility of the chemiluminescent surface, non-linearity of the response and effects of relative humidity changes in the sample gas.

A good deal of development work has been carried out with a view to utilizing the rhodamine B system for continuous automatic detection of O_3. Hodgeson *et al.* [48] eliminated the desensitizing effect of water vapour on the chemiluminescent surface by treatment of the silica-gel with a hydrophobic agent, silicone resin, prior to adsorbing the dye. The response was then independent of relative humidities between 0 and 80% r.h. The lifetime of the reactive surface was limited by the slow oxidation of rhodamine B by O_3. Since sensitivity could be sacrificed in the system, the analyser was run over an extended period with a 1 : 6 dilution of the incoming sample with dry air. In this way the useful lifetime of a surface would be increased to 2 to 3 months. However, frequent calibration using a known O_3 source is necessary to allow for the small daily decrease in sensitivity.

Guicherit [49] increased the stability of the chemiluminescent surface by protecting the rhodamine B from oxidation, with another compound, which reacts with O_3 more easily i.e. gallic acid. Following the reaction with O_3, the gallic acid or one of its oxidation products transfers energy to the dye, giving chemiluminescence. A stable surface was produced with no serious loss in sensitivity. The stability was further increased using an intermittent sampling procedure in which sample air was admitted for 15 s followed by the introduction of dry O_3 free air for about 45 s. The dry air admission eliminated moisture effects and also gave a 'zero-point' reading. Monitoring outdoor air over a period of 52 days showed no decay in sensitivity of the surface, but after this time the response fell off rapidly. The minimum detectable concentration of O_3 using the dye method is < 1 ppb which is more than adequate for monitoring ambient levels.

The gas-phase chemiluminescent reaction between O_3 and ethylene (C_2H_4), which was originally utilized by Nederbragt et al. [50] for the detection of O_3 in air, is now widely used for ambient O_3 measurement. A large number of commercial instruments based on this method are now available. Air containing O_3 is mixed with a slow flow of C_2H_4 in an injector positioned close to the end window of a photomultiplier, and the low-level light emission resulting from the O_3–C_2H_4 reaction observed. Kummer et al. [51] have investigated the nature of the light-emitting products of this reaction. Emission from excicited formal-dehyde in the 350–550 nm region and also from vibrationally excited hydroxyl radicals has been detected. Other aliphatic olefins also give chemiluminescence with O_3 and with a much higher emission intensity, at least at low pressures. However an advantage of the O_3–C_2H_4 system is that sufficient light emission for measurement is obtained at pressures near 1 atm, and therefore elaborate pumping and pressure control facilities are not required. The photocurrent is a linear function of the O_3 concentration in the sample gas and the detection limit is of the order of 1 ppb of O_3. As with all chemiluminescence methods, the electrical response must be calibrated with a known O_3 source.

Other chemiluminescence methods have been used for O_3 analysis. For example, the chemiluminescent reaction of NO with O_3, which is commonly used for the analysis of NO (see Chapter 4, Section 2.3) can be used for analysis of O_3 if NO is used as the reactant gas instead of O_3. Recently Finlayson et al. [52] have described a chemiluminescent reaction between O_3 and triethylamine. The latter compound also reacts with peroxyacetylnitrate to give luminescence but at a different wavelength to the emission following reaction with O_3.

Ultraviolet absorption

The O_3 absorption of the 253.7 nm mercury resonance line was the first spectro-metric concentrations approach employed for monitoring pollutant concentration in the lower atmosphere. However, early instruments had insufficient sensitivity and stability for routine measurements of ambient concentrations of O_3. Sensitivity and stability has been greatly improved in a recently developed commercial O_3-photometer offered by DASIBI Corporation of Glendale, California. In this instrument the absorption by O_3 of the 253.7 nm line is measured in a 70 cm folded path cell which is alternatively filled with unfiltered and filtered (O_3 free) ambient air. A schematic diagram of the system is shown in Fig. 5.4. The light source is a stabilized miniature low pressure mercury arc. The light beam is split, a fraction being directed onto the incident light detector photocell and the remainder passing through the cell to a second detector photocell. The output of the photocells is digitized and stored and the detectors are linked electronically so that the second photocell measures the integrated transmitted light for a fixed

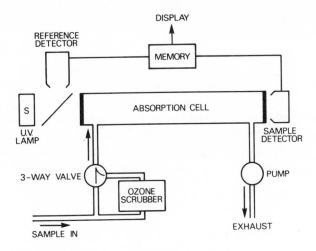

Fig. 5.4 Schematic diagram of O_3 photometer (Dasibi).

amount of incident radiation. This operation is carried out alternatively on the direct and filtered air sample, the latter providing a unique internal reference system, allowing direct measurement of the fraction of the incident light absorbed by O_3. The O_3 concentration can then be determined using the Beer–Lambert relationship for weak absorption:

$$\log \frac{I_0}{I} = \epsilon_{253.7} \times l \times [O_3]$$

The absorption data is processed electronically to provide a direct reading of O_3 concentration in ppb. Minimum detectable concentration is of the order of 1 to 2 ppb.

Two features of the system may be disadvantageous under some circumstances. Firstly the sequential mode of operation and signal integration system give the instrument a relatively long time-constant for measuring a change in O_3 concentrations. In practice a digital output, updated every 10 or 22 s is obtained, giving a 90% rise or fall time of 30 s. Whilst this is adequate for most ambient monitoring applications, it is not satisfactory for many laboratory applications. Furthermore, the sample flow requirement (1 to 4 litres min⁻¹) is quite large, which again may be a disadvantage in laboratory work. A more serious disadvantage is the potential interferences from some orgainc compounds e.g. carbonyl and aromatic compounds which also absorb in the u.v. region employed. Mercury vapour also interferes by virtue of its extremely strong absorption of the resonance 253.7 nm line. These interferences are, however, unlikely to be troublesome when making measurements in outdoor air.

Infrared absorption

Infrared absorption spectrometry has been used for the measurement of O_3 under laboratory conditions e.g. in smog-chambers. Long path lengths have to be employed since the extinction coefficient of O_3 in the i.r. ($\epsilon = 3.80 \times 10^{-4}\,ppm^{-1}\,m^{-1}$ at $1054\,cm^{-1}$) is lower than that in the u.v. ($\epsilon = 1.3 \times 10^{-2}\,ppm^{-1}\,m^{-1}$ at $254\,nm$). The advent of i.r. laser sources offers possibilities for remote sensing of O_3 in ambient air by i.r. absorption spectrometry. Hanst [29] discusses the possibility of using a CO_2 laser, wavelength shifted to $9500\,nm$ with propane, for O_3 measurement. An estimated detection limit of $0.05\,ppm$ in a 1 km path is given.

5.3.2.3 Measurement of O_3 by the C_2H_4-chemiluminescence method

Principle

O_3 reacts rapidly in the gas phase with C_2H_4 with an accompanying chemiluminescence emission in the 350 to 600 nm wavelength region. By monitoring this light emission with a sensitive photomultiplier, when air containing O_3 is mixed with C_2H_4 in a flow cell at atmospheric pressure, a signal is obtained which is proportional to the O_3 concentration.

The method is suitable for the measurement of O_3 concentrations in the range 0.001 to 100 ppm and the response is linear in this range provided the sample and C_2H_4 flow rates are maintained constant. At the lower end of the concentration range the response is limited by the thermal noise and drift of the photomultiplier tube. The latter may be minimized by thermoelectric cooling of the photomultiplier.

There are no known interferences with this O_3 method. An interfering substance would either have to produce chemiluminescence by reaction with C_2H_4 at atmospheric pressure or interfere with the reactions giving rise to the chemiluminescence from O_3 reaction. No known compounds present in atmospheric air have been demonstrated to produce these effects.

The precision of the method relies on maintaining constant gas flow rates and suppression and compensation for the 'dark' current of the photomultiplier tube. A precision of $\pm 2\%$ at a 50 ppb O_3 concentration can be readily achieved. The accuracy is dependent on the validity of the calibration which is carried out using a source of known O_3 concentration. Standard O_3 sources are usually calibrated using the neutral KI method (see Section 5.3.1.2) and are, therefore, subject to the uncertainty of $\pm 5\%$ inherent in this chemical method.

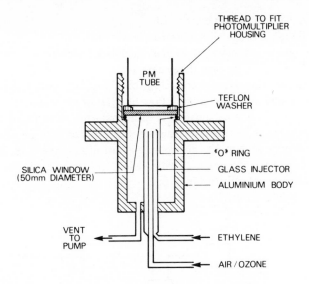

Fig. 5.5 Detector assembly for chemiluminescence O_3 analyser.

Apparatus

The apparatus consists of a reaction cell-photomultiplier detector assembly, a high voltage supply, amplifier and recorder, and ancillary equipment for providing controlled flows of sample gas and C_2H_4.

Detector assembly: The arrangement for a suitable detector assembly is illustrated in Fig. 5.5. The body of the reaction cell may be either of glass (as in the design of Warren and Babcock [53]) or of aluminium, with the quartz end window sealed on [either with cement or an 'O' ring] to give a gas tight fit. The pyrex-glass injector is sealed into the cell so that the tip is between 5 and 10 mm from the end window. It is necessary to exclude all stray light from the detector and this is most conveniently achieved by adapting the reaction cell assembly to fit onto a standard photomultiplier housing. Alternatively, the whole assembly may be enclosed in a light-tight box. The inlet and outlet gas lines should be covered with black slieving to prevent light entry.

A suitable photomultiplier is EMI 9635 QD (low dark current 13 dynode tube with quartz window and S2 cathode). This tube has a dark current of approximately 0.5 nA at 1000 volts (20 °C) and gives a response of 50 nA (ppm O_3^{-1}). The photocurrent can be amplified with a conventional d.c. amplifier with an input range of 10^{-9} to 10^{-5} A. The high voltage supply unit should be capable of giving a constant voltage of up to -1000 V d.c. The output can be displayed on a recorder.

The ancillary equipment consists of:

(a) a PTFE sample probe with a filter to prevent entry of extraneous dust;

(b) an air pump and metering system capable of drawing a constant flow of 1.0 ± 0.02 litre min^{-1} through the system and

(c) a C_2H_4 metering system consisting of a capillary orifice flow meter and rotameter capable of maintaining a flow of C_2H_4 of approximately 10 ml min^{-1} constant to $\pm 2\%$. A good quality pressure reducing valve fitted to the C_2H_4 supply cylinder can be used to maintain the required pressure at the capillary orifice. It may be advisable to incorporate a dust filter in the C_2H_4 line upstream from the capillary. Commercial purity grade C_2H_4 can be used. The flow system is shown diagrammatically in Fig. 5.6.

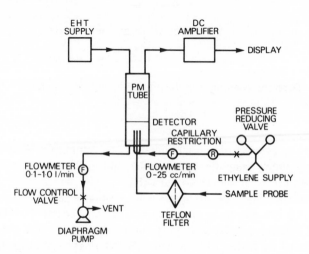

Fig. 5.6   Flow diagram for chemiluminescence O_3 analyser.

Operation

Before the instrument can be used for measurement of O_3 concentration it is necessary to establish and compensate for the photomultiplier 'dark' current and also to calibrate the electrical response as a function of O_3 concentration.

The 'dark' current is measured by either supplying the instrument with O_3 free air or by operating with the pump off so that no air enters the reaction cell. It is inadvisable to check the 'dark' current by stopping the C_2H_4 flow as a low level irreproducible response can be observed when O_3/air mixtures are present in the reaction cell even if C_2H_4 is nominally absent. A suitable EHT of, say, 800 V, is chosen and the 'dark' current measured after the photomultiplier has reached temperature equilibrium. The instrument is then supplied with air containing a constant known O_3 concentration of approximately 0.2 to 1.0 ppm, supplied from a suitable O_3 generator (see below) and the signal allowed to stabilize. The response may then be adjusted up or down to give the desired sensitivity by adjusting the EHT. The dark current must be remeasured after adjusting the EHT. The calibration factor F is then given by

$$F = \frac{I - I^0}{C_{O_3}} \, nA \, ppm^{-1}$$

where I^0, I are respectively the dark current (or equivalent recorder deflection) and the total observed photocurrent (or recorder deflection) with the known concentration of O_3 (C_{O_3}) entering the instrument.

If the instrument is equipped with a thermostatted photomultiplier, the dark current may be backed off electrically to give a zero recorder reading when no O_3 is present. If no temperature control is used, it will be necessary to check the dark current periodically to compensate for zero drift due to temperature fluctuations. For continuous monitoring applications, this may be achieved conveniently by incorporating a programmed solenoid valve which vents the pumping line for a short period (e.g. 2 min) at intervals (e.g. every 30 min) thereby stopping the flow of O_3/air into the reaction cell. If it is proposed to operate in an environment with large ambient temperature variations $> 10°$ C, a thermostatted, cooled detector assembly is desirable.

When the instrument is calibrated initially, the linearity should be checked by measuring the response at different O_3 concentrations within the operating range. For atmospheric measurements, response over the range 0.02 to 1.0 ppm should be checked; at the lowest concentrations some O_3 losses in the sample line may occur giving rise to a fall-off in response and/or longer periods to reach a steady reading. This is often most noticeable when a dust filter is incorporated in the sample level. It is advisable to condition the system with higher O_3 concentrations for a few hours before calibration to minimize any effects of this kind. After this initial check for linearity, subsequent routine calibration need only be carried out at a single O_3 concentration. For accurate monitoring, a weekly check on the response should be carried out.

C_2H_4 chemiluminescence O_3 detectors have been operated continuously for long periods. When operating in urban locations, the filter should be replaced fairly frequently and it may be necessary to clean the reaction cell periodically to remove any accumulated deposits.

5.3.2.4 Preparation of O_3/air mixtures for calibration purposes

Preparation of mixtures containing ppm concentrations of O_3 in air is conveniently achieved by exposing air or O_2 to a mercury vapour lamp emitting short

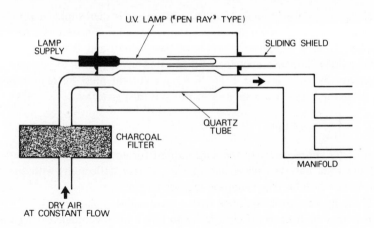

Fig. 5.7 O$_3$ generator.

wavelength (185 nm) radiation. O$_3$ formation results from the photodissociation of oxygen: $O_2 + h\nu \, (\lambda = 185\,\text{nm}) = 20$; $O + O_2 = O_3$. Two suitable types of light sources are available commercially. The filament type operates on a low d.c. voltage (e.g. Philips OZ4) and the low pressure arc type (e.g. 'Pen ray' lamp, Ultraviolet Products, Inc.) is operated on a.c. supplied from a high voltage leak transformer.

Fig. 5.7 shows a schematic diagram of a simple form of O$_3$ generator suitable for producing constant O$_3$ concentrations in the range 0.1 to 1.0 ppm in an air stream. Pure, dry air is passed continuously, at a constant flow rate, through a quartz tube which is positioned alongside a miniature mercury arc lamp. The ozonized air passes into a sampling manifold. Tubes downstream from the generator should preferably be of Teflon but clean borosilicate glass may be used. The O$_3$ concentration can be varied by adjusting either the lamp current, the air flow rate, or the length of the arc tube exposed. A sliding shield may be used to alter the length of the arc tube.

The O$_3$ concentration from the generator can be determined using the neutral KI method. After allowing the generator to stabilize by operating under constant conditions for several hours bubbler samples are taken from the manifold and analysed according to the procedure given in Section 5.6.3. The initial calibration should be carried out at several different concentrations, so that the characteristics of the generator can be established. Provided operating conditions i.e. temperature, flow rate, supply voltage are carefully controlled, reproducible O$_3$ concentrations can be obtained with this type of generator. In fact, it has been shown that an O$_3$ source of this type, once initially calibrated, provides a more reliable routine calibration for O$_3$ detectors than manual iodometric analysis [54].

As an alternative to the above dynamic method, a static system may be used to obtain O_3/air mixtures for calibration. In this case the O_3/air mixture from the generator is collected in large bags fabricated of inert plastic film (e.g. Teflon or Tedlar) where it can be diluted with clean air if required. It may be necessary to precondition the bag by exposure to higher O_3 concentrations to prevent surface destruction of the O_3. The O_3 concentration in the bag should be determined using neutral KI manual analysis prior to use for calibration purposes. The general procedures for the preparation of gas mixtures in static systems are described in Chapter 4, Section 3.8.1.

5.3.3 Analysis of H_2O_2

Although it has been suggested that H_2O_2 may constitute a significant fraction of the oxidants present in polluted air very little progress has been made in the development of techniques for the measurements of air concentrations of H_2O_2. Gaseous H_2O_2 is well known for its instability, particularly in the presence of metal surfaces. This, combined with the high affinity for H_2O of H_2O_2, makes quantitative sampling by conventional techniques very difficult.

5.3.3.1 Chemical methods

H_2O_2 is an oxidant and can be measured by the neutral KI technique (Section 5.3.1) if ammonium molybdate is added as a catalyst to assist rapid colour development [34]. Methods specific for H_2O_2 and organic peroxides which yield H_2O_2 on acid hydrolysis, based on the formation of a titanium (4^+) peroxysulphate complex have been reported [34, 55]. Sensitivity is hardly adequate for the measurement of H_2O_2 in polluted air, however. (The molar absorbitivity of the Ti (4^+)–peroxide complex at 407 nm is only 777 compared to 2.4×10^4 for the I_3^- complex formed when O_3 reacts with neutral KI [34].) Cohen and Purcell [56] have reported another method for the determination of microgram quantities of H_2O_2. The analytical procedure uses the coloured complex obtained upon extraction of titanium–H_2O_2 mixtures with 8-quinolinol in chloroform. Air containing H_2O_2 is sampled through a fritted bubbler containing 10 ml of a solution of titanous sulphate. The absorbing solution is then extracted with an 0.1% solution of 8-quinolinol in chloroform and the absorbance of the complex measured at 450 mμ. The method is claimed to be very specific and to have a sensitivity at least 4 times greater than that of the titanous sulphate method. (Molar absorbitivity $\simeq 3 \times 10^3$ at 450 mμ.) Neither method appears to have been widely utilized, although Bufalini et al. [57] report measurement of H_2O_2 concentrations of up to 150 ppb in polluted air, using the Ti (4^+)-8-quinolinol method.

5.3.3.2 Physical methods

The only physical method which has been applied to the measurement of H_2O_2 in air is i.r. absorption spectroscopy. Hanst *et al.* [28] have used a 417 m path length Fourier transform i.r. instrument to determine H_2O_2 by its adsorption at 1250 cm^{-1} ($\epsilon = 9 \pm 3 \, atm^{-1} \, cm^{-1}$). In atmospheric spectra, interference from water vapour and methane resulted in a rather high detection limit of the order of 100 ppb. In polluted air in Pasadena, California, no H_2O_2 could be detected at this concentration but by comparison of spectra recorded at different times during the day, evidence was found on one occasion for the accumulation of approximately 70 ppb H_2O_2 during the development of smog. This method seems to offer the best prospects at the present time for selective measurement of this difficult gas in polluted air.

5.3.4 Analysis of aliphatic aldehydes and oxygenated compounds

5.3.4.1 Chemical methods

Chemical methods for the analysis of aldehydes are based on the chemical reactivity of the carbonyl group. The results from analyses by these methods are often reported as 'total aldehydes' with the response to ketones ignored. Formation of the bisulphite addition complex, followed by titration of the 'trapped' bisulphite, has been widely used for industrial hygiene applications and source measurements, but the method is not sensitive enough for ambient air analysis. Aqueous sodium bisulphite is a useful trapping reagent for collecting low molecular weight aldehydes for measurement using more sensitive colorimetric techniques. However, trapping efficiency for higher molecular weight aldehydes and ketones is less satisfactory.

A useful method for the determination of total aliphatic aldehydes in air was first described by Sawacki *et al.* [58]. The aldehydes are determined colorimetrically by their reaction with 3 methyl-2-benzothiazolone hydrazone hydrochloride (MBTH) in the presence of ferric chloride to form a blue cationic dye in acid media. Air is sampled directly into the MBTH reagent and the sensitivity is sufficient for measurement of ppb concentrations of aldehydes in a 100 litre air sample. Because the contribution to the measured extinction, and also the sampling efficiency for the higher aldehydes is less than for formaldehyde (HCHO), the overall molar response depends on the relative proportions of the individual aldehydes present. For normal outdoor air samples, in which 60 to 80% of the aldehydes are HCHO, Altshuller and Leng [60] have suggested that the measured concentration calculated as HCHO should be multiplied by 1.125 ± 0.10 to obtain the real concentration of total aliphatic aldehydes in outdoor air. This is based on the molar absorbivities of 65 000, 50 000 and 23 000 for HCHO, straight chain aliphatic aldehydes and branched or unsaturated aldehydes respectively [58].

Specific colorimetric methods have been developed for HCHO and acrolien. Procedures involving the use of chromotropic acid for the analysis of HCHO have been widely investigated [61–64]. There is no significant interference from higher aliphatic aldehydes, ketones, alcohols etc. and the interference from olefins and aromatic hydrocarbons can be largely eliminated by the use of appropriate sampling conditions. Thus, for higher concentrations of HCHO as found for example in combustion effluent, etc., collection in aqueous bisulphite is recommended. For measurements in the ambient atmosphere, where HCHO concentrations are unlikely to exceed a few tenths of a part per million, sampling directly into chromotropic acid/H_2SO_4 reagent is most advantageous. The sensitivity is sufficient for the detection of 10 ppb HCHO in a 100 litre air sample, sampled directly into the reagent. A slight interference from aromatic hydrocarbons and olefins may occur if the concentration of these compounds exceeds that of HCHO by a factor of 5 to 10.

Acrolien can be measured colorimetrically using 4-hexylresorcinol [65]. The reaction between these two compounds in an ethyl alcohol–trichloro-acetic acid solvent medium in the presence of mercuric chloride results in a blue coloured product with a strong absorption maximum at 605 nm. The sensitivity is adequate for the determination of ppb concentrations of acrolien in a 60 litre air sample, with no significant interference from common inorganic and organic pollutants. Air is sampled directly into the mixed sampling reagent contained in 2 fritted bubblers in series and the colour developed by heating to $60°$ C for 15 min.

5.3.4.2 A colorimetric analysis of total aliphatic aldehydes in air (MBTH method) [58, 59]

Principle

The aldehydes in ambient air are collected in an 0.05% aqueous solution of 3-methyl-2-benzothiazolone hydrozone hydrochloride (MBTH). The resulting azine is then oxidized by ferric chloride–sulphamic acid solution to form a blue cationic dye in acid solution which can be measured at 628 nm. The concentration of total aldehydes is calculated in terms of HCHO. Normally between 60 to 80% of the aldehydes occurring in outdoor air are HCHO and acrolein. Other aldehydes which may be present are higher aliphatic aldehydes and, to a lesser extent, aromatic aldehydes. Because the contribution to the measured extinction and also the sampling efficiency for the higher aldehydes is less than that for

HCHO, the calculated concentration should be multiplied by 1.25 ± 0.10 to obtain a real concentration of total aliphatic aldehydes in ambient air [60].

The method is relatively free from interferences. Thus, none of the variety of organic and inorganic materials present in photochemical smog, generated by laboratory irradiation of diluted vehicle exhaust (e.g. hydrocarbons, ketones, nitrogen oxides, O_3, PAN) gave detectable interference [60].

From 0.03 to $0.7\,\mu g\,ml^{-1}$ of HCHO can be measured in the colour developed solution (12 ml). This corresponds to a minimum detectable concentration of 0.03 ppm aldehyde (as HCHO) in a 25 litre air sample. The reproducibility of the method is to within $\pm 5\%$.

Apparatus

The samples are collected in all glass bubblers with a coarse fritted inlet. It is normally necessary to sample ambient air for quite long periods, to obtain sufficient aldehyde for analysis and a pump capable of drawing at least $0.5\,litre\,min^{-1}$ for 24 h is required. A critical orifice, a rotameter or a gas meter can be used to meter the flow.

Reagents

Analytical reagent grade chemicals should be used.

(a) 3 Methyl-2-benzothiazolone hydrazone hydrochloride absorbing solution (0.05%): Dissolve 0.5 g MBTH in distilled water and dilute to 1 litre. The reagent may become turbid either in storage or during sampling. If this occurs, filter by gravity. The solution is stable for a week or more if stored in the cold in a dark bottle.

(b) Oxidizing reagent: Dissolve 1.6 g sulphamic acid and 1.0 g ferric chloride in distilled water and dilute to 100 ml.

(c) Formaldehyde stock solution $(1\,mg\,ml^{-1})$: Dilute 2.7 ml of 37% formalin solution to 1 litre with distilled water. This solution must be standardized by the addition of excess sodium bisulphite to an aliquot of the solution, followed by iodometric titration of the formaldehyde bisulphite addition product after the excess bisulphite has been oxidized by the addition of I_2. Alternatively, sodium formaldehyde bisulphite can be used as a primary standard. Dissolve 4.470 g in distilled water and dilute to 1 litre. Stabilize for at least 3 months.

(d) Dilute standard formaldehyde solution $(10\,\mu g\,ml^{-1})$: dilute 1 ml of standard stock solution to 100 ml with distilled water. Prepare a fresh solution daily.

Procedure

Sample a measured volume of ambient air at a rate of 0.5 litre min^{-1} through 35 ml of MBTH solution in the absorber until sufficient aldehyde for analysis has been collected. In clean air a period of up to 24 h may be required. An average collection efficiency of 84% has been determined for sampling under these conditions although higher efficiencies have been reported for slightly modified conditions. Altshuller and Leng [60] obtained efficiencies of between 90 and 95% using a 10 ml of 0.2% aqueous MBTH in a bubbler aspirated at 1 litre min^{-1}.

After sampling, make up the volume of the absorbing solution to exactly 35 ml with distilled water (to compensate for evaporation losses) and allow to stand for 1 h. For analysis, pipette 10 ml of the sample solution to a glass stoppered tube and an equal volume of unexposed reagent to a second tube to serve as a blank. To each, add 2 ml of oxidizing solution and mix well. After standing for at least 12 min, determine the absorbance of the sample at 628 nm against the reagent blank in 1 cm cells. The aldehyde content (expressed as μg HCHO ml^{-1}) can be determined from the calibration plot prepared as described below.

Calibration

The dilute standard HCHO solution, freshly prepared, is used to calibrate the method. Pipette 0, 0.5, 1.0, 3.0, 5.0 and 7.0 of this solution into a series of 100 ml volumetric flasks and dilute to volume with 0.05% MBTH solution. Allow these solutions to stand for 1 h and then transfer 10 ml aliquots of each solution to a stoppered test tube, add 2 ml of oxidizing solution and mix well. After 12 min determine the absorbance of each solution at 628 nm against the blank in 1 cm cells. Plot the absorbance against μg HCHO ml^{-1} of solution.

The air concentration of total aliphatic aldehydes (as HCHO) is given by

$$\text{ppm (volume)} = \mu\text{g ml}^{-1}\ \text{HCHO} \times \frac{35 \times 24.45}{V \times M \times E}$$

where V = volume of air sampled (at 760 Torr and 25° C); M is the molecular weight of HCHO (= 30.03) and E is the collection efficiency of the bubbler which can be established by using 2 bubblers in series. If a course fritted bubbler is used, E may be taken to be 0.84. In order to obtain the real concentration of aldehydes in ambient air the above result should be multiplied by 1.25 to allow for the lower molar extinction of the higher aldehydes when determined using MBTH.

5.3.4.3 Colorimetric analysis of HCHO (chromotropic acid method)

Principle

HCHO reacts with chromotropic acid (1,8 dihydroxynaphthalene-3,6-disulphonic acid) in concentrated H_2SO_4 solution to form a purple coloured dye, which is determined spectrophotometrically at 580 nm. The HCHO may be collected either in water, in aqueous bisulphite or directly in the chromotropic acid– H_2SO_4 solution.

Using the procedure given below, involving collection in water, from 0.1 to $2.0\,\mu g\,ml^{-1}$ HCHO can be measured in the final solution. This corresponds to a minimum detectable quantity of 0.1 ppm in a 40 litre air sample collected in 20 ml water. The sensitivity can be increased ten-fold by sampling directly into chromotropic acid–sulphuric acid solution (see below).

The chromotropic acid method has very little interference from other aldehydes ($<0.01\%$ from saturated aldehydes and a few % from acrolien). Alcohols, phenols, aromatic hydrocarbons and olefins all show a negative interference but the concentrations of these classes of compound in ambient air are too low to lead to serious error. Possible interference from this source should be considered when analysing for HCHO in combustion effluent, etc., however. The reproducibility of the method is to within ± 5%.

Apparatus

2 course-fritted bubblers in series are used to collect the samples. A pump capable of maintaining a metered flow of 1 litre min^{-1} through the sampling train for up to 24 h is required.

Reagents

(a) Chromotropic acid reagent: Dissolve 0.10 g of chromotropic acid disodium salt in water and dilute to 10 ml. Filter if necessary and store in the dark. Make up a fresh solution weekly.
(b) Concentrated H_2SO_4 (sp. gr. = 1.86).
(c) Standard HCHO solutions–prepare as for MBTH method, Section 5.3.4.2.

Procedure

Place 20 ml distilled water in each bubbler, connect up sampling train, and aspirate at 1 litre min^{-1} for a suitable time to collect enough HCHO for analysis. The collection efficiency in a single bubbler is approximately 80%, i.e. 95% of the HCHO is trapped in the 2 bubblers. For long period sampling (e.g. 24 h) evaporative losses may be considerable. In this case, it may be preferable to use larger volumes of water (e.g. 35 ml and 25 ml respectively in the two bubblers).

After sampling adjust volumes to 20 ml and pipette a 4 ml aliquot of each, to stoppered test tubes. A blank is also prepared using 4 ml distilled water. Add 0.1 ml of 1% chromotropic acid solution and mix. Then add, cautiously, 6 ml concentrated H_2SO_4 from a pipette. Allow the solution to cool and read the absorbance at 580 nm in a 1 cm cell. The colour is remarkably stable but there is a small increase in absorbance of the solutions over a period of a few days. The amount of HCHO present in each aliquot taken can be determined from the calibration curve obtained as described below.

Calibration

A freshly prepared standard HCHO solution containing $10 \mu g$ HCHO ml^{-1} is used to prepare calibration solutions. Pipette 0, 0.1, 0.3, 0.5, 0.7, 1.0 and 2.0 ml of this solution into a series of glass stoppered graduated tubes. Dilute each to 4 ml with distilled water and develop colour as described above. Plot the absorbance against HCHO concentration (μg ml^{-1}) in the colour developed solution.

The air concentration of HCHO is given by

$$\text{ppm (volume)} = \text{total } \mu g \text{ HCHO} \times \frac{24.47}{V \times M}$$

where the total HCHO is the sum of the amounts collected in each bubbler; V is the volume of air sampled (at 760 Torr and $25°$ C) and $M =$ mol. wt. of HCHO ($= 30.03$).

Variation

Altshuller *et al.* [63] have described a more sensitive procedure in which air is sampled directly at 1 litre min^{-1} through an impinger containing 10 ml 0.1% chromotropic acid in concentrated H_2SO_4. Collection efficiency was essentially 100% and, therefore, only a single collection vessel is required. The disadvantages of this method for some applications are the increased absorption efficiency of interfering organic compounds and the handling of the strong acid solutions during sampling and analysis.

5.3.4.4 Physical methods

Although GC has been widely used for the measurement of hydrocarbons, relatively little progress has been made in applying this technique to the analysis of oxygenates in emissions and particularly in atmospheric samples. One of the difficulties arises from the tendency of these compounds to adsorb on containing surfaces, particularly in moist environments. Thus, losses in sampling, storage and transfer can be serious. However, a useful procedure for the GC measurement of aliphatic aldehydes in air has been developed by Levaggi and Feldstern. Air samples are collected in 1% sodium bisulphite solution which serves to separate the aldehydes from hydrocarbons (which are not trapped at all in the aqueous solution) and other oxygenated compounds (alcohols, esters or ketones) which are only trapped with poor efficiency. An aliquot of the absorbing solution is injected onto a GC fitted with a heated injection port containing a short column of solid sodium carbonate. The aldehyde—bisulphite addition compounds are decomposed on this column and the volatile aldehydes carried onto the analytical column where they are separated and detected on a flame ionization detector. HCHO is not detected by this method. Sensitivity is sufficient for the determination of 0.02 to 0.03 ppm acetaldehyde in a 100 litre air sample. Large volume air samples are required for the measurement of aldehydes in relatively clean air by this method.

Although the carbonyl C—H stretching vibration at 3.5 to 3.7 μm should be useful for i.r. analysis of aldehydes, no studies appear to have been reported on the quantitative analysis of aldehydes, either in combustion effluents or in the atmosphere by this technique. Aldehydes have been determined in laboratory 'smog chamber' studies using long-path i.r. spectroscopy, and Fourier transform i.r. spectroscopy has been successfully used to detect HCHO in polluted atmospheres [28].

5.3.5 Analysis of PAN and related compounds

5.3.5.1 Chemical methods

The PANs absorb readily in aqueous sodium hydroxide where they are hydrolysed to yield nitrite ion in quantitative yield [66]. The nitrite ion can be determined colorimetrically using a Griess type reagent (see Chapter 4, Section 2.2). This method has been used as a check on mixtures containing PAN at high concentrations which are analysed using i.r. absorption spectrometry. It is of little practical use for atmosphere measurements because of interference from NO_2, which also yields NO_2^- in alkaline solution, and which is normally present at much higher concentration than the peroxyacyl-nitrates.

5.3.5.2 Physical methods

Infrared spectroscopy has been used to identify and analyse PAN and its higher homologues in smog chamber studies and also in the atmosphere in the Los

Angeles basin using long-path length absorption methods. These compounds show characteristic bands at 5.4, 7.7 and 12.5 μm and PAN also shows a band at 8.6 μm. The 8.6 and 12.6 μm bands are normally used for analytical work, and molar absorbtivities of the various bands for PAN, peroxypropionyl nitrate and peroxybutyryl nitrate have been reported [67]. The 8.6 μm band (absorbtivity = 14.3 x 10^{-4} ppm^{-1} m^{-1}) of PAN is used as a primary standard for measurement of calibration mixtures of this gas in the 200 to 1000 ppm range. The discovery and identification of peroxybenzoyl nitrate (PBzN) in laboratory studies of photochemical reactions involving aromatic hydrocarbons has been made by means of its i.r. absorption spectrum [7], but the concentrations of PBzN in polluted air are likely to be far too low for measurement by i.r. absorption.

The minimum detectable concentration of PAN using 120 m path length i.r. absorption spectroscopy is approximately 0.03 ppm. This sensitivity is insufficient for normal atmospheric concentrations which are of the order of a few ppb or less. The very high sensitivity of the electron capture detector enables direct GC measurement of PAN and related compounds in ambient air samples of a few ml.

Techniques for the measurement of PAN by GC were first described by Darley *et al.* [68] and have subsequently been developed by other workers [69, 70]. Since the peroxyacylnitrates are rather reactive compounds and tend to absorb or decompose on column materials, only a limited number of columns have been found that are of practical use. Suitable stationary phases are Carbowax E400 [68] or PEG 400 (polyethylene glycol) [70] with silanized Chromosorb W or Gas Chrom Z as a support. Column and sampling tubes should be of Teflon although glass and stainless steel have been successfully used. Samples cannot be stored before analysis for PAN. A typical ECD chromatogram of polluted air sample using a 1.8 m x 3 mm pyrex column containing 10% PEG on Gas Chrom Z is shown in Fig. 5.8.

Columns used for the analysis of atmospheric PAN have normally been shorter in length than used in the analysis depicted above. Thus Darley *et al.* [68] used a 9 inch Carbowax E400 and Penkett *et al.* [70] a 1.3 ft PEG 400 column. These give retention times of approximately 1.5 to 2.5 min for PAN with satisfactory separation from other electron absorbing species.

The main difficulty in the quantitative analysis of PAN by ECD gas chromatography lies in calibration of the detector response. This can only be achieved using air mixtures containing known concentrations of PAN of the order of 10 ppb (at concentrations above about 50 ppb in a 2 ml sample the detector response becomes non-linear). Preparation of these mixtures requires accurate dilution of mixtures containing 200 to 1000 ppm PAN which can be standardized by i.r. absorption spectroscopy. Furthermore, high concentration PAN mixtures are not available commercially at the present time, and PAN must be

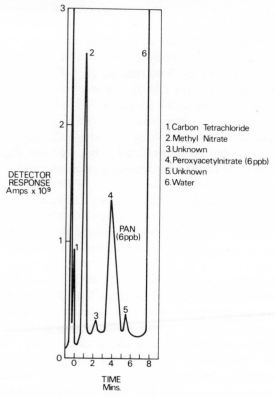

1. Carbon Tetrachloride
2. Methyl Nitrate
3. Unknown
4. Peroxyacetylnitrate (6 ppb)
5. Unknown
6. Water

Fig. 5.8   Chromatogram of electron capturing substances in Cincinnati air [13, 69]. Column: 6 ft × 3 mm pyrex tube packed with 10% PEG on 60/80 mesh Gas Chrom Z at 22° C. Carrier: Argon + 5% methane at 60 ml min^{-1}.

prepared by photolysis of ethyl nitrite in oxygen and purified by preparative GC. The techniques for preparing these mixtures, which have been described by Stephens *et al.* [71] are rather laborious and required extensive laboratory facilities. The pure PANs are violently explosive compounds and suitable precautions should be taken in handling these substances.

In addition to PAN and its higher homologues, alkyl nitrites, nitrates and nitroalkanes have good electron capture responses and can be analysed by electron capture GC. Several alkyl nitrates have been identified in irradiated motor vehicle exhaust and detection of these compounds at the ppb level in the atmosphere is possible.

Efforts to detect PBzN in the atmosphere by electron capture GC have so far been unsuccessful, possibly due to absorption losses and interference from unknown components. A new GC procedure for PBzN has recently been described [72] which involves conversion of PBzN to methyl benzoate by trapping the former in basic methanol solution at $0°$ C. The methyl benzoate is then determined by GC with flame ionization detection. With this technique, atmospheric concentrations of PBzN of less than 0.1 ppb should be detectable.

A potential chemiluminescence method for the analysis of PAN has recently been described by Pitts et al. [73]. They observed a strong chemiluminescence when PAN reacted with triethylamine, either in the gas or liquid phase. In the gas phase, PAN concentrations as low as 6 ppb could be detected. However, O_3 also gave chemiluminescence with triethylamine although the emission intensity maximum was at a different wavelength. Possible procedures for overcoming the O_3 interference were discussed with a view to developing this technique for atmospheric monitoring of PAN.

5.3.5.3 Analysis of PAN by electron capture GC

Principle

PAN is measured in a discrete air sample by an electron capture detector after GC separation from other electron absorbing components in the sample. The procedure given here is based on the techniques used by Darley et al. using modifications recommended by I.S.C. [74].

The method is extremely sensitive, the detection limit being less than 1 ppb on a 2 ml air sample. With conventional type ECD detectors, the concentration range used is limited by the non-linear response when the detector current is reduced by about 25 to 30%. The maximum concentration measurable depends on sample size, column length, temperature and carrier flow rate but under typical operating conditions this occurs at considerably less than 1 ppm PAN. A much extended linear range can be obtained using a detector operating in the pulse frequency feedback mode. There are no known interferences with the analysis. An interfering substance would have to meet three conditions:

(a) It must have a high electron capture cross-section;

(b) It must have a chromatographic retention time very close to that of PAN on the analytical column and

(c) It must be present in the sample at a concentration detectable by this procedure. These conditions eliminate virtually everything.

The detector response is calibrated by analysis of mixtures containing known concentrations of PAN and the accuracy of the method depends on the accuracy with which these mixtures are prepared. The reproducibility of measurement of a given PAN mixture is reported to be within 2% at the 50 ppb level. The

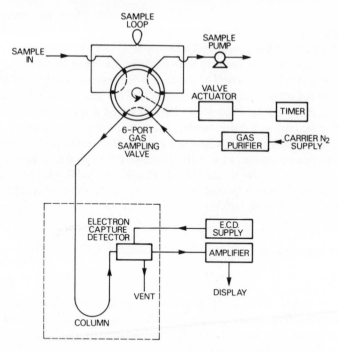

Fig. 5.9 Schematic diagram of GC system for automatic analysis of PAN.

reported overall accuracy of within 5% is probably over optimistic considering the difficulties in calibration for such a reactive substance at very low concentrations.

Apparatus

A flow diagram of the GC system for analysis of PAN is shown in Fig. 5.9. Three major components are required for the analysis.

(1) Sample injection system: A stainless steel (or Teflon) 6-way gas sampling valve (GSV) with an external sample loop (1 to 5 ml) is used to inject air samples into the GC. A manual GSV can be used for laboratory work but for atmospheric monitoring automatic operation of the GSV is necessary. A number of automatic GSVs are available commercially from chromatographic suppliers. These valves are usually operated by a compressed air supply controlled by solenoid valves. The solenoid valves are activated with a suitable cam-timer so that the GSV is switched between the 'sample' and 'inject' position at the required times. A 15 min interval between samples will give a satisfactory record of the ambient

concentration of PAN. A small pump is required to draw the air sample through the sample loop.

(2) Chromatographic column: A column is required to separate PAN from O_2 and from other electron absorbing components which may be present in the sample e.g. higher homologues of PAN, alkyl nitrites and nitrates, organic halogen compounds. This separation may be effected using a 9 inch long column of 1/8 inch i.d. Teflon tubing packed with 5% Carbowax E400 on 100-120 mesh HMDS treated Chromosorb-W. When operated at 25° C with a nitrogen carrier gas flow of 40 ml min^{-1} the retention time for PAN on this column is 60 s.

(3) Electron capture detector: The column can be mounted in any commercial chromatograph equipped with an electron capture detector, with associated amplifier and recorder. In order to avoid excessive use of recorder chart in continuous monitoring, it is preferable to arrange for the recorder chart drive to be on only during the period necessary to record the chromatogram (i.e. 3 min). A second cam-timer can be used for this purpose.

Procedure

Dry nitrogen carrier gas is passed continuously through the sample loop of the GSV, the column, and the detector at 40 ml min^{-1}. The standing current in the ECD should be checked according to the manufacturers' procedures to ensure that the detector is clean and functioning satisfactorily. The automatic sample injection system is then operated and chromatogram recorded. The large air peak appears during the first 30 s following injection followed by the PAN peak at approximately 60 s. A small shoulder on the tail of the PAN peak, assumed to be the higher homologue, peroxypropionitrate, may also appear.

Calibration

Calibration is based on comparison of the sample peak heights with those obtained by injection of mixtures containing known concentrations of PAN. The latter are prepared by dilution of mixtures containing high concentrations of PAN measured by i.r. spectrometry. Since concentrations of a few hundred ppm are required for accurate i.r. measurement, quantitative dilution by a factor of 10^4 must be carried out.

PAN is synthesized, purified as described by Stephens *et al.* [71] and stored in stainless steel cylinders. The cylinders, pressured to 100 psig with N_2 contain 500 to 1000 ppm PAN and are stored at 16° C. An i.r. spectrophotometer with a 10 cm cell is employed to determine the concentration of PAN in the cylinder gas using the absorptivity of $13.9 \times 10^{-4}\,\text{ppm}^{-1}\,\text{m}^{-1}$ at $8.6\,\text{m}$ $(1161\,\text{cm}^{-1})$.

i.e. Concn. (ppm) = Absorbance/$(0.1\,\text{m} \times 13.9 \times 10^{-4})$.

A flow dilution method can be used to reduce the concentration of PAN accurately to a level suitable for calibrating the GC system (10 to 50 ppb). Fig. 5.10 shows a suitable system involving two 100 : 1 dilutions in activated-charcoal-filtered air. The concentration of PAN in the effluent can be varied by adjusting the dilution at either stage. Samples from the diluted effluent are injected on to the chromatograph and the peak height recorded for a series of calculated concentrations.

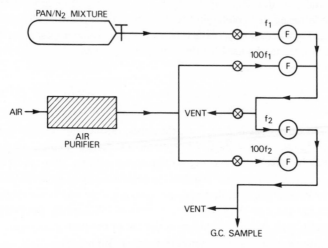

Fig. 5.10 Flow dilution system for preparing low concentrations of PAN.

Alternatively an exponential dilution system may be employed (see Chapter 4, Section 8.) In this case a single aliquot of the i.r. calibrated gas mixture is required. This enables an alternative simplified method to be used for the preparation of PAN. Small quantities of PAN are prepared directly in a 10 cm i.r. cell with a Pyrex glass body by photolysing traces of ethyl nitrite (C_2H_5ONO), in pure O_2. The PAN formed is then measured in the cell without separation from unreacted nitrite or other products. The cell is first flushed with pure O_2 and then small amounts (50 μl) of vapour from the above liquid C_2H_5ONO are added at 15 min intervals whilst the mixture is irradiated with 'Blacklight' fluorescent u.v. lamps (radiation of 300 to 400 nm is required). The C_2H_5ONO concentration must be kept low to maximize the yield. When sufficient PAN has accumulated (as measured from the i.r. absorption at 8.6 μm), an aliquot of the mixture which now contains a known concentration of PAN is injected into an exponential dilution flask connected to the appropriate flow system (Chapter 4). Since the concentration of PAN in the effluent from the flask decreases in a known manner, a calibration may be carried out by repeated injecton of samples from the effluent.

The relationship between peak height and PAN concentration should be linear up to approximately 50 ppb and, therefore, the concentration in unknown samples within this range can be calculated by multiplying the peak height by a constant. The constant varies from one instrument to another and varies inversely with changes in the detector standing current. The automatic system can be operated continuously over periods of months. Occasional cleaning of the electron capture detector is desirable since contamination of the detector causes a slow decline in standing current which leads to a reduction in sensitivity. It is desirable to calibrate the instrument about once a week to compensate for the gradual reduction in sensitivity. When the detector is thoroughly cleaned the original sensitivity is regained.

5.3.6 Analysis of oxyacids of N

Although the oxyacids of N, in the gaseous state, are widely recognized as being important products of photochemical reactions, there has been relatively little progress in developing specific analytical techniques for these compounds. A possible exception is the recent identification and measurement of HNO_3 vapour in the lower stratosphere, using both i.r. spectroscopic techniques [26, 75] and chemical analysis of filter samples [76]. Recently attempts have been made to measure HNO_3 vapour concentrations in photochemical smog using Fourier transform i.r. spectroscopy [28]. The latter technique has been successfully applied to measure HNO_3 in laboratory smog chamber systems. HNO_2 has not been detected using the new i.r. methods either in the atmosphere or in smog chambers, but its detection by these techniques should be possible.

Both HNO_2 and HNO_3 are extremely reactive; water soluble vapours and absorption losses are likely to present serious problems during sampling and analysis by conventional techniques.

5.3.6.1 Chemical methods

Laboratory measurements of low concentrations of HNO_2 were first described by Nash [77]. HNO_2 was determined as nitrite ion (NO_2^-) using a modified Griess–Saltzmann reagent, following its absorption in dilute aqueous NaOH contained in a simple impinger. By deliberately using a fast sampling flow rate ($\geqslant 1$ litre min^{-1}) absorption of NO_2, which also yields NO_2^- in aqueous alkali, was made inefficient. Less than 10% of the NO_2 was absorbed at the 0.1 ppm level whereas HNO_2 was trapped essentially quantitatively in a single bubbler. Thus HNO_2 could be measured in the presence of NO_2. However, any other nitrogen compounds which yield nitrite or alkaline hydrolysis (e.g. peroxyacyl-nitrates) can potentially interfere. Nash [78] has also reported measurements of HNO_2 in the atmosphere using this technique.

HNO_3 vapour can be measured in principle by absorption in water or dilute alkali, followed by analysis for nitrate ion (NO_3^-) using colorimetric or other techniques. Possible interference may come from particulate nitrates, NO_2 (which yields both NO_2^- and NO_3^- on alkaline hydrolysis) and organic nitrates. Filtration samples were used by Lazrus et al. [76] in their measurements of stratospheric HNO_3, which was determined by analysis of NO_3^- in the filter sample extracts. It was found that paper filters composed of cellulose fibres impregnated with dibutoxyethylphthalate (IPC 1478 paper) provided an efficient trapping medium for gaseous HNO_3. Polystyrene filters failed to trap HNO_3 vapour but would, of course, trap particulate nitrate. Measurement of the NO_3^- from samples collected on both types of filter provided a measure of gaseous HNO_3. This technique may not be applicable for measurements in the lower atmosphere because of complications arising from the much higher water vapour concentrations.

5.3.6.2 Physical methods

Commercial chemiluminescence oxides of nitrogen analysers employing carbon, molybdenum or stainless steel converters are non-specific for the determination of NO_2 (see Chapter 4). The instruments not only respond to NO_2 but also to PAN and a variety of other organic nitrates and nitrites [79]. The response of chemiluminescence oxides of nitrogen analysers to HNO_2 [80] and HNO_3 [79, 81] has also been investigated.

HNO_2 gives a quantitative response in an 'NO_x' analyser employing a stainless steel converter although some evidence of absorption in the sample lines was found giving rise to a longer than normal time constant for measurement of 'total NO_x' when HNO_2 was present. By passing the sample through a bubbler containing aqueous alkali HNO_2 could be selectively removed, and its concentration could be determined in the presence of NO and NO_2 by difference. The response of commercial NO_x analysers to HNO_3 appears to be non-quantitative. When HNO_3 vapour is sampled into a chemiluminescence analyser, a low and irreproducible response is observed. Adsorption and chemical attack on the instrument components is probably responsible for these effects.

The most promising technique at the present time for measurement of low concentrations of HNO_3 in the gas phase is i.r. absorption spectroscopy. The

HNO_3 absorption band most sensitive for atmospheric analysis is centered at about $880\,cm^{-1}$ with two distinctive peaks located at 879 and $896\,cm^{-1}$. At these frequencies water vapour interference is not serious. Using a Fourier transform i.r. spectrometer with a scanning Michelson interferometer and a folded path of 400 m, Hanst et al. [27] claim a detection limit at least as low as 10 ppb for HNO_3 in ambient air.

References

[1] Leighton, P.A. (1961). *Photochemistry of Air Pollution*, Academic Press, New York.
[2] Demerjian, K., Kerr, J.A. and Calvert J.G. (1974). *The Chemistry of Photochemical Smog* (ed. Pitts, J.N. and Metcalfe, A.) *Adv. in Envir. Sci. and Tech.* 4, 1.
[3] U.S. Dept. of Health, Education and Welfare (1970). *Air Quality Criteria for Photochemical Oxidants*, Natl. Air Pollut. Control. Admin. Publication No. AP-63.
[4] Taylor, O.C. (1969). *J. Air Pollut. Control Ass.* 19, 34.
[5] U.S. Dept. of Health, Education and Welfare (1969). *Air Quality Criteria for Particulate Matter*, Natl. Air Pollut. Control. Admin. Publication No. AP-50.
[6] U.S. Dept. of Health, Education and Welfare (1969). *Air Quality Criteria for Sulphur Oxides*, Natl. Air Pollut. Control Admin. Publication No. AP-49.
[7] Huess, J.M. and Glasson, W.A. (1968). *Environ. Sci. Technol.* 2, 1109.
[8] Hersch, P. and Deuringer, R. (1963). *Analyt. Chem.* 35, 897.
[9] Brewer, A.W. and Milford, J.R. (1960). *Proc. R. Soc. (Lond.)* A256, 470.
[10] Mast, G.M. and Saunders, H.E. (1962). *I.S.A. Trans.* 1, 325.
[11] Wartburg, A.F., Brewer, A.W. and Lodge, Jr., J.P. (1964). *Int. J. Air Water Pollut.* 8, 21.
[12] Potter, L. and Duckworth, S. (1965). *J. Air Pollut. Control Ass.* 15.
[13] Altshuller, A.P. (1966). *Atmospheric Analysis by Gas Chromatography*, in Adv. in Gas Chromatog. 5, 229.
[14] Fishbein, L. (1973). *Chromatography of Environmental Hazards*, Vol II, Elsevier Scientific, Amsterdam, The Netherlands.
[15] Levaggi, D.A. and Feldstein, M. (1970). *J. Air Pollut. Control Ass.* 20, 312.
[16] Lovelock, J.E. (1963). *Analyt. Chem.* 35, 474.
[17] Wentworth, W.E. (1971). *Recent Advances in Gas Chromatography* (ed. Domsky, I.I. and Perry, J.A.) p. 185. Marcel Decker, New York.
[18] Aue, W.A. and Kapila, S. (1973). *J. Chromatog. Sci.* 2, 255.
[19] Maggs, R.J., Jaynes, P.L., Davies, A.J. and Lovelock, J.E. (1971). *Analyt. Chem.* 43, 1966.
[20] Boettner, E.A. and Dallos, F.C. (1965). *J. Am. Ind. Hyg. Ass.* 26, 289.
[21] Sullivan, J.J. (1973). *J. Chromatog.* 87, 9.
[22] Lovelock, J.E., Adlard, E.R. and Maggs, R.J. (1971). *Analyt. Chem.* 43, 1962.

[23] Renzetti, N.A. (1956). *J. Chem. Phys.* **24**, 209.
[24] Scott, W.E., Stephens, E.R., Hanst, P.L. and Doerr, R.C. (1957). *Proc. Am. Petroleum Inst. Series III* **37**, 171.
[25] Stephens, E.R., Hanst, P.L., Doerr, R.C. and Scott, W.E. (1956). *Ind. Eng. Chem.* **48**, 1498.
[26] Harries, J.E. (1973). *Nature* **241**, 515.
[27] Hanst, P.L., Lefohn, A.S. and Gay, Jr., B.W. (1973). *Appl. Spectroscopy* **27**, 188.
[28] Hanst, P.L., Wilson, W.E., Patterson, R.K., Gay, Jr., B.W., Cheney, L.W. and Burton, C.S. (1975). *A Spectroscopy Study of California Smog*, E.P.A. Environmental Monitoring Series − EPA-650/4-75-006. U.S. Envir. Protection Agency, Research Triangle Park, N.C.
[29] Hanst, P.L. (1970). *Appl. Spectroscopy* **24**,161; also in *Adv. in Envir. Sci. and Tech.*, (eds. Pitts, Jr. J.N. and Metcalfe, R.L.) (1971) **2**, 92.
[30] Menzies, R.T. (1971). *Appl. Optics* **10**, 1532.
[31] Hodgeson, J.A., McClenny, W.A. and Hanst, P.L. (1973). *Science* **182**, 248.
[32] Byers, D.H. and Saltzmann, B.E. (1958). *J. Am. Ind. Hyg. Ass.* **19**, 251.
[33] Saltzman, B.E. and Gilbert, N. (1959). *Analyt. Chem.* **31**, 1914.
[34] Cohen, I.R., Purcell, T.C. and Altshuller, A.P. (1967). *Environ. Sci. Technol.* **1**, 247.
[35] Guicherit, R., Jeltes, R. and Lindqvist, F. (1973). *Environ. Pollut.* **3**, 91.
[36] Saltzmann, B.E. and Wartburg, A.F. (1965). *Analyt. Chem.* **37**, 779.
[37] Saltzmann, B.E. (1954). *Analyt. Chem.* **26**, 1949.
[38] Intersociety Committee (1970). *Tentative method for the manual determination of oxidising substances in the atmosphere*, 44101-02-70T Health Lab. Sci. **7**, 152.
[39] Lindqvist, F. (1972). *Analyst* **97**, 549.
[40] Byers, D.H. and Saltzmann, B.E. (1959). *Adv. Chem. Ser.* **21**, 93.
[41] Hodgeson, J.A., Baumgardner, R.E., Martin, B.E. and Rehme K.E. (1971). *Analyt. Chem.* **43**, 1123.
[42] Kopczynski, S.L. and Bufalini, J.J. (1971). *Analyt. Chem.* **43**, 1126.
[43] Dorta-Shaeppi, Y. and Treadwell, W.D. (1949). *Helv. Chim. Actu.* **32**, 356.
[44] Nash, T. (1967). *Atmos. Environ.* **1**, 679.
[45] Bravo, H.A. and Lodge, Jr. J.P. (1964). *Analyt. Chem.* **36**, 671.
[46] Hauser, T.R. and Bradley, D.W. (1966). *Analyt. Chem.* **38**, 1529.
[47] Regener, V.H. (1964). *J. Geophys. Res.* **65**, 3975.
[48] Hodgeson, J.A., Krost, K.J., O'Keefe, A.E. and Stevens, R.K. (1970). *Analyt. Chem.* **42**, 1975.
[49] Guicherit, R. (1971). *Z. Anal. Chem.* **256**, 177.
[50] Nederbragt, G.W., Van der Horst, A. and Van Duijn, J. *Nature* **206**.
[51] Kummer, W.A., Pitts, Jr., J.N. and Steer, R.P. (1971). *Environ. Sci. Technol.* **5**, 1045.
[52] Finlayson, B.J., Pitts, Jr., J.N. and Akimoto, H. (1972). *Chem. Phys. Letters* **12**, 495.
[53] Warren, G.J. and Babcock, G. (1970). *Rev. Scient. Instrum.* **41**, 280.
[54] Hodgeson, J.A., Stevens, R.K. and Martin, B.E. (1971). *I.S.A. Trans* **11**, 161.
[55] Pobiner, H. (1961). *Analyt. Chem.* **33**, 1423.
[56] Cohen, I.R. and Purcell, T.C. (1967). *Analyt. Chem.* **39**, 131.

[57] Bufalini, J.J., Gay, Jr., B.W. and Brubacker, K.L. (1972). *Environ. Sci. Technol.* **6**, 816.
[58] Sawacki, E., Hauser, T.R., Stanley, T.W. and Elbert, W. (1961). *Analyt. Chem.* **33**, 93.
[59] Intersociety Committee on Manual Methods of Air Sampling and Analysis (1972). *Methods of Air Sampling and Analysis*, American Public Health Assoc., Washington D.C. p. 199.
[60] Altshuller, A.P. and Leng, L.J. (1963). *Analyt. Chem.* **35**, 1541.
[61] Bricker, C.E. and Johnson, H.R. (1945). *Ind. Engng. Chem. Analyt. Edn.* **17**, 400.
[62] West, P.W. and Sen, B. (1956). *Z. Anal. Chem.* **153**, 177.
[63] Altshuller, A.P., Miller, D.L. and Sleva, S.F. (1961). *Analyt. Chem.* **33**, 621.
[64] Altshuller, A.P., Leng, L.J. and Wartburg, A.F. (1962). *Int. J. Air Water Pollut.* **6**, 381.
[65] Cohen, I.R. and Altshuller, A.P. (1961). *Analyt. Chem.* **33**, 726.
[66] Nicksik, S.W., Harkins, J. and Mueller, P.K. (1967). *Atmos. Envir.* **1**, 11.
[67] Stephens, E.R. (1964). *Analyt. Chem.* **36**, 929.
[68] Darley, E.F., Kettner, K.A. and Stephens, E.R. (1963). *Analyt. Chem.* **35**, 589.
[69] Bellar, T.A. and Slater, R.W. (1965). Proc. 150th Am. Chem. Soc. Meeting, Atlantic City, N.J.
[70] Penkett, S.A., Sandalls, F.J. and Lovelock, J.E. (1975). *Atmos. Envir.* **9**, 131.
[71] Stephens, E.R., Burleson, F.R. and Cardiff, E.A. (1965). *J. Air Pollut. Control Ass.* **15**, 87.
[72] Appel, B.R. (1973). *J. Air Pollut. Control Ass.* **23**, 1042.
[73] Pitts, Jr., J.N., Fuhr, H., Gaffney, J.S. and Peters, J.W. (1973). *Environ. Sci. Technol.* **7**, 550.
[74] Intersociety Committee (1971). *Tentative Method for Analysis of Peroxyacetylnitrate in the Atmosphere*, Health Lab. Sci. **8**, 1.
[75] Murcray, D.G., Kyle, T.G., Murcray, F.H. and Williams, W.J. (1968). *Nature* **218**, 78.
[76] Lazrus, A.L., Gandrud, B. and Cadle, R.D. (1972), *J. Appl. Met.* **11**, 389.
[77] Nash, T. (1968). *Ann. Occ. Hygiene* (Cambridge) **11**, 235.
[78] Nash, T. (1974). *Tellus* **XXVI**, 175.
[79] Winer, A.M., Peters, J.W., Smith, J.P. and Pitts, Jr. J.N. (1974). *Environ. Sci. Technol.* **8**, 13.
[80] Atkins, D.H.F. and Cox, R.A. (1974). A.E.R.E./R7615 HMSO, London.
[81] Stedman, D.H. and Niki, H. (1973). *J. Phys. Chem.* **77**, 2604.

CHAPTER SIX

Hydrocarbons and carbon monoxide

6.1 Introduction

All industrial processes contribute to the pollution of our environment. The greatest man-made contributions to atmospheric pollution (in terms of bulk of pollutants) are, however, processes of combustion in power generation and industrial plants, in automobile and aircraft engines, and in domestic heating. Such processes are responsible for the generation of a wide range of pollutants; amongst those pollutants which have received considerable attention in recent years are unburned and partially oxidized gaseous and particulate species. More recently, interest in this general field has been concentrated on investigations into the identification and measurement of levels for groups of compounds or even individual pollutants.

In this chapter such combustion products are examined in three sections: volatile hydrocarbons; hydrocarbon content of particulate matter; and carbon monoxide. No sampling nor analytical techniques of general applicability to all environments and the extensive range of potential pollutants have as yet been reported. In this chapter, sampling and analytical procedures are considered separately. It should be remembered, however, that the quantification of results will depend of the efficiencies of both procedures, and on their compatability.

When considering volatile hydrocarbons in the atmosphere, the analyst is concerned primarily with alkanes or with lower alkenes and aromatics as a group of unsaturated hydrocarbons. The analysis of such compounds has been the subject of a recent review [1].

Methane, which is the only hydrocarbon found naturally occurring worldwide, has a background concentration in the atmosphere of 1.3 to 1.4 ppm. Other hydrocarbons found in air are derived from a wide variety of sources, the most significant of which include: oil and petroleum refineries and storage depots; chemical production and oil burning industries; commercial and geogenic gas leaks; biological processes [2]; and agricultural and forest burning programmes [3]. In certain forest areas terpenes are emitted by the living trees [4, 5].

At the ppb* concentrations, in which they are normally encountered in the atmosphere the hydrocarbons are relatively harmless to mammals. It has been shown [6], however, that ethylene at concentrations as low as 0.01 ppm, and to a lesser extent other hydrocarbons produce deleterious effects in various species of plant. A more serious problem is the so-called 'photo-chemical smog' [7–9], which is produced under certain meteorological conditions by the reactions of nitrogen oxides with other air pollutants including hydrocarbons leading to the formation of compounds such as ozone, aldehydes, peroxyacylnitrates and alkyl nitrates. These reaction products can cause severe irritation of the eyes and mucous membrane. Visibility may also be reduced by the formation of aerosols of polymer molecules originating from photochemical reactions of hydrocarbons.

Specific hydrocarbons derived from different sources will have varying degrees of impact on the environment. Thus ideally it would be desirable to determine levels of individual hydrocarbons in the atmosphere. Many surveys, however, have been carried out measuring only total hydrocarbon levels, in for example, St. Louis [10], and comparing Paris with other cities [11]. A more significant survey was carried out measuring total hydrocarbon and calculating non-methane hydrocarbon levels in the U.S.A. [12], using an empirical relationship derived from previous measurements. Thus measurements of non-methane hydrocarbon levels are to be preferred to those of total hydrocarbon levels. More data may be obtained by monitoring specific compounds and a study of diurnal patterns for several aliphatic and aromatic hydrocarbons has resulted in their occurrence being attributed to automobile exhaust [13]. A similar range of compounds has been investigated using an alternative method of evaluating vehicular emissions [14], which involved measurements in New York and New Jersey and included calculations to demonstrate the effect of wind direction on hydrocarbon levels. A study in Los Angeles [15] has attempted to correlate hydrocarbon levels measured with possible sources of such species, and the relative significance of sources was quoted as: automobile exhausts 47%, evaporative losses of gasoline 31%, geogenic gas leaks 14%, and leaks in commercial natural gas supply systems 8%. An attempt has been made [16] to produce a distribution profile for individual hydrocarbons (benzene and toluene) over a large area (Toronto) by sampling at twelve sites simultaneously. Overall average concentrations were 13 and 30 ppb respectively but the variation in levels between individual sites reached a factor of 40. The toluene to benzene ratio has been

* 1 ppb = 1 part in 10⁹.

suggested as an indication of the proportion of hydrocarbon pollution in the atmosphere produced by automobiles. Other workers [17, 18] have found similar ratios to those recorded in Toronto. In Zurich, however, a far lower toluene to benzene ratio has been measured [19]. Benzene and toluene have been measured [20] at levels of 23 ppm and 11 ppm respectively in the vicinity of a chemical reclamation plant. A three month survey of a rural community in Eastern Pennsylvania showed that EPA limits for hydrocarbons [21] were not exceeded [22]. Other measurements have been reported covering several molecular weight ranges and include: C_1–C_5 by an automatic method [23], C_1–C_5 amongst other gases associated with peat and coal [24]; C_8–C_{18} in Paris air [25]; C_6–C_{20} in Zurich [19], C_9–C_{20} in Italy [26]; and C_6–C_{10} in the Netherlands [27].

The research outlined above has demonstrated the ubiquity of hydrocarbons in the environment. They occur in great structural variety as contributions from many sources, each source differing in the relative proportions of its component hydrocarbons. Their stabilities vary greatly but most hydrocarbons are sufficiently inert to enter into complex pathways of dispersal, involving transport by air, water, particulates, and by the food chain. Much of the research into the origins and fates of hydrocarbons in the environment has centred on the lower molecular weight alkanes and alkenes, prinicpally because of their relatively facile analysis by gas chromatography, alone or in conjunction with mass spectrometry. Combustion chemistry is notoriously complex and the combustion of fossil fuels for heat, power, and transportation produces most of the organic fraction of the atmospheric particulate matter commonly encountered in urban environments.

Some attention has been paid to the higher molecular weight aliphatic compounds in particulate matter, and compounds in the C_{11} to C_{33} range have been tentatively identified in an atmospheric extract [28]. Organic materials in airborne particulates have also been studied [29] by high resolution mass spectrometry to determine whether hydrocarbon distributions typical of major sources of pollution are readily identified. Thin layer chromatography (TLC) has been used to separate alkane fractions prior to gas chromatographic analysis [30]. The majority of such studies have involved other fractions and have concentrated on the identification of specific compounds, particularly suspected carcinogens of the polycyclic aromatic hydrocarbon (PAH) type. The increasing use of fossil fuels, whose pyrolysis produces complex mixtures of PAH, poses many environmental and public health problems. Environmental samples may contain complex PAH mixtures derived from many sources and it has been stated [31] that:

'No existing method separates and resolves adequately the entire PAH fraction on a mg to μg scale and on samples that contain a very large excess of nonhydrocarbons. Even in petroleum analysis, where large samples are available, a complete analytical resolution of the PAH fraction exceeds the capability of any existing combination of analytical techniques.'

Analyses of PAH mixtures using methods developed by Sawicki *et al.* [32] typically entail various partitioning sequences followed by column chromatography to separate the organic fraction of the particulate matter into subfractions prior to spectrophotometric examination of the subfractions. Other methods involve fluorescence techniques [33], TLC [34] and high-pressure liquid chromatography (HPLC) [35]. Specialized techniques of this type enable the separation and identification of PAH subfractions and have been extensively reported [36]. Typical examples are the analyses of groups of eleven PAH in particulate samples from Baltimore [36] and a characteristic urban area of Budapest [37].

CO is a poisonous gas, but it owes its place in the hierarchy of pollutants to different considerations. Although it is not a stable end-product of balanced combustion, unless the supply of O_2 is insufficient, it represents the penultimate stage in combustion through which all the carbon species must pass. Combustion cannot take place without the intermediate formation of CO. As a combustion-generated pollutant, CO need not be a problem under normal operating conditions except in the spark ignition engine. There, however, it is a serious problem, especially under idling and deceleration conditions. Thus, it is in the context of monitoring motor vehicle emissions that the analysis of CO has received considerable recent attention. Classically the analysis of CO has been performed by non-dispersive infrared spectroscopy (Ndir) [38], which is sensitive enough to measure the levels normally encountered in street air of 1 to 50 ppm. A gas chromatographic technique, although not allowing continuous measurement of CO, has a substantially higher sensitivity. The basis of the method is catalytic reduction of the CO to methane and detection by flame ionization. This principle is used in commercial instruments which measure methane, CO and total hydrocarbons [39]. An analyser for continuous determination of carbon monoxide at levels down to 1 ppm using an electrochemical cell has been described [40].

National air quality standards have been set for the U.S.A. by the Environmental Protection agency and are listed in Table 6.1.

6.2 Sampling of volatile hydrocarbons

6.2.1 Principles of sampling procedures

The accuracy and precision of a sampling technique is of the utmost importance and such a technique must allow one to relate readily the values measured to those which were present in the environment. These requirements are of course true of all analyses and in this application inert materials such as glass,

Table 6.1

Pollutant	Standard	Conditions
Particulate matter	$75\ \mu g\ m^{-3}$	Annual geometric mean
	$260\ \mu g\ m^{-3}$	Max. 24 h conc. not to be exceeded more than once a year
Carbon monoxide	$10\ mg^{-3}$ (9 ppm)	Max. 8 h conc. not to be exceeded more than once a year
	$40\ mg\ m^{-3}$ (35 ppm)	Max. 1 h conc. not to be exceeded more than once a year
Hydrocarbons	$160\ \mu g\ m^{-3}$ (0.24 ppm)	Max. 3 h conc. not to be exceeded more than once a year

polytetrafluorothylene (PTFE) and stainless steel should be used to reduce the adsorption effects on the walls of containers and transfer lines. Particular attention should be given to the cleaning of all apparatus and to the selection and application of any sealing or lubricating materials in order to avoid contamination.

Both the duration of sampling and the time at which samples are taken should be related to the purpose of a particular investigation. Studies on short duration pollution incidents require rapid sampling techniques, whereas in investigations into ambient levels of pollutants time-averaged results over as long a period as is practicable are desirable. German States developing legislation in respect of smog incidence are specifying half-hour sampling periods and in the U.S.A a 3 h mean sample is prescribed [41].

The requirement of gas chromatography for a small (< 10 ml), discrete injection necessitates the use of a point-sampling technique. Analytical systems employing gas chromatography are therefore not area monitors. Multiple-point sampling may be used to simulate area monitoring.

6.2.2 Apparatus and methods for gas sampling

At ambient levels normally encountered, preconcentration (of the volatile components from a highly diluted sample) prior to analysis is usually required, however some methods in which a gas sample is analysed directly have been reported [13, 23]. The high volatility of the low molecular weight hydrocarbons requires the use of cryogenic trapping techniques to ensure 100% retention, whereas for the higher molecular weight compounds ($\geqslant C_6$) adsorption on to solid adsorbents is the preferred technique [42–45]. Typical experimental arrangements for such

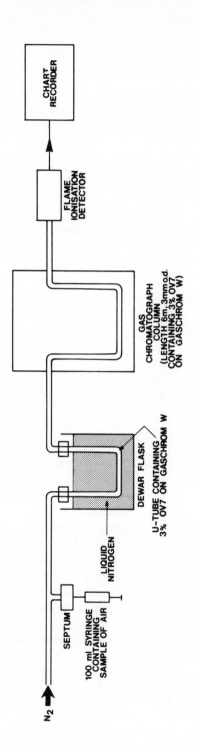

SAMPLING AND ANALYSIS OF C₁–C₅ HYDROCARBONS

Fig. 6.1   Sampling and analysis of C_1–C_5 hydrocarbons.

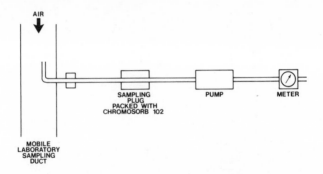

SAMPLING FOR C$_6$–C$_{10}$ HYDROCARBONS

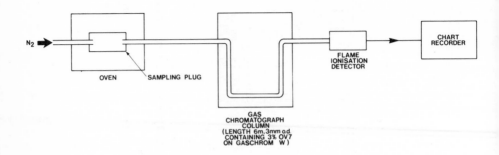

ANALYSIS OF C$_6$–C$_{10}$ HYDROCARBONS

Fig. 6.2 (a) Sampling for C$_6$–C$_{10}$ hydrocarbons; (b) analysis of C$_6$–C$_{10}$ hydrocarbons.

procedures are shown in Figs. 6.1 and 6.2 respectively. When using solid adsorbents it is essential to ensure that a valid sample is collected. Particular attention should be directed towards the determination of the 'breakthrough volume', the sampled air volume at which the compound being collected begins to elute from the tube. The breakthrough volume is dependent upon the gas chromatographic retention time of the compounds at ambient temperature, the adsorbent acting as the stationary phase and the sampled air as the carrier gas. An equilibration position will be reached where if the ambient concentration of the compound remains constant, the amount of the compound being adsorbed will be equal to the amount eluting from the tube.

The choice of adsorbent for use in a particular situation depends upon the chemical properties of the compounds to be sampled and subsequently analysed.

The following criteria must be considered: quantitative collection efficiencies and recovery of trapped vapours, high breakthrough volumes, minimal decomposition or polymerization of sample constituents during collection and recovery, low background contribution from the adsorbent and little or no affinity of the adsorbent for water. The parameters involved in determining the performance of an adsorbent can be divided into two categories: those related to the sampling environment such as sample flow rate, sampling time, air temperature and humidity and those related to the physiochemical properties of the adsorbent such as surface area, particle size and porosity, solute capacity, sorption mechanism and degree of solute affinity. Furthermore, some of the factors which influence adsorbent performance are not independent of each other.

Gas tight syringes [16, 45], glass aspirators and PTFE bags [16, 23, 46, 47] have all been used as sample containers. Ambient gas has been sampled using a modified gas pipette with 'teflon' coated rubber seals and GC/MS analysis gave a reported detection limit of $5\,\mu g\,m^{-3}$ [48]. The use of 'Saran' bags for the sampling of industrial atmospheres in the 200–1000 ppm range has been reported [46] and losses after 3 h of 15% and 40% for benzene and toluene respectively were recorded. A remote sequential sampler has been described [49], in which spring loaded syringes are activated by a timer and the samples obtained have been stored for 18 h without leakage. A small lightweight gas sampler for time integrated samples has been described [50]. An evacuated container equipped with a critical orifice is used to collect the sample, and the sampling period is dictated by the area of the orifice and the volume of the container. The system has been evaluated for the sampling of methane, ethane, benzene and hexane.

6.2.3 Cryogenic systems

A battery operated system for sampling in remote areas has been reported [51], and requires approximately 3 litres of liquid N_2. This system was originally developed for use in the tropical forests of Panama where very high relative humidities are encountered, but has exhibited variable freeze-out efficiency and requires constant attention. A portable cryocondenser has been shown [52] to have a high enrichment capacity for large air volumes and, when coupled via sub-ambient GC to a flame ionization detector, to give a detection limit at the sub-ppb level. A similar system has been used for the sampling of industrial air pollutants at various altitudes [53]. A more sophisticated application of a freeze-out technique to ultratrace analysis has been described [54], in which a temperature gradient is applied to a tube packed with a GLC stationary phase on an appropriate support. The main problem with freeze-out techniques is the presence of water vapour in the air. The application of drying agents prior to condensation has reduced this problem and systems based on this principle have been described

[55]. Separations of the contents of cryogenic traps prior to mass spectral analysis have been accomplished by thermal analysis [56] or by preparative gas chromatographic techniques [55].

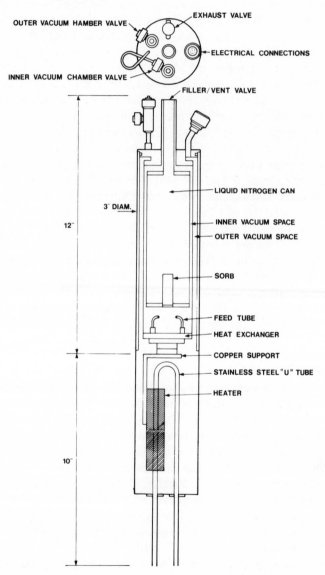

Fig. 6.3   Schematic view of cryostat capable of temperature programmed operation.

A recently developed cryostat, Fig. 6.3, is at present being used for the analysis of individual hydrocarbons in street air [57]. This instrument has a temperature range of -196 to $+300°$ C. Temperatures may be maintained within this range by the balancing of the cooling and heating effects of a liquid nitrogen reservoir and a 240 W cartridge heater.

The instrument basically consists of two independent vacuum chambers. The inner chamber contains the liquid N_2 can and an integral carbon adsorption pump. The outer vacuum space contains the 'U' tube, heat exchanger, heater and temperature sensors. The use of two independent vacuum chambers interconnected via an isolation valve facilitates the changing of the 'U' tube without removal of the refrigerant. The 'U' tube is only packed with adsorbent along the length of the cartridge heater. Once the sampling temperature has been achieved and the can refilled with liquid N_2, the supply will last for approximately 4 h, during which time the desired temperature may be maintained.

In remote areas with no electricity supply, sampling is carried out at $-196°$ C. Prior to analysis, liquid O_2 that has been condensed out in the 'U' tube, can, with controlled heating, be slowly evaporated so as to avoid interference with the subsequent chromatographic separation of the adsorbed material.

This cryostat has several advantages over other sampling techniques that are currently available. It is easily portable, which is advantageous when sampling in remote areas, and the results obtained with it are both reproducible and consistent. By selection and control of the sampling temperature it is possible to ignore certain of the more volatile compounds. If required, however, the complete range of C_1-C_{12} compounds may be analysed in a single sample. A certain amount of separation of the adsorbed material prior to introduction into the gas chromatograph is achieved by temperature programming the 'U' tube. With the cryostat it is possible to sample large volumes of air with the confidence that a truly representative sample is being collected and without the limitations of relatively small breakthrough volumes. this is an important consideration when a GC/MS analysis is to be employed and ng quantities of each compound to be identified are required. After sampling at sub-ambient temperature (to take advantage of the greater breakthrough volumes) the sample tube can be stored at ambient temperature without loss of the higher boiling hydrocarbons (C_5 and above).

Although cryogenic sampling is the only technique presently available which produces, directly, a true sample compostion in the low molecular weight range of air pollutants, procedures involving adsorption on solid adsorbents or in liquids, coated on inert supports and acting as stationary phases, are simpler and more convenient for field sampling. Adsorption techniques suffer from high percentage losses of volatile compounds. The chemical nature of a solid adsorbent has a considerable influence on the adsorption/desorption characteristics for specific compounds, and losses due to irreversible adsorption are common, particularly when a thermal desorption step is used.

6.2.4 Adsorption techniques

Activated C is particularly prone to irreversible adsorption of certain organic species. This material has, however, been used widely as an adsorbent followed by either thermal desorption [25] or by various methods for solvent extraction [19, 26, 58] of the adsorbed species. Carbon disulphide has been shown to be an effective solvent for the extraction of activated charcoal [19]. Recovery of solutes by solvent extraction is more complete than thermal desorption but is also a more time-consuming and complex process.

The usefulness of silica gel as an adsorbent for the collection of environmental samples is greatly reduced by its affinity for water and the consequent influence of relative humidity on the adsorption curves for hydrocarbons. Sampling on to silica gel followed by thermal desorption of the adsorbed species, their oxidation to CO and determination by acidmetric and coulimetric measurements has recently been reported [59]. The procedure was limited by the number of adsorption/desorption cycles that could be carried out, by inadequate desorption temperatures and by variations in retention volumes for specific organic species.

The use of support bonded chromatographic stationary phases for sampling has been investigated [60] and it was shown that high concentrations of stationary phases were required in order to maximize the retention of organic species present. This method is limited to atmospheric pollutants of relatively low volatility but could be improved by sub-ambient operation.

Porous polymers with high capacities for the sampling of organics and low affinities for water provide extremely useful adsorbent species and have been extensively investigated [2, 54, 61–7] utilizing both thermal desorption and solvent extraction. An evaluation of various organic adsorbents and their relative dynamic capacities for certain classes of compound has been reported [42]. Chromatographic packing materials studied were Poropack P, a porous polymer of styrene and divinyl-benzene; Carbosieve, prepared by thermally cracking polyvinylidine chloride and Tenax GC, a porous polymer, 2,6-diphenyl-p-phenylene oxide. In this, as in other recent publications [62, 68, 69], Tenax GC was shown to be superior as a general adsorbent, having excellent temperature stability to 380° C and proving easy to handle. In these studies relative humidity, ubiquitous background levels, repetitive use, transportation and storage were all considered. Mieure and Dietrich [66] classified the adsorption/desorption characteristics of various column packings:

Chromosorb 101 — Absorbs and desorbs acidic and neutral components.
Chromosorb 105 — Absorbs and desorbs low-boiling components.
Tenax GC — Adsorbs and desorbs basic, neutral, and high-boiling components.

6.3 Sampling of particulate matter

Most methods for the analysis of organic particulates in ambient air depend on filter collection by some form of chromatographic separation. In a recent survey at locations throughout Los Angeles County [70] conventional 'high volume' samplers as described by the Intersociety Committee [71] were used on an intermittent schedule. The full 'tentative method' [71] is a lengthy process involving particulate sample collection on glass fibre filters, soxhlet extraction, column chromatography, and u.v.-visible spectrophotometry, and owes much to the early work of Sawicki and co-workers [72–5]. Many people have attempted to apply more direct chromatographic techniques to the problem and an authorative series of papers relating to the analysis of PAH in the environment has been published by Lao and co-workers [76–8]. These workers also found the use of commercial high volume samplers, using glass fibre filters, adequate for the collection of airborne particulate samples. Filters are weighed before and after sampling to determine the weight of extrained particulate material. It has long been suspected that the volatility of PAH might prevent their complete collection by particulate filter techniques and also cause loss after collection. If equilibrium between solid and vapour (particulate on filter and vapour in passing air respectively) is established, then losses during collection depend on the equilibrium vapour concentration (EVC). These effects have been considered [77–9] and the question posed [77]: do the high volume particulate filter techniques which are presently used yield meaningful measurements of the levels of these pollutants? The authors concluded that qualititative evidence indicated that considerable losses occur during ambient air sampling of compounds having EVCs of $500 \, \mu m \, 10^{-3} \, m^{-3}$ or higher at ambient temperatures and noted that such species included pyrene and benzo(a)anthracene.

6.4 Analysis of volatile hydrocarbons

There is a need for high specificity in measuring the components of an environmental sample but it is also desirable that the analysis may be performed routinely, rapidly and with precision. The apparatus should be simple if it is required for use in a mobile laboratory; thus permitting rapid analysis in a field situation. The application of GLC to the analysis of complex environmental air samples is the only really viable system, with FID being the most commonly used detector. Conventional packed columns remain adequate in many applications, but in order to achieve the required separation of the broad range of components detectable, and to enable accurate measurement of retention times, capillary columns are gaining increasing use. There are no accepted standard methods available for the analysis of hydrocarbons; and it is difficult to envisage the establishment of any in the immediate future because of the large variation in the analytical requirements of specific applications.

Several GLC systems have been described and compared [80]; the authors concluded that there is a need for better chromatographic methods designed to resolve the 50 to 100 or more readily measurable hydrocarbons in urban atmospheres. They added that methods should be developed, standardized and made generally available in order to enable more meaningful comparative and collaborative studies. There is also a requirement to reduce experimental error in identification and estimation of individual compounds especially in the correlation of the photochemical reactivity of specific hydrocarbons with the resulting oxidant formation.

Rasmussen et al. [80] described 'some of the present-best GC-column methods used in this laboratory'. They preferred a packed capillary column for the analysis of light hydrocarbons having compared its perfomance with both 1/8 inch o.d. packed columns and support coated open tubular (SCOT) columns. The actual column used was 20 ft, 1/16 inch o.d., 0.03 inch i.d., packed with Durapak (n-octane, Porasil C). Sub-ambient temperature programming was found to give excellent separation of the C_2-C_6 hydrocarbon fraction. The construction of a suitable cryogenic GC oven has been described [81]. When considering methods for the analysis of hydrocarbons it is convenient to treat the compounds in groups specified by a range of molecular weights. Thus Rasmussen et al. [80] considered intermediate hydrocarbons, aromatic and aliphatic (C_6-C_{12}) compounds, and discussed their comparative chromatographic characteristics on several types of column: on open tubular columns of relatively large diameter capable of accepting high sample loadings, m-bis(m-phenoxyphenoxy) benzene (300 ft, 0.06 inch o.d.); on a system using three packed 1/8 inch o.d. columns in series; and on small diameter (0.01 inch i.d.) open tubular columns in conjunction with temperature programming designed to produce optimum resolution. They found a 0.02 inch i.d., 200 ft OV-101 SCOT column to be most satisfactory.

6.4.1 Low molecular weight hydrocarbons (C_1-C_5)

Jeltes and Burghardt [23, 82] have reviewed methods available for the analysis of the C_1-C_5 hydrocarbon fraction present in air and described an automatic

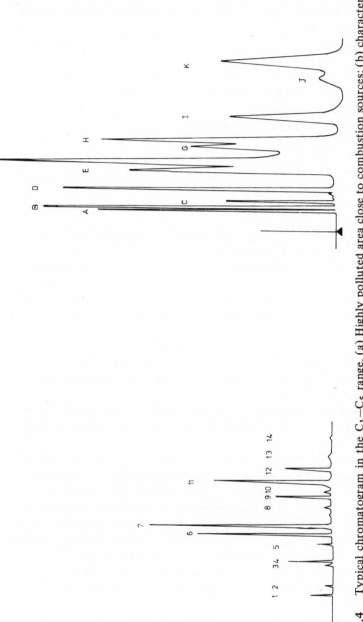

Fig. 6.4 Typical chromatogram in the C_1–C_5 range. (a) Highly polluted area close to combustion sources; (b) characteristic suburban air.
Fig. 6.4(a) (1) Ethane; (2) ethylene; (3) acetylene; (4) propane; (5) propene; (6) i-butane; (7) n-butane; (8) neopentane; (9) i-butane; (10) propyne; (11) i-pentane; (12) n-pentane; (13) cyclopentane; (14) i-pentane.
Fig. 6.4(b) (A) Methane; (B) ethane; (C) acetylene/ethylene; (D) propene/propane; (E) i-butane; (F–K) C_4–C_5 hydrocarbons.

GC method, claiming it to be 'a relatively simple, direct and inexpensive method for sensitive and quick measurement'. Having considered both the spectroscopic and chromatographic methods available they chose to develop an automatic system for measurement combining GSC and FID. They listed the requirements for a monitor for measurement of hydrocarbons:

(1) Maximum information; separated measurement of individual hydrocarbons, measurement of ethylene (phytotoxic), 'fingerprinting' of hydrocarbon fraction (source identification).

(2) Relatively comfortable automized mode.

(3) As inexpensive as possible.

(4) Simplicity of the whole set-up.

(5) Quick sampling and analysis.

(6) Quick results for guarding and 'alarms'.

(7) Specificity.

(8) Sensitivity.

The merits of methods involving preconcentration as opposed to direct analysis for the two distinct hydrocarbon fractions (C_1-C_5) and (C_5-C_9) were also discussed. A $2\,m \times 4\,mm$ i.d. column packed with $80-100$ mesh Alumina type Fl was employed for the analysis with air as carrier gas. The detection limits at a signal to noise ratio of 2 were $2-5\,ppb$ for C_1-C_3 hydrocarbons and $7-20\,ppb$ for the C_4-C_5 fraction. An 'AQUASORB' filter was used to remove water from the air samples. The cycle time for this system was $45\,min$ including $n-C_5$ compounds. Points (2) and (6) appear to lose a little clarity in the translation but the concepts of facile automation and rapid results are obviously desirable. An automatic system for C_1-C_5, total C_6 and benzene working on a $30\,min$ cycle has also been described [83].

The problems associated with the condensation of water vapour are even more significant in cryogenic sampling systems. In order to minimize such difficulties samples have been passed through a dessicant prior to freeze-out, potassium carbonate has been reported [55, 84] to be extremely effective in this application owing to its very low affinity for organics. K_2CO_3 has been used prior to the trapping of organics in a cryogenic U-tube containing glass beads [55], and the sensitivity was quoted as $0.2\,ppb$ where the compounds were concentrated from 20 litres of air with 100% efficiency. A range of dessicants has been examined [84], including both anhydrous salts such as, potassium carbonate, calcium sulphate and sodium sulphate, and adsorbents of the type Sephadex (G-100) or Linde molecular sieve 3A. The same authors considered the removal of water by its conversion to acetylene and hydrogen by calcium carbide and calcium hydride.

A cryogenic sampling system incorporating a freeze-trap containing stainless steel gauze and employing potassium carbonate as dessicant has been reported

[44] to give complete recovery of all hydrocarbons boiling above $-90°$ C with a detection limit of 0.01 part in 10^8. 50–100 litre air samples were taken and liquid O was used as coolant. The transfer of the concentrate from the trap was a two-stage process: initially by pressure equalization into evacuated containers and secondly by Toepler pumps to give a residual pressure in the trap of less than 1 Torr. Subsequent analysis was by GC using retention data from several columns or with suitable samples by GC/IR.

Light hydrocarbons have been measured [13] directly using a 5 ml sample, whilst C_2–C_5 aliphatic hydrocarbons (excluding C_4 alkenes) have been investigated [14] by a more sophisticated technique involving a silica gel packed column at liquid nitrogen temperature. The latter analysis employed an 8 ft × 1/8 inch stainless steel column packed with the same silica gel substrate and operated at $30°$ C. The same workers have measured C_4–C_8 aliphatic hydrocarbons and C_6–C_{10} aromatic hydrocarbons using different trapping and analytical columns for each group. The C_4–C_8 aliphatic hydrocarbons were trapped on a 1 ft × 1/8 inch stainless steel column packed with 10% Carbowax 1540 on 60–80 mesh Gas Chrom. Z, and separated on a 300 ft × 0.06 inch i.d. open tubular column coated with dibutylmaleate at $0°$ C. The C_6–C_{10} aromatic hydrocarbons were trapped on a 1 ft × 1/8 inch stainless steel column packed with 60–80 mesh glass beads, and analysed on a 300 ft × 0.06 inch i.d. open tubular column coated with m-bis(m-phenoxyphenoxy) benzene at $70°$ C. The CO and methane species were separated on a 8 ft × 1/8 inch stainless steel column packed with 60–80 mesh 13X molecular sieve (Wilken Instrument Co.) at $70°$ C. The three gas chromatographic procedures produced 52 measurable hydrocarbon peaks representing the C_1–C_{10} hydrocarbons. The higher molecular weight hydrocarbons (C_{11}–C_{12}) were observed on occasion but at very low concentration levels. Some peaks contained unresolved shoulders. The overlap of alkane and alkene species was not a significant problem. There were interferences between benzene and some of the C_7 and C_8 alkanes, particularly methylcyclohexane. A direct comparison has been made of the resolution obtained for similar ambient air samples on a 1/8 inch × 2 m packed analytical column and on a 0.02 inch × 100 ft SCOT column – the latter proving far superior [66].

Benzene and toluene have been analysed with other compounds [16] by trapping on a 1/8 inch i.d. × 8 inch stainless steel column packed with 20% Dow-Corning SF $\neq$ 20 on 60–80 mesh Columpak at dry ice/acetone temperature followed by analysis on a 1/8 inch i.d. × 6 ft stainless steel column packed with 8% SF96 on 60/80 Chrom. W at $50°$ C.

Acetylene is one of the most dangerous hydrocarbon traces present in air as it may be accumulated and lead to explosions. This type of hazard is at a maximum where vast amounts of air are involved in industrial processes, typically in the steel industry (particularly in the LD basic oxygen process) and obviously in the production of liquid air. Acetylene was detected and quantified colorimetrically before the advent of gas chromatography, which is a more rapid and sensitive technique and also more readily automated. Several commercial instruments similar to those described elsewhere in this chapter for the analysis of total- and non-methane have been adapted for the monitoring of acetylene levels merely by the incorporation of a chromatographic system producing separation of acetylene from other volatile hydrocarbons. The work reported by Klein [152] who used a 3.4 m × 4 mm i.d. column packed with silica gel (grain size 0.3–0.5 mm) is typical. This readily separated acetylene from other hydrocarbons at 50° C using nitrogen carrier gas and reliably detected acetylene down to 5 ppd in 10 ml air samples.

6.4.2 Higher molecular weight hydrocarbons (C_6–C_{20})

Grob and Grob [19], in agreeing with a previous worker [85] that high molecular weight hydrocarbons in the atmosphere represent a complex mixture of hundreds of compounds, considered the analysis of such mixtures as three separate components. These three components were: high resolution trace analysis on capillary columns as a necessary prerequisite; appropriate sampling of volatile organics for qualitative and/or quantitative analysis; and gas chromatographic-mass-spectrometric (GC/MS) design maintaining the full separation power of a high-resolution column and working in the ng range. They emphasized that the key to the possibility of injecting very dilute solutions on to capillary columns is the avoidance of stream splitting referencing their own previous work (see also [86]). They also illustrated that direct sampling was not viable for all but the most abundant species, as the majority of components were present in the 0.1–10 ppb range and 0.1 ng of an individual compound was required for positive identification. 108 volatile organic substances were identified in the C_6–C_{20} range the majority of which were aliphatic or aromatic hydrocarbons. The samples were trapped at ambient temperature a glass tube containing 25 mg of wood charcoal as used for cigarette filters with an average particle size of 0.08 mm. Extraction of the filters was with carbon disulphide. The analyses were carried out on two GC columns, a 120 m × 0.33 mm Ucon HB 5100 for the full spectrum (C_6–C_{20}) and an 80 m × 0.33 mm Ucon LB550 for the volatile fraction (benzene to C_3-substituted benzenes) with temperature programming from ambient to 190 and 120° C respectively. A more recent application of this type of technique (carbon adsorption) allied to semiautomatic sampling and chromatography on highly polar columns has been reported [87] for the analysis of C_6–C_{20} fractions.

Using a similar approach but substituting a thermal desorption step for liquid extraction of the sampling trap Bertsch *et al.* [62] have recognized several

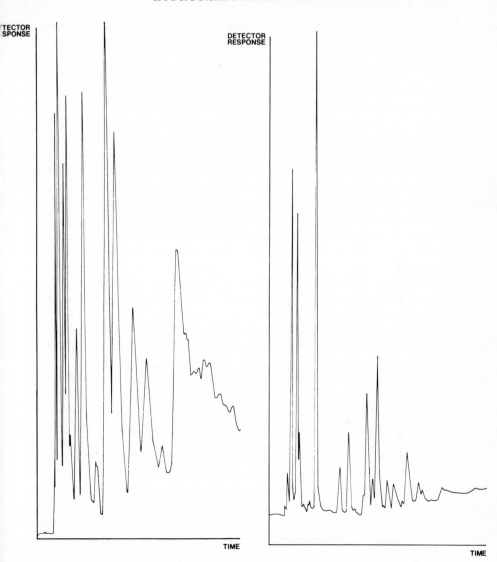

Fig. 6.5 Typical chromatogram in the C_6-C_{10} range. (a) Exhibition Road – this illustrates the large contribution made by motor vehicles being greater than (b) a highly industrialized area in Teeside.

hundred substances in the C_5-C_{16} range and identified almost 100. The sampling traps used were 110×8 mm i.d. stainless steel containing Tenax GC and the GLC was carried out on 100 m $\times 0.5$ mm i.d. nickel open tubular columns coated as described elsewhere [88] with Emulphor ON-870.

Tenax GC has been used as an adsorbent [65] with the subsequent desorption step achieved by the passage of current through a heating tape attached to the sampling tube. The tube was heated to $260°$ C for 5 min, the desorbed components being retained at the front of the analytical capillary column at ambient temperature prior to temperature programmed elution.

Raymond and Guiochon have discussed the merits of graphitized carbon black as an absorbent for a wide range of organic compounds [89] and subsequently described [25] its use as the sampling component of a GC/MS system, with which they identified more than 70 components in the air of Paris. These compounds were in the $C_8–C_{18}$ range and predominantly aliphatic or aromatic hydrocarbons. Sampling was carried out at ambient temperature using tubes 5 cm × 4 mm i.d. packed with 0.25–0.40 g of 200–250 μg sieved graphitized carbon black. Desorption was achieved by heating the trap to $400°$ C within 15 s using an electric wire (Thermocoax) soldered around it. The analytical column was maintained at ambient temperature so that the compounds eluted from the sampling tube were retained at the column inlet; desorption usually took approximately 5 min. The analytical column used was 100 m × 0.4 mm i.d. coated with OV 101, and was operated under temperature programming between ambient and $230°$ C.

A 'timed elution technique' has been described [18, 61] involving adsorption on to a solid adsorbent following by thermal desorption and analysis by GC/MS. The 'time-elution technique' is so-called because of the method of eluting the adsorbed species from the sample tube, which, after sampling, is placed in an oven and connected via a valve system to the analytical column. After the elapse of sufficient time to allow the establishment of thermal equilibrium in the sample tube a known amount of carrier gas is passed, calculated to flow for the precise time required to just displace the equilibrated gas phase from the sample tube into the gas chromatograph. The sample tubes were stainless steel 4 × 1/4 inch o.d. packed with 0.3 g Chromosorb 102, desorbed at $130°$ C on to the analytical column which was stainless steel 12 ft × 1/8 inch o.d. packed with 5% Carbowax 20m on Chromosorb W AW-DMCS at $90°$ C. Individual hydrocarbons were identified at the 20 ng level. Applications of discontinuous carrier gas flow for the elution of adsorbed species from sample tubes are at present under study [90].

A rather complex two-stage extraction/injection system has been described [62] which was designed to overcome problems of water condensation from the air sampled. The air sample was first concentrated on to Chromosorb 102 at ambient temperature; the subsequent 'thermal' desorption step was somewhat unusual in that the sampling tube was warmed only slightly above ambient but the collecting vessel (part of the injection port) was cooled to liquid N temperatures. After condensation of the organic compounds the injection port was rapidly heated and analysis was carried out on a 100 ft support-coated Carbowax 20M column temperature programmed between 60 and 180° C.

Several reports [66, 91] have appeared describing the connection of the sample collection tube directly to the analytical column. An interesting extension of this approach was presented by Mieure and Dietrich [66], who designed their sample tube such that it could be incorporated into the injection port of their chromatograph. This design has considerable potential for less volatile hydrocarbons as it enables independent and rapid heating of sample tubes to higher temperatures than those required for analytical columns.

6.4.3 Mass spectrometry

Mass spectrometry is almost mandatory if definitive identification of organic compounds in environmental samples is required. Even with this powerful technique it is often difficult to differentiate between the many similar compounds which may be present, and thus considerable effort is currently being directed towards improvement of MS techniques in this application.

GC/MS systems, many of which are interfaced to computerized data handling systems are now in widespread use [56, 58, 62, 89, 92–6]. A high resolution mass spectrometer coupled to a computer has been used for the analysis of multi-component air samples [97].

Experiments have been conducted [98] with a photoionization instrument, irradiating at specific wavelengths, selected to produce ionization of trace gases and to avoid activation of major air components. Measured sensitivities and calculated detection limits were disappointing and required improvement.

A technique [99] which haš found application in the estimation of refinery pollutant dispersion and in air quality control utilizes a mass spectrometer modified by the incorporation of a static condensation system and gives quantitative results for hydrocarbons of ± 10% at the 5 ppm level and ± 20% at 1 ppm.

A technique has been described by which it is claimed that micropollutants present in the air within the concentration range of 10^{-4} to 10^{-8}% by volume may be separated and identified routinely on commercial equipment [100]. Enrichment was carried out on a microgradient tube followed by separation on a glass thin-film open tubular column by linear-programmed low temperature gas chromatography and identified by mass spectrometry. The authors reported that significant MS data was obtained for concentrations down to 0.02 ppb.

A portable MS based vapour detection system with complete digital control has recently been described [101] which was reported to give detection limits in the ppb range with a reproducibility of approximately 5% when used with a compatible GC system. The 'advantage' of cryogenic preconcentration of volatile trace components in air samples have recently been exploited [102] in chemical ionization mass spectrometry by using the condensed water as a secondary CI reagent gas. Further refinements in the cryogenic trap and heater system should produce lower detection limits.

6.4.4 Routine instrumental methods

6.4.4.1 Gas chromatography

Several commercial instruments equipped with (FID) are available for the continuous measurement of hydrocarbons at ambient levels. The FID is capable of providing a continuous response, which is approximately linear for hydrocarbons over a wide range of concentrations. Carbon atoms bonded to hetero atoms (e.g. O, N or Cl) give a decreased FID response, and measurements of total hydrocarbon levels can be distorted by the presence of organic compounds which are not purely hydrocarbons. Comparative evaluations of the available instruments have been reported [103, 104] and their sensitivity, accuracy, portability and ease of operation discussed. Calibration techniques for such instruments have also been reported [105] using both standard calibration gases from cylinders and diffusion of permeation tubes.

FID instruments have been operated under reduced pressure [106, 107] with the sample passing through the column and detector prior to entering the pump. This method has been applied to corrosive or unstable compounds. As previously noted non-methane hydrocarbon measurements are of greater significance than methane measurements as methane contributes little to photochemical reactions.

There are now many commercial non-methane hydrocarbon monitors available, the majority of which operate by producing a value for the difference between total hydrocarbon and methane levels [108]. Methane and non-methane hydrocarbon levels may be monitored continuously and simultaneously using a dual flame instrument [109]. A system has been described [110] which measures methane and/or total hydrocarbons by selective combustion. Selective combustion has also been used, with the incorporation of a water sorption detector, to measure hydrogen, methane, reactive and non-reactive hydrocarbons. The air is first dried and then burnt, the reactive hydrocarbons being distinguished from the non-reactive species by the relative ease of their combustion. In this context the term reactive refers to the ease with which a particular hydrocarbon will undergo photochemical reaction with NO or O_3. All hydrocarbons appear to be capable of taking part in such processes, but the reaction rates are extremely dependent on molecular structure [111]. Methane is the least reactive and its estimated atmospheric residence time is up to sixteen years [112], thus in terms of urban air quality it may be considered as photochemically inert. The most reactive of such species probably only have residence times of hours in ambient air and strong sunlight [113].

Instruments have been described [114–16] for the measurement of CO and CH_4, in which the CO is first reduced to CH_4 and subsequently detected by FID and detection limits of 'a few ppb' have been quoted [117]. Interference free measurements of CO, CH_4 and total hydrocarbons in the range 0–1 ppm have been obtained automatically with a cycle time of 5 min [118], similarly an automatic timing device has been incorporated into a total hydrocarbon analyser to enable repetitive measurements of CH_4, total hydrocarbons and reactive hydrocarbons [119]. An automatic gas chromatographic system, which measured CO, CH_4 and non-methane hydrocarbons in a single sample, has recently been described [120].

Instruments are also available for measurement of CH_4, C_2H_4, C_2H_2, CO and total hydrocarbons [39, 121–3] with detection limits in the region of 200 ppb for total hydrocarbons and 1 ppm or better for individual components. An air quality chromatograph (AQC) system has been reported [124] and evaluated for the measurements of CH_4, C_2H_4, C_2H_2, propene, propane, iso- and normal-butane. CH_4 has been separated from the other hydrocarbons in air by the use of a cryogenic trap [125]. The device traps more than 95% of all non-methane hydrocarbons and when used in conjunction with a total hydrocarbon analyser makes the continuous determination of methane possible. A commercial AQC is described in Section 6.5.

6.4.4.2 Spectroscopic methods

Reactive hydrocarbons can be estimated utilizing the chemiluminescent reaction of O atoms with unsaturated hydrocarbons at pressures of ~ 1 mb which produces intense emission in the 700–900 nm region of the spectrum [126]. Owing to the high background level of methane techniques for the determination of total hydrocarbons in air do not require exceptional sensitivity (merely a capability for precise operation at the ppm level). Although only of barely adequate sensitivity for ambient measurements, non-dispersive i.r. (Ndir) [127] has been a popular technique. The sensitivity and selectivity of the method have been improved by refinements in instrumental design [128].

High resolution i.r. spectroscopy has been used to monitor methane in the Pyrennes [129]. A problem in using i.r. for the measurement of trace levels of organics has been its limited sensitivity. The sensitivity has been improved [130] by using a scanning Michelson interferometer, cooled solid state detector, fast minicomputer and multiple pass long path length cells. The spectra are analysed by the computer which minimizes the interference from H_2O and CO and has enabled the measurement of organics at ppb levels.

Laser Raman spectroscopic techniques have been employed for the remote sensing of organics [131] and in particular for levels of CH_4 [132, 133],

C_2H_4 [134] and C_2H_2 and C_2H_6 [135]. The detection limits for hydrocarbons using Laser Raman methods have been quoted as $1-7$ ppm with a range resolution of 10 m [136]. Direct current discharge emission spectra have been reported for selected organics [137]. An advanced theoretical treatment of remote air pollution measurement has been published [138], in which several techniques made possible by the advent of high energy tunable laser sources are discussed and compared.

6.4.5 Methods for specific compounds

6.4.5.1 Ethylene [139]

Air is passed through 200 ml of a 0.1% methanolic solution of mercuric acetate at 500 ml min^{-1}. The solution is concentrated to 10 ml and a 1 ml aliquot applied to a TLC plate. After separation the band is removed and reacted with sodium borohydride releasing Hg, which is detected by atomic adsorption and is directly related to C_2H_4 levels, which may be determined, it is claimed, to a detection limit of 0.1 ppm.

6.4.5.2 Benzene

It has been claimed [140] that benzene may be determined using a tube containing silica gel impregnated with paraformaldehyde and H_2SO_4. A critical appraisal [141] of another tube system showed that none of the tubes tested gave a result to within 25% in the $20-160$ ppm range.

Several other 'chemical' methods which had been in use for many years prior to the advances in GC techniques have been reported [41] and may still be of value when instrumental methods are not practicable. These methods deal mainly with alkenes and aromatic species.

6.4.6 Calibration methods

The types of analytical method described in this chapter do not determine the absolute amount of compound present but do measure a property of that compound which may be related to its concentration by an appropriate calibration procedure. Gas chromatographic techniques require particular care in calibration as the response of the many components of a complex mixture may vary nonlinearly with relatively slight changes in operating parameters.

Gas mixtures containing very low but accurately known levels of organic compounds are extremely difficult to prepare. The techniques which are available for the generation of such standard gas mixtures fall into two categories, static and dynamic procedures. Static procedures are usually necessary in the preparation of multicomponent systems while dynamic methods are applied to the introduction of trace levels of single species into carrier or diluting gas streams.

A major problem in preparing very dilute mixtures is adsorption of trace components on to the walls of cylinders and tubing; this effect may be reduced by rigorous cleaning and a careful choice of materials. Mixtures of saturated hydrocarbons in clean glass or metal containers are relatively stable. Unsaturated compounds however, especially C_2H_2 and butadiene are slowly decomposed, particularly in the presence of water vapour and O.

6.4.6.1 Static procedures

Mixtures may be prepared in glass containers or purchased as 'standard gases' in cylinders; it is often unwise to place too much reliance on commercially available standards without first checking them against their specification. Mixtures for less rigorous work can be prepared in plastic bags but these must be used rapidly to avoid excessive diffusion losses.

6.4.6.2 Dynamic procedures

Gas mixing pumps are available and the use of a two-stage apparatus has enabled the production of standards down to 1 ppm concentration levels. Another system, suitable for work not demanding the highest accuracy, employs motor driven syringes.

6.4.6.3 Permeation and diffusion tubes

A more sophisticated and extremely accurate dynamic procedure employs permeation tubes, in which the component of interest, as the liquid or gas phase, is enclosed in a cylindrical ampoule made of plastic (often PTFE). After a certain induction period the material permeates slowly through the plastic at a uniform rate by a diffusion controlled process. An apparatus based on this principle requires precise and stable control of two physical parameters, temperatures and dilution gas flow rate. These parameters are readily controlled to within 1% and thus the overall precision of the system is high. The absolute amount of a compound released has been determined in several ways; for liquid hydrocarbons gravimetric methods are accurate and simple to perform. Such techniques allow the confident production of standards at the ppb level. Diffusion tubes may also be used to produce mixtures at similar concentrations. Although not specifically relating to their application to hydrocarbons, the principles of operation and diffusion tube devices have been extensively discussed [142–6].

6.4.6.4 Capillary restricted flow

In this method the calibration component to be diluted is dosed into a carrier gas stream through a restricting capillary. The concentration is calculated in accordance with the flow laws of Hagen–Poiseuille. As the capillary radius enters these calculations raised to the 4th power it must be known precisely. This procedure has application as an absolute primary standard with an accuracy of 1–2% [147].

6.5 Analysis of CO

A major problem encountered in the analysis and quantification of atmospheric levels of individual hydrocarbons is that of separating the compound of interest from many potential interferences prior to its measurement by a non-selective detector (usually FID). When dealing with CO, however, sampling and measurement may be considerably simplified by the use of detector systems sensitive specifically to that gas or to effects produced by that gas.

CO exhibits i.r. adsorption of sufficient strength to be detected over reasonable path lengths at normal urban atmospheric concentrations. The use of a 10 m path-length i.r. cell enables the measurements of concentrations in excess of 10 ppm to within 10%. The actual measurement made is of the i.r. absorbance at $4.67\,\mu m$. The vast majority of published data on CO levels have been obtained using Ndir methods, and such measurements usually have an uncertainty of about 1 ppm.

The essence of the Ndir method is the design of the detector, which is a two-sided cell containing equal concentrations of a mixture of CO and an i.r. transparent gas. Infrared radiation is incident on both halves of the detector, one beam having passed through a sample cell and the other through a reference cell containing no CO. There is an integrated absorption of i.r. energy over all wavelengths passed by the optical system by the CO present in both halves of the detector. This absorbed energy is converted into heat producing a change in pressure or volume of the gases in the detector. If CO is present in the sample cell it will absorb radiation reducing that incident on the detector and producing a decreased heating effect. It is normally the unequal volume change between the sample and reference sides of the detector which is converted into an electrical signal directly related to the CO concentration. Principal interferences are from H_2O and CO_2 but these may be greatly reduced by the use of filter cells. The Intersociety Committee have produced a full tentative method together with a calibration procedure [148].

Several chemical methods have been developed for the estimation of CO. One method giving a detection limit of 5 ppm in a 50 ml sample involves the reaction of CO with an alkaline solution of the Ag^+ salt of p-sulphaminobenzoic acid to

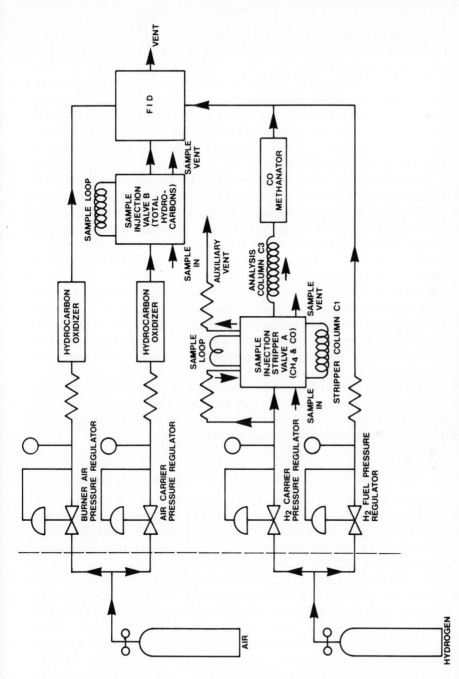

Fig. 6.6 Bechman air quality chromatograph model 6800 simplified flow diagram of three-component system.

form a colloidal suspension of Ag^+. The adsorbance of the resulting suspension is measured spectrophotometrically and related to the CO concentration [149, 150]. A more sensitive technique, useful in the range $0.025-10$ ppm, is based on the reaction of CO with hot mercuric oxide to produce CO_2 and Hg vapour. The Hg vapour is sensed by its strong absorption of light from a Hg lamp. Interference from H_2 and some organics can be overcome [151]. The oxidation of CO to CO_2 over an oxidation catalyst (Hopcalite) has also been used for its quantification. The measurement made is that of the increase in temperature of the air stream produced by the heat of oxidation.

Commercial instruments which measure CH_4, CO and total hydrocarbons have been referred to above and discussed in the literature [39]. An analyser capable of sequential measurements of CH_4, CO and total hydrocarbons is shown in Fig. 6.6. H_2 carrier gas flushes a small air sample from a sample loop through a stripper column, (Cl) packed with an adsorbent porous polymer, for sufficient time to allow passage of CH_4 and CO but not the heavier compounds which are subsequently removed by back-flushing. Passage of the CH_4 and CO through a chromatographic column containing molecular sieves (C3) causes separation of these compounds which are passed to the FID via a catalytic methanator. Sub-ppm levels are readily determined. Modification of the instrument allows passage of C_2 hydrocarbons through the stripper column, and these are separated on a second column of porous polymer and passed directly to the detector. Total hydrocarbons are determined by passage of a $10\,cm^3$ air sample directly into the FID (valve B). The carrier gas is air, purified by catalytic oxidation of impurities and this, rather than more conventional carrier gases, eliminates problems resulting from radical changes in flame characteristics upon introduction of the polluted air sample into the detector.

6.6 Analysis of the organic fraction of airborne particulate matter

Methods for the analysis of particulate organic air pollutants have been reviewed [153]. Such analyses are not simple – over 100 PAH species alone have been separated and identified by GC/MS techniques in a single study [76]. In this study airborne particulate samples were collected on glass fibre filters using a high volume sampler; the filters were Soxhlet-extracted using cyclohexane and extractable matter was subjected to a Rosen separation. The PAH fraction was concentrated and injected into a GC/MS system. Samples of less than $100\,\mu g$ produced good MS data for the individual components emerging from the GC column in ng quantities. A number of other procedures have been used for the analysis of PAH in air. Most accepted methods including the ASTM Standard (1971) consist of an extraction process, a fractionation by column chromatography and the spectrophotometric analysis of the eluted fractions. The complexity of the samples involved however continues to promote research interest, often directed towards achieving a more convenient analytical procedure for an extended range of compounds. GC and LC have provided avenues of approach to this problem.

6.6.1 Separation of organic fractions and analysis by liquid chromatography

PAH are soluble in many organic solvents including cyclohexane, benzene, chloroform, acetone and alcohols, but the efficiency of such solvents in the extraction of PAH is largely dependent on the nature of the particulate matter under investigation. So strongly are PAH adsorbed by carbon black [154], for example, that many hours are required to solvent extract samples of airborne particulates and other source materials containing carbon black. There is, by no means, agreement on the length of time for complete extraction, and the various workers in the field have suggested 2 h [155], 12 h [156] and 20—30 h [157] using cyclohexane. The Intersociety Committee has adopted 24 h with cyclo-hexane for one method [33] and 6—8 h for another [34], and also 6—8 h with benzene [32]. A quantitative investigation into the extraction efficiencies of these solvents has been carried out [158], in which 76% and 95% recoveries of benzo(a)pyrene from enriched air samples were obtained after 6 h refluxing with cyclohexane and benzene respectively. Possible losses of PAH as the result of decomposition during long periods of Soxhlet extraction have been investigated [159]. The possibility of decomposition and the length of time involved in sol-vent extraction methods makes the use of alternative extraction procedures very attractive. Such a procedure has been reported [160] using ultrasonic vibration at ambient temperature and requiring only 30 min. Having extracted the organic matter from a particulate sample the aromatic fraction is usually separated by liquid—liquid partition followed by benzene elution from a silica gel chromato-graphic column [161]. Tabor *et al.* [162] chromatographed the neutral fraction of the organic material in air pollution samples on silica gel, eluting the aliphatics with isoctane and the aromatics with benzene. A quantitative method for PAH based on this procedure required additional chromatography, on alumina [163]. These types of procedure enable the production of simplified mixtures of PAH species from the more complex arrays of organics present in the particulate extracts.

Quantification of the PAH has generally been based on u.v. absorption or fluorescence spectrophotometry. Many PAH display characteristic u.v. absorp-tion and/or fluorescence spectra in the 250—435 and 340—500 nm regions respectively. Their identification in the sub-fractions of environmental extracts have normally involved comparison of such spectra with reference data derived for pure PAH compounds. It has been shown that fluorescence can be at least an order of magnitude more sensitive than absorption spectroscopy for PAH analysis [164, 165]. It is also more sensitive than mass spectral detection, which normally has a ng limit of detection [76, 166, 167], although integrated ion current techniques have been reported [168, 169] to reduce this limit to the sub-pg 1. Fluorescence is a non-destructive detection technique permitting further examination of samples and is also highly selective for certain PAH. The selectivity of fluorescence measurements has been successfully employed in the analysis of PAH mixtures [36, 170, 171]. Fox and Staley [36] described the use of high pressure liquid chromatography on octadecylsilyl stationary phases of micro particle ($5\,\mu$m) size with on-line fluorescence detection of the PAH fractions in atmospheric particulate matter.

6.6.2 Analysis by high pressure liquid chromatography (HPLC)

The distinctive features of gas chromatography are the use of long pressurized columns, a long-lived adsorbent, a single mobile phase throughout the procedure, and the highly reproducible retention volumes that the system provides. Many of the criteria applied to GC have been adapted to various forms of liquid chromatography, liquid—solid, liquid—liquid, gel and ion exchange. GC has often been considered inappropriate to the analysis of PAH because of the thermal instability of that class of compounds and the low volatility of the higher molecular weight species. During the relatively short and rapidly advancing history of modern liquid chromatography many workers have been of the opinion that some variation of this technique will provide the most convenient procedure for the rapid analysis of an extensive range of PAH, and particularly for the separation of high molecular weight compounds. Since Karr *et al.* [172] used a 25 ft column of 0.25 inch i.d. copper (providing a length: diameter ratio of 1200) packed with 80/100 mesh alumina (4% water content), and cyclohexane at $50\,\mathrm{lb\,inch^{-2}}$ separate a synthetic mixture of PAH applications of LC techniques to such analyses have been numerous.

A significant advance in the application of HPLC to PAH analysis was reported by Schmit *et al.* [173], who permanently bonded octadecyl silicone to spherical siliceous particles with a porous surface of specific thickness and pore size [174]. Such columns have been used to give, for example, near baseline separation of the carcinogenic benzo(a)pyrene and its non-carcinogenic isomer benzo(e)pyrene. 'Pressure-assisted reverse-phase LC' has been described for the analysis of PAH in engine oil [175], using Corasil/C_{18} and methanol—water solvents. It seems probable that such a system would be effective for the analysis of particulate air samples. The separation of benzo(a)pyrene from its isomers has been reported [176] using cellulose acetate columns and ethanol/dichloromethane solvent (2:1).

6.6.3 Analysis by thin layer chromatography (TLC)

In general thin layer chromatography as applied to air pollutants has been concerned with qualitative analysis, although some semi-quantitative work has been done. Most analyses have been carried out on benzene extracts of particulate fractions, often following clean-up on a chromatographic column as discussed in Section 6.6.1. Identification of separate species has normally involved fluorescence spectrophotometry or occasionally absorption spectrophotometry.

The use and relative merits of three TLC systems for the analysis of 20 PAH extracted from airborne particulate matter has been described [177]. The systems described were: (i) aluminium oxide with pentane-ether (19 : 1, v/v); (ii) cellulose with N,N-dimethylformamide–water (1 : 1, v/v); and (iii) cellulose acetate with ethanol–toluene–water (17 : 4 : 4, v/v). The third system has enabled the separation of benzo(a)pyrene and benzo(k)fluoranthene [178]. Quantitative determinations of benzo(a)pyrene have been made using aluminium oxide–cellulose (2 : 1, w/w) with ethanol–toluene–water (17 : 4 : 4, v/v) [159]. A two-dimensional system has proved satisfactory in the separation of benzo(a)-pyrene and benzo(k)fluoranthene employing aluminium oxide–cellulose acetate (2 : 1, w/w) with pentane (first dimension) and ethanol–toluene–water (17 : 4 : 4, v/v) (second dimension) [179]. The significance of PAH in the environment is their carcinogenic health hazard. The hazards posed by individual members of this group of compounds vary over several orders of magnitude and most researchers are seeking a convenient quantitative technique, particularly for the analysis of the most potent carcinogens. It is unlikely that TLC will provide such a technique in the near future.

6.6.4 Analysis by gas liquid chromatography

The analysis of the C_{15}–C_{36} n-alkane range in benzene-soluble fractions from airborne particulates has been achieved on a 20 ft × 1/8 inch o.d. stainless steel column packed with 3% SE30 on Chromosorb W (100/200 mesh) [180]. This study demonstrated that petrol and diesel fuel combustion processes can generate this range of aliphatic compounds in particulate matter, although they are not present in the fuels. A similar range of n-alkanes (C_{11}–C_{33}) was analysed on a glass column, 5.4 m × 6 mm o.d. packed with OV-7, 1% on Chromosorb W (80/100 mesh) by Lane et al. [28] who also separated 20 PAH.

Lao et al. [76] have concluded a feasibility study into the quantitative measurement of sub-µg quantities of PAH in polluted air using GC/MS by stating that such techniques 'can provide the definitive approach to complete PAH analysis which to date has been impossible'. The same authors have subsequently [78] added that 'the GC-FID-Quadrupole-MS-Computer system is, at present, the method of choice for PAH analysis in all types of environmental samples. In their first study both packed and SCOT Dexsil-300 columns were used, with the packed columns proving more satisfactory. In the second study packed columns of Dexsil-400 and 410 were used. The claims made by these workers may appear extravagant but they have recorded some very impressive chromatograms. Despite such claims it seems unlikely that any one system will provide adequate separation of all the PAH species present in airborne particulates – but specialized techniques have enabled resolution of some extremely stubborn fractions. For example [181] GLC in the nematic region of N,N'-bis (p-methoxybenzylidene)-α,α'-bis-p-toluidine, utilizing the unique selectivity of this liquid phase, based upon differences in the molecular length to breadth ratio of solute geometric isomers, has enabled the separation of 16 3–5 ring PAH compounds.

References

[1] Leinster, P., Perry, R. and Young, R.J. (1977). *Talanta*, in press.
[2] Leggett, D.C. Murrmann, R.P., Jenkins, T.J. and Barriers, R. (1972). *U.S. Nat. Techn. ADRep.* 745125.
[3] Darley, E.F., Biswell, H.H., Miller, G. and Goss, J. (1973). *J. Fire Flammability* 4, 74.
[4] Rasmussen, R.A (1970). *Environ. Sci. Technol.* 4, 667.
[5] *Idem* (1972). *J. Air Pollut. Control Ass.* 22, 537.
[6] Abeles, F.B. and Heggestad, H.F. (1973). *ibid* 23, 517.
[7] Perry, R. and Slater, D.H. (1975). *'Chemistry and Pollution'* F.R.Benn and C.A. McAuliffe (eds.) Macmillan, London.
[8] Huess, J.M. and Glasson, W.A. (1968). *Environ. Sci. Technol.* 2, 1109.
[9] Glasson, W.A. and Tuesday, C.S. (1971). *ibid* 5, 151.
[10] Breeding, R.J., Klonis, H.B., Lodge, J.P., Pate, J.B., Sheesley, D.C., Englert, T.R. and Sears, D.R. (1976). *Atmos. Environ.* 10, 181.
[11] Nerat, G. (1973). *Ingenieursblad* 42, 109.
[12] Altshuller, A.P., Lonneman, W.A. and Kopczynski, S.L. (1973). *J. Air Pollut. Control Ass.* 23, 597.
[13] Altshuller, A.P., Lonneman, W.A., Sutterfield, F.D. and Kopczynski, S.L. (1971). *Environ. Sci. Technol.* 5, 1009.
[14] Lonneman, W.A., Kopczynski, S.L., Darley, P.E. and Sutterfield, F.D. (1974). *ibid* 8, 229.
[15] Mayrsohn, H. and Crabtree, J.M. (1976). *Atmos. Environ.* 10, 137.
[16] Pilar, S. and Graydon, W.F. (1973). *Environ. Sci. Technol.* 7, 628.
[17] Lonneman, W.A., Bellar, T.A. and Altshuller, A.P. (1968). *ibid* 2, 1017.
[18] Perry, R. and Twibell, J.D. (1974). *Biomed. Mass Spec.* 1, 73.
[19] Grob, K. and Grob, G. (1971). *J. Chromatogr.* 62, 1.
[20] Smoyer, J.C., Shaffer, D.E. and De Witt, I.L. (1971). *Inst. Environ. Sci. Technol. Meeting Proc.* 17, 339.
[21] Miller, S.S. (1971). *Environ. Sci. Technol.* 5, 503.
[22] Grob, R.L., Schuster, J.L. and Kaiser, M.A. (1974). *Environ. Lett.* 6, 303.
[23] Jeltes, R. and Burghardt, E. (1972). *Atmos. Environ.* 6, 793.

[24] Kim, A.G. and Douglas, L.J. (1973). *J. Chromatog. Sci.* **11**, 615.
[25] Raymond, A. and Guiochon, G. (1974). *Environ. Sci. Technol.* **8**, 143.
[26] Ciccoli, P., Garetti, G., Liberti, A. and Passanzini, L. (1974). *Ann. Chim.* **64**, 753.
[27] Burghardt, E. and Jeltes, R. (1975). *Atmos. Environ.* **9**, 935.
[28] Lane, D.A., Moe, H.K. and Katz, M. (1973). *Analyt. Chem.* **45**, 1776.
[29] Shultz, J.L., Sharkey, A.G., Friedel, R.A. and Nathanson, B. (1974). *Biomed. Mass Spec.* **1**, 137.
[30] Brocco, D., Dipalo, V. and Possanzini, M. (1973). *J. Chromatog.* **86**, 234.
[31] Giger, W. and Blumer, M. (1974). *Analyt. Chem.* **46**, 1663.
[32] Sawicki, E., Carey, R.C., Dooley, A.E., Giselard, J.B., Monkman, J.L., Neligan, R.E. and Ripperton, L.A. (1970). *Health Lab. Sci.* **7**, 31.
[33] *Idem* (1970). *ibid* **7**, 45.
[34] *Idem* (1970). *ibid* **7**, 60.
[35] Fallick, G.J. and Walters, J.L (1972). *Am. Lab.* **4(8)**, 21.
[36] Fox, M.A. and Staley, S.W. (1976). *Analyt. Chem.* **48**, 993 and references therein.
[37] Kertesz-Saringer, M. and Morlin, Z. (1975). *Atmos. Environ.* **9**, 831.
[38] U.S. Federal Registry (1971). **36** (228) 22384.
[39] Villalobos, R. and Chapman, R.L. (1971). *Analyt. Instr.* **9**, D-6, 1.
[40] Perry, R. and Harrison, R.M. (1976). *Chem Brit.* **12**, 185.
[41] Bongers, W. (1976). *Stichting Cowcawe, Hague*, 3/76.
[42] Zlatkis, A., Lichtenstein, H.A. and Tishbee, A. (1973). *Chromatographia* **6**, 67.
[43] Rasmussen, R.A. and Holden, M.W. (1972). *Chromatog. Newslett.* **1**, 31.
[44] Narain, C., Marron, P.J. and Glover, J.H. (1972). *Gas Chromatog. Proc. Int. Symp.* **9**, 1.
[45] Westberg, H.H., Rasmussen, R.A. and Holdren, M.W. (1974). *Analyt. Chem.* **46**, 1852.
[46] Desbaumes, E. and Imhoff, C. (1971). *Staub-Reinhalt Luft.* **31**, 257.
[47] Dimitriades, B. and Seizinger, D.E. (1971). *Environ. Sci. Technol.* **5**, 223.
[48] Schneider, W. and Fronhne, J.C. (1975). *Staub-Reinhalt Luft.* **35**, 275.
[49] Griffith, G.A., Drivas, P.J. and Shair, F.H. (1974). *J. Air Pollut. Control Ass.* **24**, 776.
[50] Williams, F.W., Stone, J.P. and Eaton, H.G. (1976). *Analyt. Chem.* **48**, 442.
[51] Rasmussen, R.A. and Hutton, R.S. (1972). *Bio. Science* **22**, 294.
[52] Rasmussen, R.A. (1972). *Am. Lab.* **4(7)**, 19.
[53] Rohrscheider, L., Jaeschke, A. and Kubik, W. (1971). *Chem. Ing. Tech.* **43**, 1010.
[54] Kaiser, R.E. (1973). *Analyt. Chem.* **45**, 965.
[55] Tyson, B.J. and Carle, G.C. (1974). *ibid* **46**, 610.
[56] Schubert, R. (1972). *ibid* **44**, 2084.
[57] Leinster, P., Perry, R. and Young, R.J. (1977). in preparation.
[58] Rollet, M. and Moisson, M. (1972). *Rev. Inst. Pasteur Lyon* **5**, 439.
[59] Crecelius, H.J. and Forweg, W. (1975). *Staub-Reinhalt Luft* **35**, 330.
[60] Aue, W.A. and Teli, P.M. (1971). *J. Chromatog.* **62**, 15.

[61]   Perry, R. and Twibell, J.D. (1973) *Atmos. Environ.* **7**, 929.
[62]   Bertsch, W., Chang, R.C. and Zlatkis, A. (1974). *J. Chromatog. Sci.* **12**, 175.
[63]   Dravnieks, A., Kiotoszynski, B.K., Whitfield, J., O'Donnell, A. and Burgwald, T. (1971). *Environ. Sci. Technol.* **5**, 1220.
[64]   Bourdin, M., Badre, R. and Dumas, C. (1975). *Analusis* **3**, 34.
[65]   Russell, J.W. (1975). *Environ. Sci. Technol.* **9**, 1175.
[66]   Mieure, J.P. and Dietrich, M.W. (1973). *J. Chromatog. Sci.* **11**, 559.
[67]   Pellizzari, E.D., Bunch, J.E., Berkley, R.E. and McRae, J. (1976). *Analyt. Lett.* **9**, 45.
[68]   *Idem* (1976). *Analyt. Chem.* **48**, 803.
[69]   Novotny, M., Lee, M.L. and Bartle, K.D. (1974). *Chromatographia* **7**, 333.
[70]   Gordon, R.J. (1976). *Environ. Sci. Technol.* **10**, 370.
[71]   Intersociety Committee (1972). *Methods of Air Sampling and Analysis*, Am. Pub. Health Assoc., Washington D.C. 11104-01-69T.
[72]   Sawicki, E., Elbeit, W.C., Stanley, T.W., Hauser, T.R. and Fox, F.T. (1960). *Int. J. Air Poll.* **2**, 273.
[73]   *Idem* (1960). *Analyt. Chem.* **32**, 810.
[74]   Sawicki, E., Fox, F.T., Elbert, W.C., Hauser, T.R. and Meeker, J.E. (1962). *Am. Ind. Hyg. Assoc. J.* **23**, 482.
[75]   Sawicki, E. (1964). *Chem. Anal.* **53**, 24, 28, 56 and 88.
[76]   Lao, R.C., Thomas, R.S., Oja, H. and Dubois, L. (1973). *Analyt. Chem.* **45**, 908.
[77]   Pupp, C., Lao, R.C., Murray, J.J. and Pottie, R.F. (1974). *Atmos. Environ.* **8**, 915.
[78]   Lao, R.C., Thomas, R.S. and Monkman, J.L. (1975). *J. Chromatog.* **112**, 681.
[79]   Murray, J.J., Pottie, R.F. and Pupp, C. (1974). *Can. J. Chem.* **52**, 557.
[80]   Rasmussen, R.A., Westberg, H.H. and Holdren, M. (1974). *J. Chromatog. Sci.* **12**, 80.
[81]   Giannovaria, J.A., Gondek, R.J. and Grob, R.L. (1974). *J. Chromatog.* **89**, 1.
[82]   Jeltes, R. and Burghardt, E. (1972). *Chemisch Weekblad* **68**, 16.
[83]   Siegel, D., Mueller, F. and Neuschwander, K. (1974). *Chromatographia* **7**, 399.
[84]   Heatherbell, D.A., Wrolstad, R.E. and Libbey, L.M. (1971). *J. Agric. Food Chem.* **19**, 1069.
[85]   Altshuller, A.P. (1968). *Adv. Chromatog.* **5**, 229.
[86]   Grob, K. and Grob, G. (1970). *J. Chromatog. Sci.* **8**, 635.
[87]   Burghard, E. and Jeltes, R. (1975). *Atmos. Environ.* **9**, 935.
[88]   Bertsch, W., Shunbo, F., Chang, R.C. and Zlatkis, A. (1974). *Chromatographia* **7**, 128.
[89]   Raymond, A. and Guiochon, G. (1973). *Analusis* **2**, 357.
[90]   Leinster, P., Perry, R. and Young, R.J. (1977). unpublished results.
[91]   Shadoff, L., Kallos, G. and Woods, J. (1973). *Analyt. Chem.* **45**, 2341.
[92]   Snyder, R.E. (1971). *J. Chromatog. Sci.* **9**, 638.
[93]   Mignano, M.J., Rony, P.R., Grenoble, D. and Purcel, J.E. (1972). *ibid* **10**, 637.

[94] Nicotra, C., Cornu, A., Massot, R. and Perilhon, P. (1972). *Gas Chromatog., Proc. Int. Symp. (Eur)* **9**, 9.

[95] Lao, R.C., Oja, H., Thomas, R.S. and Monkman, J.L. (1973). *Sci. Total Environ.* **2**, 223.

[96] Pebler, A. and Hicham, W.M. (1973). *Analyt. Chem.* **45**, 315.

[97] Schuetzle, D., Crittenden, A.L. and Charlson, R.J. (1973). *J. Air Pollut. Control Ass.* **23**, 704.

[98] Driscoll, J.N. and Warnech, P. (1973). *ibid* **23**, 858.

[99] Rasmussen, D.V., Fisher, T.P. and Rowan, J.R. (1972). *Can. J. Spectrosc.* **17**, 79.

[100] Bergert, K.H., Betz, V. and Pruggmayer, D. (1974). *Chromatographia* **7**, 115.

[101] Evans, J.E. and Arnold, J.T. (1975). *Environ. Sci. Technol.* **9**, 1134.

[102] Wang, I.C., Swafford, H.S., Price, P.C., Martinsen, D.P. and Butrill, S.E. (1976). *Analyt. Chem.* **48**, 491.

[103] Dennison, J.E., Viscon, R.E. and Broyde, B. (1973). *West Elect. Eng.* **17**, 3.

[104] Derwent, R.G. and Stewart, H.N.M. (1974). *Meas. Control.* **7**, 101.

[105] Decker, C.E., Royal, T.M. and Tammerdahl, J.B. (1974). *Govt. Rept. Announce (US)* **74**, 68.

[106] Frostling, H. and Brantte, A. (1972). *J. Phys. (A)* **5**, 251.

[107] Fischer, H., Neafelder, Koh, G. and Pruggmayer, D. (1974). *Fachz. Lab.* **18**, 214.

[108] Saena, O., Boldt, C.A. and Tarazi, D.S. (1972). *Govt. Rept. Announce (US)* **72**, 64.

[109] Poli, A.A. and Zinn, T.L. (1973). *Analyt. Instr.* **11**, 135.

[110] King, W.H. (1974). *Environ. Sci. Technol.* **4**, 1136.

[111] Altshuller, A.P. and Bufalini, J.J. (1971). *Environ. Sci. Technol.* **5**, 39.

[112] Kamens, R.M. and Stern, A.C. (1973). *J. Air Pollut. Control. Ass.* **23**, 592.

[113] Kopczynski, S.L., Lonneman, W.A., Sutterfield, F.D. and Darley, P.E. (1972). *Environ. Sci. Technol.* **6**, 342.

[114] Stevens, R.K. (1971). *Govt. Rept. Announce (US)* **71**, 184.

[115] Todd, T.M. (1971). *Am. Lab.* **3(10)** 51.

[116] Smith, R.C., Bryan, R.J., Feldstein, M., Levadie, B., Miller, F.A., Stephens, E.R. and White, N.G. (1972). *Health Lab. Sci.* **9**, 58.

[117] Stevens, R.K., O'Keeffe, A.E. and Ortman, G.C. (1972). *Air Qual. Instrum.* **1**, 26.

[118] Fee, G.G. (1971). *Analyt. Instr.* **9**, D-4 1.

[119] McCann, R.B. (1971). *J. Air Pollut. Control Ass.* **21**, 502.

[120] Burgett, C.A. and Green, L.E. (1976). *Am. Lab.* **8(1)**, 79.

[121] Villalobos, R. and Chapman, R.L. (1974). *Chimia* **28**, 411.

[122] *Idem* (1971). *ISA Trans.* **10**, 356.

[123] *Idem* (1972). *Air Qual. Instr.* **1**, 114.

[124] Karten, M.P.H. (1972). *Chemisch Weekblad* **68**, L19.

[125] Cooper, J.C., Birdseye, H.E. and Donnelly, R.J. (1974). *Environ. Sci. Technol.* **8**, 671.

[126] Krieger, B., Malki, M. and Kummler, R. (1972). *ibid* **6**, 742.

[127] Houben, W.P. (1971). *Analyt. Instr.* **9**, D-3, 1.

[128]   Hodgeson, J.A., McClenny, W.A. and Hanst, P.L. (1973). *Science* **182**, 248.
[129]   Bargues, P. (1973). *Off. Nat. Etud. Rech. Aerosp. Note Tech.* 213.
[130]   Hanst, P.L., Liefoh, A.S. and Gray, B.W. (1973). *Appl. Spectrosc.* **27**, 188.
[131]   Robinson, J.W. and Guagliardo, J.L. (1974). *Spectrosc. Lett.* **7**, 121.
[132]   Kobayasi, T. and Inaba, H. (1971). *Rec. Symp. Electron Ion Laser Beam Technol.* **11th**, 385.
[133]   Lidholt, L.R. (1972). *Opto-electronics* **4**, 133.
[134]   Katayana, N. and Robinson, J.W. (1975). *Spectrosc. Lett.* **8**, 61.
[135]   Loane, J. and Krishman, K. (1974) *Govt. Rept. Announce (US)* **74**, 72.
[136]   Hirschfeld, T., Schildkraut, E.R., Tannenbaum, H. and Tannenbaum, P. (1973). *Appl. Phys. Lett.* **22**, 38.
[137]   Braman, R.S. (1971). *Atmos. Environ.* **5**, 669.
[138]   Byer, R.L. (1975). *Opto-electronics* **7**, 147.
[139]   Jennen, A., Alaerts, G. and Ronsmans, G. (1975). *Analusis* **3**, 427.
[140]   Pop, C.S. (1973). *Rev. Chim. (Buch).* **24**, 44.
[141]   Ash, R.M. and Lynch, J.R. (1971). *J. Am. Ind. Hyg. Ass.* **32**, 410.
[142]   Scarengelli, F.P., O'Keeffe, A.E., Rosenberg, E. and Bell, J.P. (1970). *Analyt. Chem.* **42**, 871.
[143]   Blacker, J.H. and Brief, R.S. (1971). *J. Am. Ind. Hyg. Ass.* **32**, 668.
[144]   Saltzmann, B.E., Burg, W.R. and Ramasaway, G. (1971). *Environ. Sci. Technol.* **5**, 1121.
[145]   Homolya, J.B. and Bachman, J.P. (1971). *Int. Lab.* **(5)**, 37.
[146]   Dietz, R.M., Cote, E.A. and Smith, J.D. (1974). *Analyt. Chem.* **46**, 315.
[147]   Savitsky, A.C. and Siggia, S. (1972). *ibid* **44**, 1712.
[148]   Intersociety Committee (1972). *Methods of Air Sampling and Analysis*, Am. Pub. Health Ass., Washington, D.C. 42101-04-69T and 42101-01-69T.
[149]   *ibid*, 42101-02-69T.
[150]   Levaggi, D.A. and Feldstein, M. (1964). *Am. Ind. Hyg. Ass. J.* **25**, 64.
[151]   Robbins, R.C., Borg, K.M. and Robinson, E. (1968). *J. Air Pollut. Control Ass.* **18**, 106.
[152]   Klein, G. (1964). Linde-Reports **17**, 24.
[153]   Sawicki, E. (1970). *Crit. Rev. Anal. Chem.* **1**, 275.
[154]   Falk, H.L. and Steiner, P.E. (1952). *Cancer Res.* **12**, 30 and 40.
[155]   Lindsey, A.J. and Stanbury, J.R. (1962). *Int. J. Air Water Poll.* **6**, 387.
[156]   Clearly, G.J. and Sullivan, J.L. (1965). *Med. J. Australia* **52**, 758.
[157]   Del Vecchio, V., Valori, P., Melchiorri, C. and Grella, A. (1970). *Pure and Appl. Chem.* **24**, 739.
[158]   Stanley, T.W., Meeker, J.E. and Morgan, M.J. (1967). *Environ. Sci. Technol.* **1**, 927.
[159]   Sawicki, E., Stanley, T.W., Elbert, W.C., Meeker, J. and McPherson, S. (1967). *Atmos. Environ.* **1**, 131.
[160]   Chatot, G., Castegnaro, M, Roche, J.L., Fontagnes, R. and Obaton, P. (1971). *Analyt. Chim Acta* **53**(2), 259.
[161]   Rosen, A.A. and Middleton, F.M. (1955). *Analyt. Chem.* **27**, 790.
[162]   Tabor, E.C., Hauser, T.R., Lodge, J.P. and Burtschell, R.H. (1958). *Arch. Ind. Health* **17**, 58.

[163] Moore, G.E. and Katz, M. (1960). *Int. J. Air. Poll.* **2**, 221.
[164] Conlon, R.D. (1969). *Analyt. Chem.* **41**, 107A.
[165] Stubert, W. (1973). *Chromatographia* **6**, 205.
[166] Horning, E.C., Carroll, D.I., Dzidic, I., Haegele, K.D., Horning, M.C. and Stilwell, R.N. (1974). *J. Chromatog.* **99**, 13.
[167] Popl, M., Stejskal, M. and Mostecky, J. (1975). *Analyt. Chem.* **47**, 1947.
[168] Majer, J.R. and Perry, R. (1970). *Pure and Appl. Chem.* **24**, 685.
[169] Perry, R. (1971). *Int. Symp. Ident. Meas. Environ Pollut.* (Proc.) 130, (Westley, B. ed.), Natl. Res. Council Canada, Ottawa.
[170] Mulik, J., Cooke, M., Guyer, M.F., Semenink, G.M. and Sawicki, E. (1975). *Analyt. Lett.* **8**, 511.
[171] Wheals, B.B., Vaughn, C.G. and Whitehouse, M.J. (1975). *J. Chromatog.* **106**, 109.
[172] Karr, C., Childers, E.E. and Warner, W.C. (1963). *ibid* **35**, 1290.
[173] Schmit, J.A., Henry, R.A., Williams, R.C. and Dieckman, J.F. (1971). *J. Chromatog. Sci.* **9**, 645.
[174] Kirkland, J.J. (1969). *Analyt. Chem.* **41**, 218.
[175] Vaughan, C.G., Wheals, B.B. and Whitehouse, M.J. (1973). *J. Chromatog.* **78**, 203.
[176] Klimisch, H.J. (1973). *Analyt. Chem.* **45**, 1960.
[177] Sawicki, E., Stanley, T.W., Elbert, W.C. and Pfaff, J.D. (1964). *ibid* **36**, 497.
[178] Sawicki, E., Stanley, T.W., Pfaff, J.D. and Elbert, W.C. (1974). *Chem. Anal.* **53**, 6.
[179] Sawicki, E., Stanley, T.W., McPherson, S. and Morgan, M. (1966). *Talanta* **13**, 619.
[180] Hauser, T.R. and Pattison, J.N. (1972). *Environ. Sci. Technol.* **6**, 549.
[181] Janini, G.M., Johnston, K. and Zielinski, W.L. (1975). *Analyt. Chem.* **47**, 670.

CHAPTER SEVEN

Halogen containing compounds

7.1 Fluorides

7.1.1 Occurrence

Fluorine is present in waste gases from several manufacturing processes and occurs in the gaseous form as HF or less commonly as silicon tetrafluoride and in particulate form as metal fluorides. Elemental F is only found in the atmosphere in trace amounts.

Apart from those industries in which F_2 or fluorides are specifically produced, a variety of other industrial processes use fluorine compounds as catalysts or fluxes with fluorine also occurring as an impurity in raw materials used.

The most significant sources of emissions are from such heavy manufacturing processes as aluminium and steel smelting, the fertilizer industry, and cement, brick, tile and ceramic industries. Less significant sources of emission include combustion of coal, glass manufacture and production of high octane motor fuels.

Fluorine compounds in the atmosphere are generally considered to be toxic to plants and animals, the present Threshold Limit Value (TLV) for total fluorides as F^{θ} in the air is $2.5\,mg\,m^{-3}$. Fluorides can affect man directly through air pollution and indirectly through contaminated plants and through animals feeding on contaminated vegetation. The highest levels tend to occur in green vegetables and fruits grown in the vicinity of emissions. Unlike SO_2, fluorides do not take part in plant metabolic reactions, but tend to accumulate in concentrations over 50 ppm both on the inside and outside of leaves. Thus, atmospheric concentrations as low as 0.001 ppm fluoride may cause appreciable damage to vegetation. Plants most sensitive to fluorides are tulips, gladiolas, conifers, and stone fruits such as apricot and prune. In the presence of SO_2, HF produces synergistic effects on these plants. Symptoms of chronic fluoride poisoning in man include abnormal hardening of bones and spots on teeth, caused by disturbances in Ca metabolism and a lowered resistance to illness, especially amongst young children. These effects occur after prolonged exposure to levels as low as $0.2\,\text{mg}\,\text{m}^{-3}$ of fluorides. At higher concentrations, acute effects include bronchitis, pneumonia, laryngitis and tracheitis, again, predominately amongst the young and old. Fluorides also affect, animals especially grazing stock through plants rather than by direct inhalation causing mainly chronic and in some instances, acute fluorosis.

7.1.2 Analytical methods for particulate matter

Despite the TLV for fluorides of $2.5\,\text{mg}\,\text{m}^{-3}$, concentrations in the μg range are known to cause damage to vegetation. Therefore, several analytical methods are required for the determination of fluorides (as F^{θ}) in both the μg and mg ranges, the former concentrations generally being found in ambient air, whilst the latter are to be encountered in stack gases and from other emission sources.

Most analytical methods involve the formation of coloured complexes, the intensity of which is proportional to the concentration of fluoride ions in the sample. The analysis is achieved by colorimetric fluoride titration, the end point being detected by a conductimetric, photometric or potentiometric procedure. Another more widely adopted method involves the use of a spectrophotometer in observing the intensity of the coloured complex.

Table 7.1 Effect of interfering substances on fluoride determinations

Substance	Concentration (mg litre^{-1})	Effect on $1.0\,\text{mg}\,\text{litre}^{-1}$ Fluoride reading	
		Increase (mg litre^{-1})	Decrease (mg litre^{-1})
Acidity	3000	0.1	
Alkalinity	5000		0.1
Aluminium (3^+)	0.1		0.1
Arsenite	1300	0.1	
Calcium (2^+)	800	0.1	
Chloride ($^-$)	7000	0.1	
Chlorine	(should be completely absent)		
Colour	(should be low or compensated)		
Iron (3^+)	10		0.1
Magnesium	1250	0.1	
Manganese (2^+)	40	0.1	
Phosphate (2^-)	16	0.1	
Sodium	1	0.1	
Sulphate (2^-)	200	0.1	

Several cations and anions interfere with the determination of fluorides, particularly when colorimetric methods of analysis are used. These interfering substances are shown in Table 7.1, together with their effects on fluoride determination. Cl_2 is easily the most powerful interfering agent, but at the same time, it is also a most unlikely substance to be found as a gas in the sample prepared for titration or colorimetric determination. The effect of interfering agents varies a great deal with the method employed and in most instances, their removal is required prior to determination and for this, several methods are available. The most commonly utilized method is steam distillation, originally developed by Willard and Winter [1], which has since been modified [2]. If the distillation is carried out with adequate care, recoveries of up to 95% of the amount of fluorides present may be achieved. Fluorides may also be recovered from interfering substances by preferential adsorption on to an ion exchange resin [3, 4] particularly when the interfering agents are cationic. The fluoride is then recovered as hydrofluoric acid on an anionic exchange resin which is then removed as sodium fluoride by elution with NaOH solution.

Newman [4] used De-Acidite FF to remove both cationic and anionic interferences successfully. The efficiency of the method depends on the resin used, and overall use of these resins is limited to samples containing low concentrations of interfering ions. Several authors have reported that samples containing high concentrations of Al, phosphate or Fe (3^+) ions may not be suitable for treatment by ion exchange.

Pyrohydrolysis of fluorides in samples has been successfully carried out to recover fluorides as HF [5] Newman [6] has further developed the method so that the HF evolved during the process is absorbed into an alkali. Isolation of fluorides can be achieved by the microdiffusion technique [7, 8]. The sample is heated with a diffusion acid in an unsealed container. The HF evolved is trapped in an alkaline coating on the lid of the vessel. On the whole, steam distillation is the most satisfactory method for the elimination of interfering agents. Although it is tedious, time consuming and complicated, it is the most frequently used pretreatment process.

7.1.2.1 Sampling

Fluorine compounds can exist in the atmosphere in either gaseous or particulate forms. The chosen sampling technique will therefore depend on the nature and source of the fluorine-containing species.

Sampling for total fluorides is relatively easy and several methods exist which can be usefully employed. The most commonly used method is the wet impingement technique, where either water or an alkaline solution is used to collect or trap fluorides. Farrah [9] has shown that water is preferable to alkaline solution, due to its ready availability in a purified form, stability and easy dissolution of HF gas. Collection efficiencies for this method of 97–100% are reported. The impinger used may be of Greenburg-Smith or similar design, the efficiencies of which have been reported by Hill *et al.* [10] to be about 90–98%. These impingers are designed to operate at flow rates of 20–30 litres min^{-1} for gaseous sampling; and for particulates, flow rates of 50 litres min^{-1} should be adopted. The efficiency of the collection process is improved by collecting several impingers in series. Glass fibre filters have been successfully used by Benedict *et al.* [11] for the collection and separation of gaseous and particulate fluorides. The filters were more efficient in collecting particulates, especially in the lower size range, than the standard impinger, which is more efficient in connecting any gaseous fluorides.

Brek and Stratmann [12a] have developed a method where fluorides are absorbed into tubes, packed with 3 mm diameter silver beads, which have previously been drenched with 20% sodium carbonate solution and dried at 200° C. Sampling is carried out at a rate of 60–70 litres min^{-1} for 10–15 min, achieving collection efficiencies of up to 99% for air containing between 5 and 500 $g\,m^{-3}$ fluorides.

Sampling for total fluorides has very little significance in air pollution studies generally. Several techniques are necessary to sample gases and particulates separately. Hill *et al.* [10] used an electrostatic precipitator to remove particulates prior to sampling for HF by absorption. Losses of gas in the particulate removal stage and the instability of HF reduced the overall collection efficiency of this method.

Good separation of gases from particulates can be achieved by drawing the air sample through a molecular membrane filter, which under low dust loading can retain only as little as 7% of the HF content, but retains up to 37% at higher dust levels, as found in stack gases. Collection of the gas can be performed by adsorption on to a solid material or absorption into a suitable liquid.

Collection efficiencies of the solid-adsorption method for gaseous fluorides are governed by the type of material used, surface area per unit mass of solid, rate of flow through the adsorbent and temperature of the gas.

Benedict *et al.* [11] have used concentric alumina tubes for the removal of HF from air samples and have achieved collection efficiencies of 80–96% for sampling rates of between 8 and 35 litres min^{-1} [10]. However, the use of sodium carbonate-coated glass tubes has given better recovery levels by washing than alumina, with the sacrifice of good separation of gas and particulates. These methods are only applicable for air sampling where fluoride levels are relatively low. For higher concentrations, such as are to be found in stack gases, alternative methods must be used.

Dorsey and Kemnitz [12b] and Smith and Martin [13] have developed a method for this purpose, in which HF is converted to the less reactive silicon tetrafluoride, followed by the separation of particulates by filtration and collection of the gaseous phase in two impingers containing solutions of NaOH. Filters used to trap the particulates may be of sodium carbonate-impregnated paper or membrane type [14], but silicon tetrafluoride may be retained unless the filter is maintained at a temperature high enough to prevent moisture condensation. According to Finkelstein *et al.* [15], the retention of gaseous fluorides on filters may be minimized by having sampling rates greater than 2 litres min^{-1} and a sample humidity below 90%.

7.1.2.2 Analytical procedures — chemical

Chemical methods

Colorimetric methods are the most widely used in the determination of fluorides, based on either the removal of free dyestuff from a metal dye lake as in zirconium–alizarin or on the formation of a coloured complex between fluoride and alizarin complexan. The earlier methods were based on the former principle and modifications of these are now used in the analysis of fluorides. When fluoride ions react with alizarin–thorium or alizarin–zirconium colour lakes, the metals are removed from the lake to form stable fluorides, initiating a change in colour, which can be measured with a spectrophotometer at low concentrations and measured visually at high concentrations. The determination of fluorides using the latter principle involves the titration of fluorides against a zirconium or thorium salt using alizarin as the indicator, until the yellow alizarin changes to the complex red colour. A revision of this method has been published by the American Society of Testing Materials [2] to cover a wide range of weights in the final solution from 10 mg down to $1\,\mu$g.

There are other modifications and revisions of these basic methods, notably those of Bellack and Schoboe [16], who introduced the use of SPADNS (sodium-2-(p-sulfophenylazo)-1,8-di-hydrocynapthalen-3,6-dilulphate) instead of alazarin, applicable for concentrations between 0.05–1.5 mg litre^{-1}. The sensitivity of SPADNS to interference is minimal. Instead of SPADNS, zirconium-erichrome cyanine R may be used [17, 18]. Here the colour of the reagent is bleached out by the fluoride ions in solution and the intensity of the colour is measured using a spectrophotometer.

Barney and Hensley [19] developed a spectrophotometric method which is based on the reaction between fluoride and thorium chloroanilate in aqueous methyl cellulose at pH 4.5. For solutions of low fluoride concentrations (0–2 mg litre^{-1}) a methyl cellulose to water ratio of $1:9$ is recommended and optical density measured at 330 nm. For higher concentrations, a $1:3$ ratio is used at 540 nm. However, interferences by both cations and anions are common. Belcher *et al.* [20, 21] proposed a method where an alizarin blue complex is produced by reaction between fluoride ions and a red complex consisting of a lanthanum salt and alizarin complexan (3-aminomethyl-(dicarboxymethyl)-1,2-dihydroxyanthranquinone). The reaction occurs at pH 4.3 to form a red metal lake which gives a measurable blue colour under the specific action of fluoride ions. The most sensitive pH range for the spectrophotometer is 5.0–5.2 [22] and at this pH, the lanthanum complex is most sensitive to fluoride ions. The absorbance may be measured in the u.v. range at 281 nm or in the visible range at 618 nm, the sensitivity in the u.v. range being 200% greater than that in the visible [23]. However, a marked increase in sensitivity at 618 nm can be achieved by performing the reaction in a 25% acetone solution [24]. The method can be used for a concentration range of 5 to 50 μg per unit volume of final solution, and can be extended down to 1.0 μg by further modification [22].

This method has been further developed by West *et al.* [25] and Liiv and Luiga [26]. They adopted a lanthanum to alizarin complexan molar ratio of 2 : 1 for the visible range, which reportedly gives a considerable increase in sensitivity towards fluoride ions. Liiv and Luiga [26] have further shown that for determinations in the u.v. range, a lanthanum to alizarin molar ratio of 1 : 1 gave an increased sensitivity towards fluoride. The optimum concentration of the reagent was 6×10^{-5} M and the absorbance was measured at 280 nm.

In the past few years, automation of colorimetric techniques for fluorides has been affected by the development of an auto-analyser by Technicon (U.S.A.). A typical arrangement consists of a sampler, proportioning pump, gas/liquid separator, a colorimeter and a recorder. In the automated procedure, fluorides react with the red lanthanum alizarin complexan to form a lilac-blue species and absorbance measured by a colorimeter at a suitable wavelength. The equipment may either be used for continuous monitoring of fluorides or for discrete samples in the laboratory [27–29].

The continuous monitoring mode uses NaOH to absorb HF only, particulates being removed by filtration at the sampling stage and cannot take account of effects produced by interfering agents. For discrete samples, the auto-analyser can be linked to a distillation apparatus, which removes interfering ions. The method has mostly been used in the determination of fluorides in plant tissues, water and impinger solutions [27, 28, 30, 31].

The potentiometric method as described by O'Donnel and Stewart [32] is based on the formation of a cerium (4$^+$)-fluoride complex causing a lowering of the cerium (4$^+$)-cerium (3$^+$) redox potential in a system consisting of two half-cells, each containing cerium (3$^+$) and cerium (4$^+$) ions. When a sample containing fluoride ions is added to one half-cell and is subsequently compensated for by addition of a standard fluoride solution to the other half-cell, the volume required for compensation is directly related to the concentration of fluoride ions in the sample.

A semi-micro determination of fluoride ions by a conductimetric method has been developed by Kubota and Surak [33], in which fluoride in the sample are recovered as fluorosilic acid by steam distillation and estimated by the conductimetric titration of the acid with lanthanum acetate, causing a decrease in conductance. Fluorides may also be determined by a complexometric titration, as suggested by Leonard [34]. Fluoride in solution is precipitated as lead chlorofluoride in the presence of chloride ions, and ethyl alcohol by a standard solution of lead nitrate. The excess lead nitrate is then titrated against EDTA using xylenol orange as an indicator.

Chemical methods in general are time consuming and in most cases require pretreatment of the sample for the removal of interfering ions, however this does not detract from the simplicity and effectiveness of the procedures.

7.1.2.3 Analytical procedures — physical

Physical methods

Only a very few physical methods are presently available for use in the determination of fluorides and of these, the recently developed specific ion electrode, is the most widely used [17, 35, 36]. The apparatus consists of a lanthanum fluoride crystal doped with europium [37], one face of which is in contact with a reference solution of chloride and fluoride ions, the other face coming into contact with the sample solution. A potential difference is generated when a sample containing fluoride ions F^θ is introduced, and its magnitude is related to the ratio of the fluoride ion activities of the sample to that in the reference solution. The potential developed is measured against a standard single junction reference electrode on an extended scale pH meter.

The electrode can be used down to concentrations of $20 \, \mu g \, litre^{-1}$, but at this concentration, it is necessary to remove any interfering cations. This can be achieved by the addition of a sodium citrate buffer solution, which complexes interfering cations such as aluminium and iron (3+), buffers the solution and controls the total ionic strength [35].

Both hydroxyl and hydrogen ions are known to affect the determination of fluorides by this method, however hydroxyl OH^θ ions only have a marked effect at low fluoride concentrations in the sample solution, whilst at pH's below 5.0, hydrogen ions can complex fluoride ions to give HF_2^θ, but this can be avoided by adding a suitable buffer solution. The most suitable pH for use of the electrode is in the range 5.0—8.0 for ambient air and stack gas concentrations, where greatest accuracy and reliability is achieved. At other pH values, sodium citrate buffer can be used to good advantage.

Table 7.2

Location	Sampling volume (m³)
Inside manufacturing plant or emitting sources, stack gases after pretreatment	0.1
Outside emission sources	1.0
Ambient air-away from emission sources	10

Table 7.3

Method	Rate (litre min⁻¹)	Particle size (μm)	Collection efficiency %
Electrostatic precipitator	50–100	all sizes	99+
Filter paper, dose texture	25– 50	min. 0.5	99+
Similar grade 125 mm	25– 50	max. 0.5	90
Membrane filters (Millipore or equivalent)			
AA 47–50 mm	25	min. 0.3	99+
HA 47–50 mm	25	min. 0.3	99+
Standard impinger	20– 30	min. 0.5	99
	20– 30	min. 0.5	50
	40– 55	min. 0.1	99+

7.1.2.4 Experimental methods

Sampling

The volume of air sampled depends generally upon the location selected for sampling and the type of sample required. The sampling volume needed to give a sufficient quantity of fluoride for analysis may be estimated as in Table 7.2. The sampling rate adopted depends on the system used and ASTM [2] gives the guidelines as in Table 7.3. The sampling train generally consists of a filter holder and three impingers connected in series, followed by a control valve, vacuum pump and a gas meter. The first two impingers are two-thirds full with 0.1 M NaOH and the third filled with silica gel to act as a moisture trap. If desired, an electrostatic precipitator may be substituted for the filter apparatus and the sampling train modified accordingly (see Chapter 1).

The particulate matter collected usually contains fluorides as metallic compounds in an insoluble form, which require conversion to a soluble form by fusion with NaOH prior to the isolation process. This is achieved by soaking the filter paper in a slurry of lime water prepared from fluoride free lime, until the mixture is alkaline to phenolphthalein, then evaporating to dryness and ashing in a muffle furnace at 550–600° C for 1 h, until all the organic matter is completely oxidized.

The contents of the crucible are then fused with 2 g of NaOH and dissolved in 10–15 ml water, followed by the addition of H_2O_2 solution (30%) to convert sulphites to sulphates and boiling to remove excess H_2O_2.

The collected particles from an electrostatic precipitator are removed from the apparatus and made into a slurry with 0.1 g calcium oxide and 2 g NaOH. The dried contents are then fused and dissolved in 5 ml water and treated as previously described with H_2O_2.

Impinger samples will almost certainly contain no insoluble fluorides so all that is necessary is to adjust the pH to alkaline phenolphthalein, evaporate down to between 10 and 15 ml and treat with H_2O_2.

Ion exchange

Apparatus is set up as shown in Fig. 7.1; the ratio of the height of the column to the diameter should be about 20:1 and the outlet capillary end of the column is set about 5 mm higher than the top of the resin bed. This procedure is used for the removal of interfering cations.

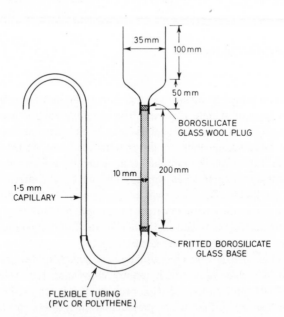

Fig. 7.1   Ion exchange column.

Reagents

Intermediate-base of the granular aliphatic polyamine type Anolite A41, A43; Permutit A; Rexyn 204 (OH) (− 60 to + 100 B.S. sieve); 2.0 and 1.0 M HCl; and 2.0, 0.1, 0.01 M NaOH.

The resin should first be treated with 200 ml 2 M HCl followed by 200 ml distilled water, then 200 ml 2 M NaOH and a final rinse of 200 ml de-ionized water.

The treated sample is acidified with 1 ml 1 M HCl and filtered to remove suspended matter. It is then added to the column and allowed to pass through at 5 ml min^{-1}. The column is then washed with 10 ml distilled water before elution of the fluoride ions by adding 25 ml of 0.1 M NaOH followed by 25 ml of 0.01 M NaOH.

When interferences are present in both cationic and anionic forms, a modification of the previous procedure is used [4]. The procedure and apparatus is basically the same, but with the following changes.

Reagents

Strongly basic anion exchange resin; Polystyrene quaternary ammonium type; De-Acidite FF (Chloride form) size-60-+ 100 B.S. sieve; 1 M HCl; 1 M NaOH (Analar); 1 litre 0.025 M ammonium chloride, pH 9.2; and EDTA solution.

The column is prepared as follows; wash the resin with 100 ml 1 M HCl, 100 ml water rinse, 100 ml NaOH and a final rinse with 100 ml de-ionized water, followed by 50 ml of the eluting solution, ammonium chloride.

After treating the sample with 2 ml EDTA solution and filtering to remove suspended matter, add it to the column, allowing it to pass through at 5 litres min^{-1} and wash the column with 10 ml de-ionized water. Elute the fluoride ions with 50 ml 0.025 M ammonium chloride solution.

Note. For the general theory of ion exchange methods see [38].

Steam distillation

Apparatus is set up as shown in Fig. 7.2 and the sample is distilled from either perchloric or sulphuric acid solution in the presence of silica as fluorosilicic acid.

Reagents

Concentrated H_2SO_4 (98% by weight); concentrated perchloric acid (72% by weight); 50% silver perchlorate solution; and 0.2 M NaOH solution.

Transfer the prepared sample to the distilling flask, limiting the total volume to 50 ml. Rinse the sample container with 50 ml perchloric acid, add 1 ml silver

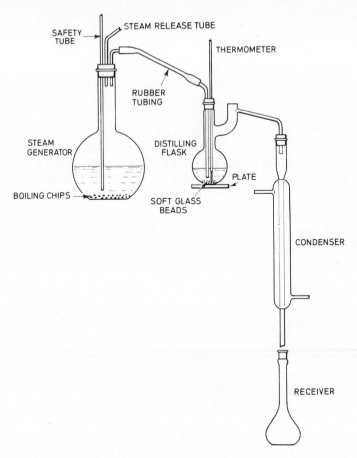

Fig. 7.2   Steam distillation apparatus for fluoride determination.

perchlorate solution and transfer to the distilling flask plus washings. Commence distillation, heat the contents of the distilling flask to 135° C and maintain for about 1 h, until 250 ml of distillate is collected for analysis. If all interferences are not removed, a double distillation may be adopted.

Microdiffusion technique

Apparatus

Polystyrene Petri dish, 50 mm diameter, 8 mm deep; polythene micropipette 0.1 ml capacity; and thermostatically controlled oven, 55–60° C range.

Reagents

Perchloric acid solution (45%); and methanolic NaOH solution. Coat the under-side of the Petri dish lid with approximately 0.1 ml methanolic NaOH solution to an area about 3 cm in diameter and dry in a clean, fluoride-free atmosphere. Transfer 2 ml of the sample solution to the dish bottom, add 3 ml perchloric acid solution, replace the lid immediately and place in oven at $60°$ C for 16–20 min. Remove the Petri dish and allow to cool. Transfer the alkaline coating from the lid quantitatively to a 25 ml flask and make up with distilled water.

Note. A simplified version of this method has been devised by Marshell and Wood [18]. Recoveries with this method are reported to be about 99% in the μg range and 98% for the sub-μg range.

7.1.2.5 Analytical methods

Titrimetric method

Here, the sample solution is titrated against thorium nitrate using alizarin sul-phuric acid as the indicator.

Reagents

Alizarin solution; 0.1 N thorium nitrate solution; 0.25 M HCl; 0.25 M NaOH; and chloroacetate buffer solution.

Pipette an aliquot (10–15 ml) of the sample solution into a 250 ml flask and dilute to 100 ml with de-ionized water. Add 2 drops of indicator followed by NaOH dropwise until a pink colour appears. Carefully add HCl until the pink colour just disappears, add 5 drops buffer solution and titrate against thorium nitrate solution to a permanent faint pink end-point. A blank should be carried out using de-ionized water in place of the sample, and a calibration curve should be constructed for the determination of F^{θ} ion concentration. By determining the concentration of fluoride ions in the sample aliquot and working backwards the original atmospheric concentration can be determined, and expressed as either $mgF^{\theta} m^{-3}$ or ppm by volume of F or HF.

Spectrophotometric methods

Alizarin complexan forms a red lake with lanthanum which in the acetate buffer solution at pH 4.3 or 5.2 yields a blue water-soluble complex with fluoride ions. The principle is suitable for samples of low fluoride concentration.

Interfering cations may be removed by the use of masking agents such as pot-assium cyanide and sodium sulphide or by extraction with 8-hydroxy quinoline.

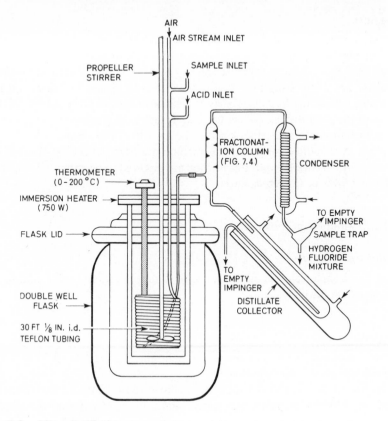

Fig. 7.3 Microdistillation apparatus.

Reagents

5×10^{-4} M alizarin complexan; 0.02 M lanthanum nitrate; 5×10^{-4} M lanthanum nitrate; buffer solution (pH 4.3); buffer solution (pH 5.2); acetone (Analar); glacial acetic acid; and stock fluoride solution.

Visible region determination

Transfer an aliquot of the sample solution to a 100 ml flask, add 10 ml 5×10^{-4} M alizarin complexan and 2 ml buffer solution (pH 5.2), followed by 10 ml 5×10^{-4} M lanthanum nitrate solution and shake. Finally add 25 ml acetone and dilute to 100 ml with water and mix. Allow to stand for 90 min and measure the absorbance against a reagent blank at 618 nm in a 1 cm cell. Determine the fluoride concentration in the sample by reference to a suitably constructed calibration curve.

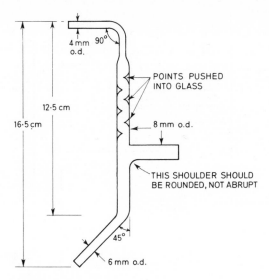

Fig. 7.4 Microdistillation column.

The modification of this method previously described gives an increase in senisitivity and is basically the same, but uses a lanthanum nitrate to alizarin complexan ratio of $2:1$.

Ultra violet region determination

Repeat the procedure as described for the visible region determination above, but with the following variations:

(a) Use 2 ml pH 4.3 buffer solution;

(b) Do not add acetone;

(c) Measure the absorbance at 281 nm in a 1 cm cell.

There are a number of variations on the basic spectrophotometric determinations, notably those of Belcher and West [22], Bellaek and Schouboc [16], Dixon [39] and Megregian [18]. It is not intended to go into detail concerning these methods as the basic principles are similar, variations occurring in the use of reagents and wavelength measurements only.

Semi-automated and automated analysis

As previously mentioned in Section 7.1.2.2, these methods involve the use of colorimetric determination of an alizarin-fluorine blue complex. In the semi-automated mode, the analysers can be used with a microdistillation unit to

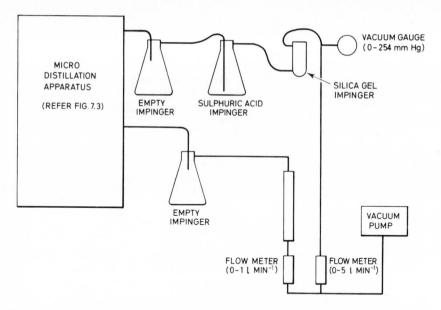

Fig. 7.5   Schematic drawing of air flow system.

remove any interferences for discrete, operator-supplied samples, see Figs. 7.3–7.6. In the automated mode, the microdistillation unit is not used and therefore autoanalysers should only be used when interfering substances are low or absent, if accurate results are required.

Physical methods

Specific ion electrode

This instrument has been described in Section 7.1.2.3 and readers should refer to [1] for details of the setting up, operation and calibration of these instruments.

7.2 Chlorine

7.2.1 Occurrence

Cl_2 is emitted to the atmosphere mainly as a result of its own manufacturing process, manufacture of associated compounds or as a bi-product or waste from other industrial activities. Less significant emission sources include sewage and potable water treatment plants, where it is used as a disinfectant.

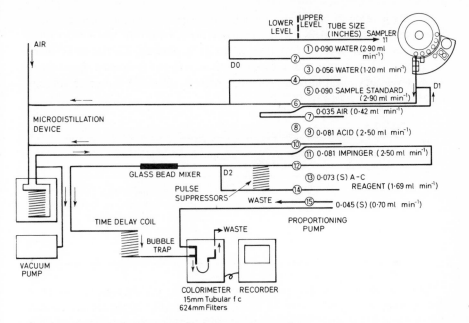

Fig. 7.6   Flow diagram of semi-automated procedure.

Table 7.4   Typical emissions from a chlorine-producing plant (U.S.A.)

Source	Chlorine (kg $(100\text{ t})^{-1}$ of liquefied chlorine)
Mercury cell plant (uncontrolled)	17.87−71.50
Diaphragm cell (uncontrolled)	8.94−44.68
Water absorber	1.79
Carbon tetrachloride absorber	0.40
Sulphur monochloride	0.13
Caustic/lime scrubber	0.0004
Tank car vents	2.01
Storage car vents	5.36
Air blowing of mercury cell	2.23

During the manufacture of Cl_2 gas by electrolysis of brine in a mercury or diaphragm cell, the product is subjected to a number of repurification steps in various plant, and the relevant emissions are shown in Table 7.4 [40, 41]. Where Cl_2 is used in the pulp and paper industry for bleaching purposes, emissions of up to $0.08-0.19$ kg t^{-1} chlorine used have been reported [42].

Cl_2 gas is toxic to plants and animals, the suggested maximum concentration in working areas in the U.K. being 1 ppm or $3 \, \text{mg m}^{-3}$ [43]. At low levels, typical symptoms in man include irritation of the eyes, nose and throat, whilst acute symptoms are permanent lung damage and bronchitis. Prolonged exposure to severe levels may result in fatal pulmonary oestema. The effects on man are summarized in Table 7.5 [33].

Table 7.5

Concentration of chlorine		
ppm v/v in air	mg m⁻³ (20° C)	Effects
1	3	Mild smarting of eyes and
4	12	irritation of nose and throat
10	29	Severe coughing and eye irritation within 1 min
> 10	> 29	Immediate and delayed serious effects

There is little conclusive data available regarding the damage to vegetation due to gaseous Cl_2, reflecting its infrequent occurrence in significant concentrations in ambient air. This is in part due to a relatively low number of possible emission sources and the fact that most waste gases are recovered for economic reasons. The most commonly reported damage to plants includes marginal and bifacial neurosis and interveinal lesions, with the most sensitive species being Eastern cotton wood, silver maple, and Bartlett pears. Defoliation was evident in walnut, willow, sycamore and peach with premature fruit drop in apple, peach and pear [44]. Injury may occur at Cl_2 concentrations of greater than 0.3 ppm in the atmosphere, the lower levels in the ppb range not causing damage because chlorine is not accumulated in the leaves, as reported by Brennan et al. [45].

7.2.2 Analytical methods

Although Cl_2 is rarely found in toxic concentrations in the atmosphere, determination in the 0.1–1.0 ppm range is important as toxic effects occur both in plants and man at these levels. There are several methods of analysis available, but none sufficiently sensitive nor specific for accurately determining atmospheric concentrations. Almost all the available methods rely upon the strong oxidizing properties of Cl_2 and are therefore not selective for Cl_2, and are subject to interference from other strong oxidizing agents such as O_3, NO_2, bromine and chlorine dioxide.

The majority of the methods in use are based upon colorimetric reactions, the most frequently used being the o-tolidene procedure, where Cl_2 reacts under acid conditions to form the yellow haloquinone of o-tolidene dihydrochloride. Air is drawn through impingers containing NaOH, which is later acidified and analysed for free chlorine colorimetrically at 435 or 490 nm by the addition of o-tolidene [46, 47].

Alternatively, air may be drawn directly through impingers containing an acid solution of o-tolidene and the colour change is measured spectrophotometrically as before. The method is suitable for concentrations down to about 1 ppm or $3 \, mg \, m^{-3}$ which is the TLV for Cl_2, requiring a sample volume of at least 3 litres. However, Technicon Industrial Systems claim that the autoanalyser using the same reagent can detect down to about 0.024 ppm Cl_2 in ambient air [47]. More recently, o-tolidene has been replaced by 3, 3',-dimethyl-napthidine, which is more sensitive [48], but also subject to interference from oxidizing agents.

It has been reported in the Russian literature [49] that the oxidation of arsenious anhydride in alkaline solution in the presence of potassium iodide and starch has been used to measure chlorine gas concentrations down to $1 \, mg \, m^{-3}$. As before, the colour change is measured spectrophotometrically, but other oxidizing agents do not interfere with the determination.

A method using N,N' dimethyl-p-phenylene diamine, was originally used for measurements in the ambient air [50]. The reagent is oxidized by Cl_2 and other oxidants to a mauve coloration, the intensity of which is measured by a spectrophotometer at 555 nm. The procedure is only applicable for low Cl_2 concentrations in the presence of HCl, as high levels of Cl_2 tend to bleach out the colour completely giving erroneous readings. At higher concentrations, Cl_2 is absorbed into 1% KI solution in 1% sodium acetate and the iodine replaced by chlorine produces a red colour with the reagent.

Methyl orange is also used as the detecting reagent [28, 51] which at pH 3.0 is decolorized, and monitored spectrophotometrically. The Intersociety Committee [28] claim that this method can be used to measure chlorine in the range $0.05-1.0 \, ppm$ $(145-2900 \, \mu g \, m^{-3})$. Other oxidizing agents interfere with this analytical procedure.

7.2.2.1 Sampling

Unlike HF, Cl_2 gas poses few problems in sampling due to its low interaction with particulate matter in the air. Therefore, the two can easily be separated by use of a suitable filter or glass wool plugs. The gas is then drawn through two midget impingers containing an absorbing medium. Midget impingers are used as they give better recovery efficiencies than the normal types. Several types are available, their use having been discussed in detail in Chapter 1.

The choice of absorbing medium is governed by the type of analytical method to be employed, but the most commonly used medium is NaOH solution, in which Cl_2 is converted to sodium hypochlorite and NaCl. The other solutions used react with Cl_2 gas to produce a colour change which is measured spectrophotometrically.

The sampling rates adopted depend on the reactivity of the absorbing media and the type of media used. These will be discussed in more detail in the specific analysis sections.

7.2.2.2 Analytical procedures

Methyl orange method

The reaction is carried out at pH3, which is the most sensitive level for Cl_2. Substances such as manganese (3^+), bromine, iron (3^+), nitrates and SO_2 interfere in the determination, of which SO_2 is critical, being generally found in similar or greater concentrations than Cl_2 in the atmosphere. Presence of SO_2 can cause up to 30% reduction in the Cl_2 reading. Nitrates give possible interferences of the order of 20% of the true Cl_2 concentration.

Reagents

Stock Methyl Orange solution; sample solution; and stock Cl_2 solution.

The air sample is drawn through a 250 ml fritted tip bubbler containing 100 ml of sample solution at a rate of $1-2$ litres min^{-1} over a suitable time period to give at least $5 \mu g$ free Cl_2 in the solution. The solution is then transferred to a 5 cm cell and the absorbance measured at 505 nm using distilled water as a reference solution. The amount of Cl_2 in the aliquot can be determined by reference to a suitably constructed calibration curve. By working back to the original sample volume, the Cl_2 concentration in the air can be calculated.

o-tolidene method

This reagent reacts with chlorine at pH less than 1.3 to give a yellow-coloured solution, which is measured with a spectrophotometer. An air sample is drawn through an acid solution of o-tolidene or through an impinger containing NaOH solution. In the latter, the hypochlorite formed is then treated with o-tolidene, the absorbance being measured in both in the range 440—490 nm. The methods are applicable for concentrations of Cl_2 around $10 \mu g$ in the sample solution.

o-tolidene solution

Reagents

o-tolidene solution and calibration solution.

The air sample is drawn through an impinger containing 100 ml o-tolidene solution at a rate of 1 litre min^{-1} until the contents of the bubbler turn distinctly yellow. An aliquot is transferred to a 1 cm cell and the absorbance is measured at 440 nm against a distilled water blank. The concentration of Cl in the sample solution is then calculated from the calibration curve previously constructed.

NaOH solution

Reagents

o-tolidene solution; NaOH – 0.0125 M for sub-ppm range; NaOH – 1–0.1 M for stack gases and higher levels; 8 M nitric acid; calibration solutions; and Cl$_2$ stock solution.

Air sample is drawn through a solution of 0.0125 M NaOH at 3 litres min^{-1}, and should be at least 1 m^3 in volume. Transfer an aliquot of the sample solution to a 100 ml flask and heat to 30° C in a water bath. Neutralize the contents with nitric acid, add 2 ml o-tolidene reagent and dilute to the mark with distilled water. Transfer an aliquot to a 1 cm cell and measure the absorbance of 490 nm using a reagent blank as a reference. The Cl$_2$ concentration can be determined in the usual way.

For the method using *N,N′*-dimethyl-*p*-phenylene diamine, the above procedures are repeated, with the substitution of this reagent for o-tolidene.

7.3 HCl

7.3.1 Occurrence

HCl is found in the atmosphere in three different forms; as a gas, acid mist or in chloride-containing particulates. HCl gas emitted to the atmosphere will react with water droplets to produce a HCl$_{aq}$ mist which will in turn adsorb on to particulate matter. Sometimes, adsorbtion is followed by a reaction which produces metal chlorides, depending on the nature of the particulates. HCl is rarely found in the atmosphere as a gas. It is emitted to the atmosphere from metallurgical and chemical production processes and combustion processes. In general, emissions are very small or rare, due to its ease of removal from waste gases by scrubbing and due to economic factors.

In the manufacture of HCl, Cl$_2$ is burned in an excess of H$_2$ and the atmospheric emissions from this process are usually less than 0.5% of total flue gas volume. It is also a by-product of the chlorination of organic compounds generally and specifically from the production of chlorinated hydrocarbons, but emissions from these sources have been virtually eliminated out of the necessity for process efficiency. Emissions from the combustion of coal containing 0.2% Cl by weight can result in 230 ppm HCl in the air surrounding the source [52]. Stack gases from refuse incinerators have been reported to give levels as high as 2000 ppm [53]. Other sources of emissions include steel pickling plants and glass manufacture.

Like, Cl_2, HCl is a phytotoxic pollutant, causing injury by acid reaction in plant tissues resulting in necrosis, strong decolorization and whitening of leaves. The most sensitive plants include sugar beet, cherry, larch, maple and tomato. Injury has been caused to plants by concentrations of greater than 5 ppm, particularly at high humidities. (The present TLV for HCl is 5 ppm or $7 \, \mathrm{mg \, m^{-3}}$.)

The strong acidity of this compound causes irritation of the eye membrane and upper respiratory tract in man, chronic effects are manifested as lowered resistance to illness and bad general health whilst severe, acute exposures can result in pulmonary edemia and laryngeal spasm.

7.3.2 Analytical methods

Several classical analytical methods are available for the determination of HCl as chloride ions Cl^θ and in air pollution studies, the method employed depends upon the quantity of Cl^θ collected by the absorbing media. For the determination of high Cl^θ ion concentrations, such as in stack gases, and for daily averages, the more common precipitation titration methods are suitable. The method involves precipitation of Cl^θ ions by $AgNO_3$ as $AgCl_2$, the end-point of the titration being determined by potassium chromate (Mohr), thiocyanate (Volhard) or fluorescence (Fajan). Of these, Volhard's back titration using thiocyanate is more accurate and is generally preferred to the other two methods. Volhard's method has been used successfully to determine Cl^θ concentrations down to about $25 \, \mu g \, Cl^\theta \, ml^{-1}$ of solution. Leithe [54] modified the Mohr method and is reported to give a detection limit of $1 \, \mu g \, Cl^\theta$ ions in the total sample analysed.

Chlorides can also be determined by titration of mercuric nitrate in the presence of a mixture of diphenyl carbazone and bromophenol blue [55]. In this method, chlorides are precipitated as mercuric chloride and bromophenol blue, yellow at pH 3.5, gives a blue–violet coloration due to reaction of excess mercuric ions with diphenyl carbazone. The procedure is applicable for absorber solutions containing at least $2 \, \mu g \, Cl^\theta$ ion ml^{-1} [28].

Chlorides in the lower concentration range can be effectively determined by spectrophotometric methods and a method has been developed by Hagino et al. and Leithe [56, 57] which depends upon the displacement of thiocyanate ions from mercuric thiocyanate by Cl^θ ions. In the presence of ferric ions, a highly coloured ferric thiocyanate complex is formed and the intensity is measured spectrophotometrically. The method is applicable for quantities of Cl^θ in the range 0–100 g in final solution. Torrance [58] has reported a lower detection limit of $0.015 \, \mu g \, ml^{-1}$ at the 95% confidence level for this method.

These methods for chlorides are severely interfered with by other halides, sulphides and thiocyanates. The effect of sulphides can be eliminated to a great extent by treating the sample with H_2O_2. The other interferences cannot be eliminated and are usually reported as chlorides.

The original method of West and Coll [59] was later adopted to measure chlorides in the atmosphere by drawing air through a solution of iron (3^+) perchlorate in perchloric acid [60]. The absorbance of the chloro complex of iron (3^+) formed is then measured at 340 nm. Bromides, iodides and sulphides do not interfere with this method. Cl in small concentrations can also be measured by a nephelometric method [61–63] where the absorbing liquid is treated with silver nitrate and the turbidity of the precipitate is observed. Bromides, iodides and sulphides interfere with this method.

Russian literature [61, 64] refers to the frequent use of the nephelometric method to determine chloride ions in ambient air in the presence of HCl_{aq} aerosol. The HCl_{aq} is determined by micro-titration or spectrophotometry, where methyl red in ethyl alcohol is used as the indicator. Total chlorides are measured nephelometrically.

Few physical methods exist for the determination of chlorides, but the method developed by Belcher et al. [65] involves the use of atomic absorption spectrophotometry. Chloride ions are converted to phenyl mercury (2^+) chloride which is quantitatively extracted into chloroform and the determination is carried out in ethyl acetate. Belcher et al. [66] have used this method for the analysis of chloride by gas chromatography with flame ionization detection (FID) using 2.5% diethylene glycol adipate as the stationary or liquid phase.

In recent years, silver chloride-type solid state membrane electrodes have been widely used for the measurement of Cl^{θ} ion activities in water. This can also be adapted for use in determining Cl^{θ} levels in absorber solutions [67, 68]. The common interfering anions (nitrates, sulphates, hydroxyl, fluorides) have little or no effect on the measurement. Ions that form stable silver complexes or insoluble silver salts, however, interfere with the response to chloride ions. Table 7.6 shows the maximum allowable ratio of interfering ions to chloride ions.

Table 7.6

Interfering ions	Maximum allowable ratio
Hydroxyl	80
Ammonium	0.12
Thiosulphate	0.01
Bromide	3×10^{-3}
Iodide	5×10^{-2}
Cyanide	2×10^{-2}
Sulphide	Must be absent

The electrode can be used to measure Cl^θ ion concentrations in the range 2 to 35 000 ppm and has been used effectively by Florence [67] to detect chloride ions in boiler water in the range 0 to $0.355\,g\,ml^{-1}$.

Torrance [69] has used a silver—silver chloride wire electrode with mercury (I) sulphate reference electrode to determine chloride ions in boiler water in the range 0.01 to $10\,\mu g\,ml^{-1}$ and has produced a linear range for the instrument from 0.1 to $15\,\mu g\,Cl\,ml^{-1}$ by plotting C against log chloride ion activity, where C is the concentration of chloride ions in the sample.

7.3.2.1 Sampling

HCl presents problems in sampling, as it may exist in either gaseous, particulate or aerosol form, as shown by Russian workers. Mists pose a special problem in sampling because it is often difficult to separate them from particulate and gaseous HCl. For general air pollution studies it is often considered sufficient to differentiate between so-called particulates (solid particles and mists) and gaseous HCl. The separation step involves filtration, usually with a membrane or fibre glass filter, followed by wet impingement of the HCl gas. Sampling rates of $10\,litres\,min^{-1}$ or greater should be adopted, depending on the type of filter used.

The filter must be heated to prevent condensation of moisture, as this would retain gases, but heating may cause loss of HCl as gas from the mist. Midget impingers are normally used in ambient air sampling, with 0.01 or 0.001 M NaOH as the absorbing medium, which rapidly achieves better than 99% recovery of gaseous HCl. The sampling rate is not particularly critical the ambient air sampling rate normally used being between 15 and 35 $litres\,min^{-1}$.

7.3.2.2 Analytical procedures (experimental)

Mercuric thiocyanate method

Cl^θ ions displace thiocyanate ions from mercuric thiocyanate which in the presence of ferric ions forms a red-coloured complex.

Reagents

2 M HNO_3; absorber solution — 0.01 M NaOH; 0.1 M ferric ammonium indicator; saturated solution of mercuric thiocyanate in methanol; stock chloride solution; and standard chloride solution.

A known volume of air is drawn through an impinger or bubbler containing 50 ml 0.01 M NaOH solution at a rate of about 30 $litres\,min^{-1}$. Pipette an aliquot (20 ml) of the sample solution into a 25 ml flask, add 2–3 drops 2 M HNO_3, 2 ml 0.1 M ferric alum sulphate followed by 2 ml mercuric thiocyanate in methanol.

Mix the contents well and measure the absorbance of 460 nm using a 1 or 5 cm cell, depending on the intensity of the colour, and a reagent blank as reference. Compare the results obtained with a suitably constructed calibration curve.

Iron (3^+) perchlorate method

Dilute solutions of this reagent in perchloric acid produce an intense yellow coloured complex when chloride ions are present. Optimum concentration for the acid is 8.5 M. If the moisture content of the air is high, 5 M acid must be used to eliminate error due to the hygroscopic properties of perchloric acid.

Reagents

Sampling solution -0.1 M iron (3^+) perchlorate in 5 M perchloric acid; and standard chloride solution $(10 \mu g \, Cl^\theta \, ml^{-1})$.

An air sample is drawn through a micro-bubbler enclosed in an opaque vessel containing 10 ml of the sampling solution, for a period of 10 min at a rate of 30 litres min^{-1}. Transfer an aliquot of the solution into a 1 cm cell and measure the absorbance at 350 nm using clean solution as a reference. Results are then compared with the calibration curve and the concentration of chloride ions as HCl in the original air sample may then be calculated.

Potentiometric method using the chloride ion electrode

Reagents

Acetic acid—ammonium acetate buffer solution; stock chloride solution; standard solution A – 1 $\mu g \, Cl^\theta$ ml; standard solution B – 10 $\mu g \, Cl^\theta$ ml; and sample solution – 0.001 M NaOH.

Air is drawn through a suitable volume of absorber solution at a predetermined rate and an aliquot (100 ml) is transferred to a 200 ml beaker followed by the addition of 4 ml buffer solution. The beaker is then stabilized at a temperature of 20—25° C for 10 min in a thermostatically controlled water bath, using a magnetic stirrer to mix the contents well. Immerse the indicator and reference electrodes in the solution and clamp them 1.5 cm above the magnetic stirrer, which should be kept running during measurement. Calculate the concentration of the chloride ions in the aliquot by reference to a suitably constructed calibration curve of e.m.f. plotted against log $(C+1)$, where C is the concentration of chloride ions ml^{-1}.

7.4 Halogenated hydrocarbons

7.4.1 Occurrence

The halogenated hydrocarbon gases include a wide range of compounds containing fluorine, chlorine and, to a far lesser extent, bromine. The great variations in the nature, properties and uses of these compounds, and in the scale and mode of their production ensure that no standard methods are applicable to their identification and measurement in ambient air. In recent years certain groups within this general class of compounds have been the subjects of intense interest and particular study.

The 'fluorocarbon* problem' came to prominence in mid-1974 with the publication by Molina and Rowland [70] of a paper, in which they suggested a possible adverse impact of fluorocarbon release on the steady-state ozone concentration in the atmosphere. Many far-reaching and conflicting publications on this topic have followed and the controversy has not, as yet, been resolved. Chlorofluorocarbons are the working fluids responsible for the safety and efficiency of almost all air-conditioning and refrigeration equipment and are used as propellants in many types of aerosol products.

For quite some time there has been concern about the potential toxic hazards of chlorinated organic compounds used as insecticides. More recently some degree of attention has been transferred to the lower molecular weight organo-halogen species which are produced and used industrially on a vast scale. A recent publication [71] has discussed the problems associated with the chloro-derivatives of methane, ethane and ethylene which were found to be widely distributed in the atmosphere at a background level in ppb range (1 part in 10^9).

Investigations into the presence of low molecular weight brominated compounds in the atmosphere have been limited; bromoform [72], methyl bromide [73] and 1.2-dibromoethane [74] being the only compounds detected.

7.4.2 Analytical methods for flurocarbons

The first direct measurements of halocarbon species in the atmosphere were made using gas chromatography and electron capture detection (ECD). Grimsrud and Rasmussen [73] have reviewed this work and noted that, although the ECD was of adequate sensitivity to measure $CFCl_3$, CF_2Cl_2 could not be monitored adequately in non-urban air masses. These workers described a gas chromatographic-mass spectrometric method which gave a detection limit for CF_2Cl_2 and $CFCl_3$ of 5 ppt (1 ppt is 1 part in 10^{12} by volume) in $20\,cm^3$ air samples. Fluorocarbons-11 and -12 have also been measured by long path i.r. absorption spectroscopy.

The system used [75] had a detection limit at partial pressures of 10^{-11} atmospheres, and involved a cryogenic procedure for concentrating a wide range of trace gases.

---

* The compounds most widely studied have been $CCl_3\,F$(F-11) and CCl_2F_2(F-12) which are usually referred to as fluorcarbons or chlorofluorocarbons.

7.4.2.1 Sampling

Several systems and materials have been used for the sampling and storage of samples containing fluorocarbons. As noted above cryogenic gas concentration techniques have been employed. Concentration by adsorption of polymeric materials, similar to the techniques discussed in Chapter 6 for hydrocarbons, have also been reported [72]. The nature of the sample required and the circumstances of operation (particularly in the construction of altitude/concentration profiles) have often limited workers to grab sampling. The sensitivity of the GC–MS analytical technique [73] required only $20 \, cm^3$ samples which were conveniently obtained using $100 \, cm^3$ glass gas syringes. $500 \, cm^3$ Pyrex bottles fitted with Teflon barrelled vacuum stopcocks were used to store samples collected from ram air passing through sampling Pitot tubes mounted in the nose of aircraft performing scheduled flights [76].

7.4.2.2 Analytical procedures

Gas chromatography is the most widely used method of analysis for fluorocarbon gases using either EC or MS detection. A Hewlett-Packard research chromatograph model No. 5750, with pulsed ^{63}Ni ECD has been used to measure routinely fluorocarbons-11 and -12 at 1 and 10 ppt respectively. A 4.5 ft stainless steel column, 1/8 inch i.d., containing 10% Na_2SO_4 on Porasil A was used and temperature programmed between 55 and $155°C$. The superior resolution desirable when using mass spectrometric detection has been achieved with a 20 ft × 1/16 inch o.d. stainless steel column of Durapak n-octane/Porasil C (100–120 mesh) programmed between -60 and $100°C$. The detection of these trace concentrations has also been achieved by the use of a Fourier Transform Spectrophotometer, nitrogen-cooled photodetectors and i.r. paths of 400 m or more [75].

7.4.3 Analytical methods for chlorinated hydrocarbons

The methods described above for the analysis of fluorocarbons are also generally applicable to chlorinated species. The sensitivity of ECD increases with chlorine content and thus mass spectrometric detection is not required for the more common chlorinated solvents which have relatively high background levels as the result of extensive industrial use.

7.4.3.1 Sampling

Nineteen chlorine-containing species including vinyl chloride and substituted benzenes have been detected in ambient air samples collected in either Houston or Los Angeles by adsorption concentration with Tenax GC [72]. Carbon tetra-chloride has been cryogenically concentrated from ambient air [75]. Chlorinated hydrocarbon samples have been stored without decomposition for extensive periods in the absence of light in glass, Teflon and stainless steel containers.

7.4.3.2 Analytical procedures

Again analytical procedures for trace levels of chlorinated hydrocarbons are centered on gas chromatography with either EC or MS detection. Sixteen halo-carbons (including methyl bromide and methyl iodide) were investigated in samples from the rural North-west of the U.S.A. [73]. It was found necessary to use three chromatographic columns to ensure resolution of all these species: 20 ft × 1/16 inch i.d. stainless steel Durapak n-octane/Porasil (100–120 mesh); 50 ft Scot OV-101; and 20 ft × 1/16 inch i.d. stainless steel Durapak Carbowax 400/Porasil F.

References

[1]   Willard, H.H. and Winter, O.B. (1933). *Ind. Eng. Chem., (Analyt.)* **5**(1) 7–10.

[2]   ASTM Standards on Methods of Atmospheric Sampling and Analysis (1973). Part 23, D1606-60, Annual Book of ASTM standards, Phila-delphia.

[3]   Dangerfield, A.D. and Nielson, J.B. (1955). *A.M.A. Arch. Ind. Hyg. Occup. Med.* **11**, 61–5.

[4]   Newman, A.C.D. (1958). *Analyt. Chim. Acta* **19**, 471–6.

[5]   Clive, W.D., Warf, J.C. and Terebaugh, R.D. (1954). *Analyt. Chem.* **26**(2) 342–6.

[6]   Newman, A.C.D. (1968). *Analyst* **93**(12) 827–31.

[7]   Armstrong, W.D. and Singer, L. (1954). *Analyt. Chem.* **26**, 904.

[8]   Marshell, B.S. and Wood, R. (1969). *Analyst* **94**(6) 493–9.

[9]   Farrah, G.H. (1967). *J. Air Pollut. Control Ass.* **17**(11) 738–41.

[10] Hill, A.C., Pack, M.R., Thomas, M.D. and Transtrum, L.G. (1960). ASTM Special Technical Publication (281) 27–44, Philadelphia.

[11] Benedict, H.M., Pack, M.R. and Hill, A.C. (1960). *J. Air Pollut. Control Ass.* **13**(8) 374–7.

[12a] Brek, M. and Stratmann, H. (1965) *Brennstoff-Chem.* **46**, 231–5.

[12b] Dorsey, J.A. and Kemnitz, D.A. (1968). *J. Air Pollut. Control Ass.* **18**(1) 12–14.

[13] Smith, W.S. and Martin, R.M. (1969). Paper 86, 60th Annual Confer-ence of Air Pollution Control Association, Ohio.

[14] Monteriolo, S.C. and Pepe, A. (1970). *Pure Appl. Chem.* **24**(4) 707–14.

[15] Finkelstein, D.N., Polykovskaya, N.A. and Morozova, N.M. (1968). *Cigiena i Sanitariia* **33**(2) 54–7.

[16] Bellack, E. and Schouboe, P.J. (1958). *Analyt. Chem.* **30**(12) 2032–4.
[17] Deeker, C.E. and Elfers, L.A. (1968). *Analyt. Chem.* **40**(11) 1658–61.
[18] Megregian, S. (1954). *Analyt. Chem.* **26**(7) 1161–6.
[19] Barney, J.E. and Hensley, A.L. (1960). *Analyt. Chem.* **32**(7) 828–31.
[20] Belcher, R., Leonard, M.A. and West, T.S. (1961). *J. Chem. Soc.* B, 2390–3.
[21] Belcher, R., West, T.S. and Leonard, M.A. (1959). *Talanta* 2, 92–3.
[22] Belcher, R. and West, T.S. (1961). *Talanta* 8, 853–62.
[23] Belcher, R. and West, T.S. (1961). *Talanta* 8, 863–70.
[24] Belcher, R., Wilson, C. and West T.S. (1964). *New methods of analytical chemistry* 2nd Edn, Chapman and Hall Ltd, London.
[25] West, P.W., Lyles, G.R. and Miller, J.L. (1970). *Environ. Sci. Technol.* **4**(6) 487–91.
[26] Liiv, R. and Luiga, P. (1972). Paper 72-1. 65th Annual Meeting of the Air Pollution Control Association, Florida.
[27] Hitchcock, A.E., Jacobson, J.S., Mandl, R.H., McCune, D.C. and Weinstein, L.H. (1965). *Simplified semi-automated analysis of plant tissues*. Proceedings of the Technicon Symposium on Automation in Analytical Chemistry. New York.
[28] Intersociety Committee. (1972). *Methods of air sampling and analysis*. American Public Health Association, Washington.
[29] Technician Industrial Method No. 113-71AP. (1974). *Hydrogen fluoride in ambient air*. Techinen Industrial System, New York.
[30] Stewart, E.D. (1970) *An improved method for the semi-automated analysis of fluorine and plant tissue*. Technicon International Congress. New York.
[31] Technician Industrial Method No. 129-71W. (1972). *Fluoride in water and waste-water*. Technician Industrial System, New York.
[32] O'Donnell, T.A. and Stewart, D.F. (1961). *Analyt. Chem.* **33**(3) 337–41.
[33] Kubota, K. and Surak, J.G. (1959). *Analyt. Chem.* **31**(2) 283–6.
[34] Leonard, M.A. (1958). *Analyst* 88(5) 539–6.
[35] Sutler, E. (1973). *Staub-Reinhalt. Luft* 33(3) 113–15.
[36] Warren Spring Laboratories, Department of Trade and Industry. (1973). *The determination of fluorides in samples obtained from process gases or from contaminated atmospheres*. Method Sheet 98, Department of Trade and Industry, U.K.
[37] Belcher, R., West, T.S. and Leonard, M.A. (1959). *J. Chem. Soc.*, A, 3577–9.
[38] Vogel, A.I. (1964). *A text-book of quantitative inorganic analysis*. Third Edn, Longmans, Green and Co. Ltd, London.
[39] Dixon, E.J. (1970). *Analyst* 95(3) 272–7.
[40] National Air Pollution Control Administration (USA). (1971). *Atmospheric emissions from chlor alkali manufacture*. Publication AP-80, Environmental Protection Agency North Carolina.
[41] U.S. Public Health Service. (1968). *Compilation of air pollution factors*. US-PHS 999-AP-42.
[42] Devitt, T.W. and Gerstle, R.W. (1971). *Chlorine and hydrogen chloride emissions and control*. Paper 71–5, 64th Annual Meeting of the Air Pollution Control Association. New Jersey.

[43] Her Majesty's Stationery Office. (1966). *Methods for the detection of toxic substances in air-chlorine*. Booklet **19**, London.

[44] McCormac, B.M. Ed. (1971). *Introduction to the scientific study of atmospheric pollution*. D. Reidel Publishing Co., Dordrecht, Holland.

[45] Brennan, E., Leone, I.A. and Daines, R.H. (1965). *Int. J. Air Water Pollut.* **9**, 791—7.

[46] Cooper, H.B.H. and Rossano, A.T. (1970). *Source testing for air pollution control*. Environmental Science Services Division, Wittan, Connecticut.

[47] Technicon Industrial Method No. 145-71AP. (1973). *Free chlorine in ambient air*. Technican Industrial System, New York, March.

[48] Her Majesty's Stationery Office. *Dust and fumes in factory atmosphere*. Booklet **8**, London.

[49] Demidov, A.V. and Mokhov, L.A. (1962). Survey of USSR Literature on Air Pollution and Related Occupational Diseases, **10**, 114. (Trs. B.S. Levine) Technical translation no. TT 66-11767.

[50] Polezhaer, N.G. (1960). Survey of USSR Literature on Air Pollution and Related Occupational Diseases, **3**, 29—30. (Trs. B.S. Levine) Techical Translation No. TT-60 21475.

[51] Rayazanov, V.A. (1962). Survey of USSR Literature on Air Pollution and Related Occupational Diseases, **9**, 1—8, (Trs. B.S. Levine).

[52] Biestock, D., Deniski, R.J. and Lapalncei, T.L. (1969). *Chlorine in coal combustion*. U.S. Bureau of Mines, R.I. 7260, May.

[53] Hiyayama, N., Kazuo, H., Sadao, K. and Toshio, O. (1968). *Bull. Japanese Soc. Mech. Eng.* **47**(11) 902—12.

[54] Leithe, W. (1947). *Mikrochem-Michrochim Acta* **33**, 167—75.

[55] Clarke, F.E. (1950). *Analyt. Chem.* **22**(4) 553—55 and 1458.

[56] Hagino, K., Iwasaki, J., Ozawa, T. and Utsuiur, S. (1956). *Bull. Chem. Soc.* (Japan) **29**(8) 860—4.

[57] Leithe, W. (1971). *Analysis of air pollutants*. (Trs. R. Kondor) Ann Arbor Science Publishers, Ann Arbor.

[58] Torrance, K. (1971). *Analyt. Chim. Acta* **54**, 373—7.

[59] West, P.W. and Coll, H. (1957). *Spectrophotometric determination of chloride in air*. Proc. Symp. Atmospheric Chemistry of Chloride and Sulphur Compounds. Cincinatti, Ohio.

[60] Coll, H. and West, P.W. (1956). *Analyt. Chem.* **28**(12) 1834—8.

[61] Alexseyeva, M.V. and Elfimova, E.V. (1960). Survey of U.S.S.R. Literature on Air Pollution and Related Occupational Diseases. **3**, 31—3. (Trs. B.S. Levine) Technical Translation no. TT-60-21475.

[62] Katz, M. (1968). *Air pollution*. Chapter 17, (Ed. A.C. Stern) 2nd Edn. Academic Press, New York.

[63] Denice, E.C., Luce, E.N. and Akerland, F.E. (1943). *Ind. Eng. Chem.* **28**, 365—6.

[64] Manita, M.D. and Melekhina, V.P. (1964). *Hyg. Sanit.* **29**(3) 62—6.

[65] Belcher, R., Najafi, A., Rodriguez-Vazgne, J.A. and Stephen, W.I. (1972). *Analyst* **97**(12) 993—7.

[66] Belcher, R., Major, J.R., Rodriguez-Vazgne, J.A., Stephen, W.I. and Uden, P.C. (1971). *Dral. Chim. Acta*. **57**, 73—80.

[67] Florence, T.M. (1971). *J. Electroanal Chem.* **31**, 77—86.

[68] Van Loon, J.C. (1968). *Analyst* **93**(12) 788—91.

[69] Torranee, K. (1974). *Analyst* **99**(4) 203–10.
[70] Molina, M.J. and Rowland, F.S. (1974). Stratospheric sink for chloro-fluoromethanes: chlorine atom-catalysed destruction of ozone. *Nature, Lond.* **245**, 27.
[71] McConnell, G., Ferguson, D.M. and Pearson, C.R. (1975). Chlorinated hydrocarbons in the environment. *Endeavour* **34**, 13.
[72] Pellizzari, E.D., Bunch, T.E., Berkley, R.E. and McRae, J. (1976). Determination of trace hazardous organic vapour pollutants in ambient atmospheres by gas chromatography-mass spectrometry-computer. *Analyt. Chem.* **48**, 803.
[73] Grimsrud, E.P. and Rasmussen, R.A. (1975). Survey and analysis of halocarbons in the atmosphere by gas chromatography-mass spectrometry. *Atmos. Environ.* **9**, 1014, 1015.
[74] Going, J. and Long, S. (1975). Sampling and analysis of selected toxic substances. Task 11. Ethylene dibromide. *EPA* 560/6-75-001.
[75] Hanst, P.L., Spiller, L.L., Watts, D.M., Spence, J.W. and Miller, M.F. (1975). Infrared measurement of fluorocarbons, carbon tetrachloride, carbonyl sulfide and other atmospheric trace gases. *J. Air Pollut. Control Ass.* **25**, 1220.
[76] Zafonte, L., Hester, N.E., Stephens, E.R. and Taylor, O.C. (1975). Background and vertical atmospheric measurements of fluorocarbon-11 and fluorocarbon-12 over Southern California. *Atmos. Environ.* **9**, 1007.

CHAPTER EIGHT

Remote monitoring techniques

8.1 Introduction

8.1.1 Overview of laser monitoring

Remote monitoring of the atmosphere with laser sources has now been actively pursued for over a decade. However, only recently has the development of tunable lasers with adequate energy and spectral purity reached a state that meets the remote monitoring source requirements. Thus, in the past few years tunable lasers have allowed an expanding range of applications and measurements to be made by remote monitoring methods.

The laser as a source offers four unique features unavailable in incoherent light sources: high power, spatial collimation, spectral purity and coherence. Remote monitoring uses all of these laser properties in the measurement. In addition to being generally useful, the remote monitoring laser transmitter must also be tunable so that interesting atomic or molecular absorption transitions can be probed to yield information about species concentration and local environment.

It is not appropriate to review the laser remote monitoring field in this chapter. However, previous review articles and a recent book should provide an adequate survey of remote monitoring up to 1975 [1-4].

Remote monitoring measurements are conducted in the atmosphere. Although the normal atmosphere is relatively transparent in the visible, a number of complex scattering phenomena occur for a transmitted optical frequency beam. These include elastic Rayleigh (molecular) and Mie (particulate) scattering, inelastic Raman, resonance fluorescence and fluorescence scattering and absorption. Information about the atmosphere can be obtained from all these scattering phenomena. However, for detection of molecular species in the atmosphere

Table 8.1 Representative cross-sections

Process	Cross-section $\dfrac{d\sigma}{d\Omega}$ (cm^2)	Lifetime (s)	Scatterer	Spectral range (μm)
Elastic Scattering				
Rayleigh	10^{-27}		N_2, O_2	$0.20 \approx 1.0$
Mie	10^{-21} to 10^{-24}		Particulates, aerosols	$0.20 \approx 10$
Inelastic scattering				
Raman	10^{-30}		All molecules	u.v., visible
Resonance fluorescence and fluorescence				
u.v., visible	10^{-14} to 10^{-17}	10^{-6} to 10^{-8}	Atoms, molecules	u.v., visible
i.r.	10^{-18} to 10^{-23}	10^{-1} to 10^{-6}	Molecules	2 to 20
Absorption				
u.v., visible	10^{-14} to 10^{-17}		Atoms, molecules	u.v., visible
i.r.	10^{-17} to 10^{-21}		Molecules	2 to 20

Raman scattering, fluorescence scattering and absorption are molecular or atomic specific.

A comparison of the above scattering and absorption processes is shown in Table 8.1. From Table 8.1 it is evident that Raman scattering has the smallest optical cross-section of 10^{-30} cm^2. Resonance fluorescence is better with cross-sections between 10^{-14}–10^{-23} from atomic transitions in the visible to molecular transitions in the i.r. Unfortunately, in the troposphere resonance fluorescence cross-sections are reduced by quenching thus reducing the measurement sensitivity. However, resonance fluorescence has been used to measure stratospheric sodium at 90 km altitude on a regular basis, since the first experiment of Bowman *et al.*, in 1969 [5].

The absorption cross-sections are the largest of the optical interactions and are not affected by quenching. Thus the most sensitive measurement methods use absorption to determine the atomic or molecular species and their concentrations. Only homonuclear diatomic molecules such as N_2, O_2, H_2 and atoms such as He, A and Kr are not detectable by absorption measurements.

Fig. 8.1 shows a schematic of a remote monotoring transmitter—receiver system. A laser source is used as a high power transmitter. It operates at a selected frequency and in general transmits a short, high peak power optical pulse. The received radiation is collected by a telescope and focussed on a

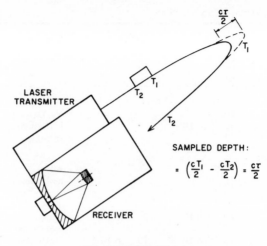

SAMPLED DEPTH:

$$= \left(\frac{cT_1}{2} - \frac{cT_2}{2}\right) = \frac{c\tau}{2}$$

MIE BACKSCATTERING

Fig. 8.1 Schematic diagram of a single ended, depth resolved, absorption measurement using Mie backscattering as a distributed reflector. The sampled depth is $\Delta R = c\tau/2$ where τ is the laser pulse width or detection integration time.

detector where it generates an electrical signal which is amplified and processed by appropriate electronics.

There are three types of absorption measurement methods. The first is long path absorption using a remote detector or retro-reflector. This is a double ended measurement method which requires the least amount of laser power but has the disadvantage of requiring a remote detector. It measures the integrated absorption along the closed path between the transmitter and receiver. In general only μW to mW of laser power are required for long path, double ended absorption measurements.

The second method is single ended long path absorption measurement. This method utilizes a remote topographical or natural target to reflect energy back toward the receiving telescope. Thus it is a single ended long path absorption measurement but without the need for a remote detector. Typically topographical targets scatter incident radiation in a Lambertian manner with a reflectivity near 20%. The result is that significantly more laser energy is needed for the single ended measurement, typically 1 to 10^5 W for 1 to 10 km ranges.

The third method of absorption measurement utilizes Mie scattering in the atmosphere as a distributed reflector. This method offers the advantage of allowing a depth resolved absorption measurement and is schematically illustrated in Fig. 8.1. The sampled depth of the atmosphere is $\Delta R = c\tau/2$ where c is the

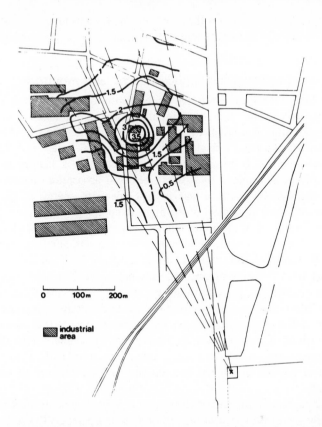

Fig. 8.2 NO_2 distribution over a chemical factory as derived from differential absorption measurements at the directions at an altitude of 45 m. The concentrations of NO_2 are given in ppm (after Rothe *et al.* [6]).

velocity of light and $\tau = \tau_p + \tau_d + \tau_e$ is the sum of the laser pulse length τ_p, the detector response time τ_d and the electrical integration time τ_e. Thus a 330 ns total response time gives a 50 m depth resolution. The location of the 50 m sample is determined by the round trip time of flight of the light pulse in typical radar fashion. For example, a range element located at 1 km appears 6.6 μs after the laser pulse is transmitted.

The unique advantages of a single ended depth resolved absorption measurement more than offset its disadvantage for requiring the highest transmitted energy of the three absorption methods. Typically transmitted powers of between 1–10 MW are required for measurement ranges from 1–3 km in the troposphere. Only recently have tunable laser sources become available with the desired output powers.

Before proceeding to discuss the measurement systems in more detail it is useful to briefly consider the limitations of laser monitoring in the atmosphere.

8.1.2 Future prospects and applications of laser monitoring

From the brief description given above, the future of remote monitoring looks very bright indeed. The method potentially allows one to make a three-dimensional concentration measurement of a specified pollutant over a 3 km range to ppb sensitivities and from a single location in near real time. Fig. 8.2 illustrates such a measurement. It shows isopleths of NO_2 over a factory in Cologne, Germany [6] as determined by a depth resolved absorption measurement. In this case the dye laser source was of just adequate energy to perform the measurements, which took most of the night to complete.

What then are some of the limitations of remote monitoring using laser sources and how do they effect future applications of these very powerful measurement techniques? The major limitation is the atmosphere itself. In dealing with the atmosphere from a remote monitoring view point, one is struck by its relatively opaque appearance [7]. In order to make a remote measurement the transmitted laser beam must penetrate the atmosphere out to the desired range, be scattered and return to the receiving telescope. The surprising fact is that even in absorption free regions of the spectrum the atmosphere is characterized on the average by an optical depth of 0.1 to 1 km^{-1} over a horizontal path. Thus horizontal ranges much beyond a few km become very difficult against an exponential decay length.

The atmosphere is not free of absorption and measurements must be made in atmospheric windows or spectral regions of high transparency. The visible window extends from near 3000 Å to about $1 \mu m$. On the u.v. end, O_3 and O_2 absorption cause the cutoff. In the i.r. the overtone bands of water vapour at 1.4 and $1.9 \mu m$ become important. Further toward the i.r. both H_2O and CO_2 limit the transparency of the atmosphere to windows that lie between 1.9–2.6, 3–4.2, 4.4–5.4 and $8–13 \mu m$. Fortunately, a large number of interesting pollutant molecules have absorption bands that lie within these windows. However, care must be taken to determine possible interferences with H_2O and CO_2 if calibration errors are to be avoided. Fig. 8.3 shows the i.r. transmittance of the atmosphere over a horizontal path. Also shown are band centres for various pollutant molecules.

Finally, the atmosphere is not a quiescent optical medium but is characterized by fluctuations in space and time which lead to scintillation. The atmospheric scintillation in turn limits the accuracy of an optical absorption measurement by introducing signal fluctuations. The measurement errors introduced by scintillation have been discussed [8, 9]. Fortunately, methods such as double pulsing the transmitted laser output in a short time compared with the 0.1–0.01 s

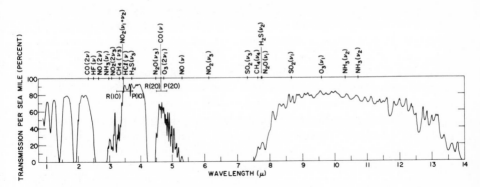

Fig. 8.3   Transparency of the atmosphere in the i.r. over a horizontal path. Absorption band centres of some pollutant molecules are indicated.

scintillation time can be used to normalize the signal fluctuations introduced by atmospheric scintillation.

The above discussion is intended to project future capabilities of laser remote monitoring with some caution.

Remote monitoring using laser sources can be applied to global, stratospheric, tropospheric, ambient and source emission problems. Because of the wide range of remote monitoring applications, research measurement are being made at an increasing rate. Here we can only give a sample of the present and future remote monitoring measurement potential.

Lidar systems [10] using fixed frequency laser sources have been used extensively for measurements of atmospheric turbidity, particulate distribution, [11] atmospheric mountain waves, [12] cloud heights and water-ice composition, [13] upper atmosphere dust, [14] smoke plumes, [15] and city haze distribution [16]. A recent review of Lidar applications has been written by Hall [17].

With tunable laser sources the application areas broaden considerably. Meteorological parameters [18] such as temperature [19, 20] density, humidity [21] and wind velocity [22] can be measured with a single ended remote monitoring system. A knowledge of these parameters is essential for the interpretation of gaseous pollutant interactions and for modelling pollutant sources and sinks.

Tunable u.v. and visible sources allow the measurement of SO_2 and O_3 [23] and NO_2 [6, 24] by the differential absorption method. Furthermore, fluorescence emission form water surfaces can be probed using the available high peak power laser sources. These measurements allow tracking and identification of waterborne contaminants such as oil [25]. Ambient levels of OH radical have been monitored in a closed path absorption measurement by Wang and co-workers [26–28].

It is in the i.r. spectral region, where molecules have characteristic vibrational rotational 'fingerprint' spectra, that remote monitoring capability has the greatest potential. Double ended absorption measurements [29] using i.r. lasers have been, made using diode [30, 31] laser sources and heterodyne detection techniques [32–5]. Single ended measurements have been made of CO [36] and differential absorption measurements have been made on a number of pollutant molecules using step tuned i.r. laser sources [45].

In addition to a large number of molecular absorption lines that can be probed in the i.r., i.r. systems operate in the eye safe region of the spectrum and can be operated in full daylight. The remote monitoring capabilities of two operational i.r. systems will be discussed in Section 8.9.3.

8.2 Brief theoretical discussion

8.2.1 Lidar equation

The remote pollutant measurement methods are listed in Table 8.1. In this section a general formulation is presented for the analysis of the sensitivity, range, and S/N for the pollutant detection methods.

The Lidar equation is a statement of conservation of energy. In its simplest form it involves keeping track of the laser energy from the transmitter out into the atmosphere and back into the receiving telescope after scattering. Since the laser beam is collimated the received signal has a $1/R^2$ dependence due to the Lambertian scattering characteristics of the effective reflector.

The backscattered signal from a target at range R is given by:

$$P_r(R) = \left(\frac{\rho}{\pi}\right) KP_o \frac{A}{R^2} \exp\left(-2\int_0^R \alpha_A(r)dr\right) \tag{1}$$

where P_r and P_o are the received and transmitted powers, K is the optical system efficiency, A the area of the receiving telescope and (ρ/π) the effective reflectivity of the remote target. The transmitted radiation interacts with the pollutant molecules through the atmospheric volume extinction coefficient α_A defined by

$$\alpha_A = \alpha_R + \alpha_{Mie} + \alpha_{ABS} \tag{2}$$

where α_R, α_{Mie} and α_{ABS} are the Rayleigh, Mie and absorption extinction coefficients. In general it is assumed that the transmitted beam intensity obeys an exponential decay law of the form

$$I = I_o \exp\left(-\alpha_A r\right) \tag{3}$$

where r is the distance.

Returned signals due to Raman, resonant fluorescence and fluorescence are included in the effective target reflectivity ρ/π. The strength of the returned signals by the inelastic scattering processes is related to the scattering cross-section for the process. The depth of the atmosphere probed by backscatter methods is shown schematically in Fig. 8.1. The sampled depth ΔR is given by

$$\Delta R = \frac{c\tau}{2}, \tag{4}$$

where $\tau = \tau_p + \tau_d + \tau_F$ is the sum of the laser pulse width τ_p, the detector integration time τ_d, and the molecular fluorescence time τ_F. For Raman, u.v. fluorescence and differential absorption measurements τ_F is negligible and the sampled depth is set by either the laser pulse width or the detector integration time. Of course the particular sampled depth at range R is determined by the standard Radar time of flight measurement.

Table 8.1 lists the scattering and absorption cross-sections for the measurement processes of interest. It is readily apparent that absorption and fluorescent cross-sections are about ten orders of magnitude larger than Raman scattering cross-sections. This physical fact leads to the immediate conclusion that the most sensitive remote detection methods are by absorption. Resonance fluorescence and fluorescence also have cross-sections that are nearly that of absorption. However, for monitoring in the troposphere, quenching reduces the fluorescence cross-section. Quenching is not a problem at high altitudes and remote monitoring by resonance fluorescence has a clear advantage in that case.

We can now return to Equation 1 and consider the effective reflectivity of the 'target' for remote sensing. Table 8.2 lists the scattering process and the corresponding effective reflectivity.

For absorption measurements with a remote sensor or retroreflector and a collimated transmitted beam, the range squared dependence is effectively cancelled. Thus all of the transmitted power is collected except for a collection efficiency factor ξ, which is near unity, and the effective target reflectivity is orders of magnitude better than for a Lambertian scatterer. Letting $\rho/\pi = R^2\xi/A$ in Equation 1 reduces it to the form commonly used to describe long path absorption.

For long path absorption by scattering from topographic targets $\rho \approx 1$ for all i.r. wavelengths and $\rho \sim 0.1$ for u.v. and visible wavelengths. Thus the effective target reflectivity is approximately $1/\pi$. For single ended long path absorption measurements from topographic targets over km ranges, 10^4 to 10^6 more transmitted power is required than for double ended measurements using a remote detector or retroreflector. Since the required power for double ended long path absorption is less than $1\,\mu W$, single ended measurements can be made with mW to W laser sources. The HeNe laser geodolite ranging systems are an example of single ended range measurements by scattering from topographic targets.

Table 8.2   Effective reflectivity

Method	(ρ/π)	Magnitude
Long path absorption with retroreflector	$\dfrac{R^2\xi}{A}$	10^4 to 10^6
Long path absorption from topographic target	ρ/π	$1/\pi$
Differential absorption	$\dfrac{c\tau}{2}\dfrac{\beta_{\text{Mie}}}{4\pi}$	10^{-4}
Resonance fluorescence and fluorescence	$\dfrac{c\tau}{2}\dfrac{N\sigma_{\text{F}}}{4\pi}$	10^{-5} (1 ppm) (u.v. and visible)
Raman	$\dfrac{c\tau}{2}\dfrac{N\sigma_{\text{RAM}}}{4\pi}$	10^{-15} (1 ppm) 10^{-9} (1 atm)

Remote detection by backscattering involves the range resolution $\Delta R = c\tau/2$ as well as the backscattering cross-section. For Mie backscattering the returned power is independent of the pollutant concentration and depends only on the Mie backscatter coefficient β_{Mie}. For resonance fluorescence and Raman backscattering the returned signal depends on the scattering cross-section and on the pollutant density. The dependence on pollutant density reduces the returned signal for weak pollutant concentrations and thus also reduces the measurement S/N ratio which further effects the detection sensitivity. For depth resolved pollutant measurements differential absorption has the advantages of a relatively large effective reflectivity which is independent of pollutant concentration and the additional advantage of the large absorption cross-section which is not affected by quenching.

To complete the general formulation of remote pollutant detection we need to introduce detector parameters and evaluate the detector S/N ratio for a given returned power. This procedure has been discussed previously, but a summary is useful here.

The voltage signal to noise ratio for a detector is:

$$S/N = P_{\text{r}} \left/ \left[4P_{\text{r}} \left(\frac{h\nu}{\eta} \right) \Delta f + 2NEP^2 \Delta f \right]^{\frac{1}{2}} \right., \tag{5}$$

where P_{r} is the received signal power and Δf is the amplifier bandwidth. The first term in the denominator is due to the shot noise generated by the signal itself. Here $h\nu$ is the photon energy and η is the detector quantum efficiency. The second term in the denominator represents the noise of the detector in the absence of the signal (dark current and background noise and pre-amplifier noise characterized by the noise equivalent power NEP). A factor of 2, added under the square root, takes account of the noise resulting from two measurements required to determine the pollutant concentration. This noise is $P_N^2 = (P_N^2)_{on} + (P_N^2)_{off}$ where on and off refer to making the measurement on resonance and off resonance of the characteristic line of the pollutant. Usually either the first or the second term in the denominator need to be considered.

For shot noise limited detection, the dark current term in the denominator can be neglected. For practical purposes this limit can be attained with photomultipliers if all background radiation is adequately filtered. At a given signal to noise ratio, the minimum detectable power from Equation 5 is given by:

$$P_r^{min}(\text{shot noise}) \geqslant 4 \left(\frac{h\nu}{\eta} \right) F(S/N)^2 \Delta f, \qquad (6)$$

where the factor F has been added (between 2 and 5) to account for the partition noise effect of the multipler dynodes.

For dark current or background limited detection, the minimum detectable power evaluated from Equation 5 is

$$P_r^{min}(\text{dark current}) = NEP(S/N)\sqrt{(2\Delta f)}. \qquad (7)$$

This type of detector noise limitation applies, in particular, to photoconductors and other detectors used in the i.r. It is characteristic for dark current limited detectors that they achieve, for constant average power, a higher signal to noise ratio with increasing peak power in pulsed operation.

Detector performance is also expressed in terms of the detectivity D^* which is related to the NEP by $D^* = \sqrt{(A)}/NEP$ where A is the detector area. Typical values of D^* are 6×10^{12} cm Hz$^{\frac{1}{2}}$W^{-1} for room temperature silicon diodes, 3×10^{11} cm Hz$^{\frac{1}{2}}$W^{-1} for 77 K operation of InAs (1 to 3.3 μm) photovoltaic detectors, 1.1×10^{11} cm Hz$^{\frac{1}{2}}$W^{-1} for 77 K InSb (1.5 to 5.5 μm) photovoltaic detectors, to 2×10^{10} cm Hz$^{\frac{1}{2}}$W^{-1} for 77 K HgCdTe (2 to 25 μm) detectors. The rise time of these detectors is less than 100 ns.

In the above expressions for detector minimum detectable power, we have assumed Δf is the electronic bandwidth required to resolve a pulse. More specifically we take $\Delta f \tau = 2$ as the condition necessary to resolve a pulse of width τ. For pulsed generation, the effective detector bandwidth decreases with the number of averaged pulses n as $\Delta f/n$.

We are now in a position to combine Equation 1 for the received power with the expressions for minimum detectable power and solve for the detection parameter of interest at specified values of the other variables. The parameters of interest are required transmitted energy versus range, S/N ratio versus range at a fixed energy, and minimum detectable pollutant concentration relative to one atmosphere, η, at a given transmitted energy and range. The above procedure is carried out and the results are discussed for the integrated path absorption methods, and for the differential absorption method in the following section.

8.2.2 Absorption measurement methods

8.2.2.1 Double ended long path absorption

The long path resonance absorption method measures the total integrated pollutant concentration over the path. It has the advantages of increasing sensitivity with range, the largest interaction cross-section which is not modified by quenching, and the least required transmitted laser power. It has the disadvantages of being double ended and of lacking depth resolution.

The transmitted intensity at the detector follows from Equation 1 if we let $(\rho/\pi) = R^2/A$, assume a round trip, and separate the beam attenuation due to pollutant absorption from the atmospheric attenuation. We then have:

$$P_r^{on}(R) = KP_0 \exp\left[-2\int_0^R \sigma_{ABS}N(R')dR'\right] \exp\left[-2\int_0^R \alpha_A(R')dR'\right], \quad (8)$$

where $N(R')$ is the pollutant density over the range R and we assume that the transmitted intensity is less than the absorption saturation intensity.

The returned power for the pump frequency tuned off the absorbing transition is:

$$P_r^{off} = KP_0 \exp\left[-2\int_0^R \alpha_A(R')dR'\right], \quad (9)$$

so that the integrated pollutant concentration is determined by the log ratio of the on and off returned power as

$$2\int_0^R N(R')dR' = \frac{1}{\sigma_{ABS}} \ln\left(\frac{P_r^{off}}{P_r^{on}}\right). \quad (10)$$

Assuming that it is possible to measure a 1% change in transmitted power when tuned on and off the absorbing line, the minimum integrated density is given by:

$$\left[2\int_0^R N(R')dr\right]_{min} = \frac{0.01}{\sigma_{ABS}}. \quad (11)$$

Table 8.3 lists representative absorption cross-sections for molecules and atoms and gives the minimum detectable pollutant concentration for an absorption length R of 100 m. Since the absorption cross-sections are the largest of the Raman, resonance or fluorescence cross-sections, these sensitivity limits are the best obtainable by optical means. Thus, there is a physical limitation for pollutant

Table 8.3   Minimum measurable concentration for an absorption length of 100 m

	CO	NO$_2$	SO$_2$	C$_6$H$_6$	Na	Hg
$\lambda(\mu m)$	4.7	0.4	0.29	0.29	0.5896	0.2537
σ^{abs}(cm^2)	1.8×10^{-18}	2.8×10^{-19}	3.4×10^{-19}	1.3×10^{-18}	4.8×10^{-13}	5.6×10^{-14}
$\eta(1\%)$	2.1×10^{-8}	1.3×10^{-7}	1.1×10^{-7}	1.5×10^{-8}	7.9×10^{-14}	6.7×10^{-14}

detection that leads to a depth resolution-sensitivity trade off. Fortunately, σ_{ABS} is large enough that an absorption distance of 100 m allows ambient pollutant levels to be measured to less than 20 ppb.

The absorption sensitivity can be improved by measuring less than 1% signal power change. The allowed measurement accuracy is discussed by Kildal and Byer [37] and by Byer and Garbuny [38]. These authors show that to detect a 0.1% change in intensity with an InSb detector requires only 10^{-7} W of incident power at the detector. There is an optimum value of absorption cross-section given by $\sigma_{opt}N = 2.22$ and $\sigma_{opt}N = 1.11$ for shot noise and dark current limited detection. In addition, the signal to noise ratio required to achieve a measurement accuracy $\Delta x/x$ is:

$$S/N = \Delta x/x, \tag{12}$$

where $x = 2N\sigma_{ABS}R$.

Other parameters which affect the measurement accuracy and limit the detection sensitivity are atmospheric turbulence and laser power fluctuations. For cw measurements the atmospheric turbulence is a significant limitation for chopping rates on and off resonance of less than 200 cps. Since the signal to noise ratio improves with higher peak power for dark current limited detectors, the use of Q-switched or pulsed laser sources is advantageous. The laser power fluctuations can be reduced by careful engineering or by ratio normalization of each laser pulse. The ultimate accuracy achieved by these steps is determined by the dynamic range and linearity of the detector. However, 0.1% ratio accuracy is routinely achieved.

We can invert Equation 8 and solve for the required transmitted laser power for a given signal to noise ratio. Thus

$$P_o = \frac{3}{K}\left(\frac{S}{N}\right) P_N \exp\left[2\int_0^R \alpha_A(R')dR'\right]$$

$$= \frac{3}{K}\left(\frac{S}{N}\right) NEP\sqrt{(2\Delta f)} \exp\left[2\int_0^R \alpha_A(R')dR'\right].$$

For an $NEP = 0.3 \times 10^{-11}$ W Hz$^{-\frac{1}{2}}$, $K = 0.1$, a bandwidth of 1 MHz, $(S/N) = 100$ which implies a 1% detection sensitivity, and assuming $\alpha_A \sim 1$ km^{-1}, the required peak transmitted power for a 1 km range is $P_o = 1$ mW in a $2\,\mu s$ pulse. This is well within the range of cw and pulsed i.r. lasers and diode lasers.

We can reverse the above procedure and ask what the maximum range is for a 1 kW peak power laser source. For $\alpha_A = 1$ km^{-1} we find $R = 16$ km at a $(S/N) = 100$. Beyond that range the signal to noise ratio decreases reaching unity at 20 km.

These examples show the advantages of long path absorption measurements which include low required transmitted power, long ranges and of course good sensitivity. However, in some situations the requirement for a double ended system is not possible to meet. In those cases the single ended topographical absorption technique is an alternative measurement method.

8.2.2.2 Single ended long path absorption

The advantage of absorption measurements using topographical targets is the ability to make a single ended measurement with the sensitivity of the closed path absorption method. The disadvantage is the requirement for increased transmitted power.

In order to find the required transmitted energy for absorption using topographical targets we must consider the measurement procedure carefully since pump depletion due to the pollutant itself may be important.

The power received off resonance is given by Equation 1 with an effective reflectivity (ρ/π), thus:

$$P_r^{off} = \left[\left(\frac{\rho}{\pi}\right) KP_o \frac{A}{R^2}\right] \exp - (2\alpha_{sc}R), \tag{14}$$

where $\alpha_{sc} = \alpha_{Mie}$ is the atmospheric extinction coefficient given by Equation 2 without the absorption term. If we assume that the atmosphere may itself contain the pollutant gas in a normal concentration N_A which is less than the pollutant concentration N_p, then the power received on resonance is given by:

$$P_r^{on} = \left[\left(\frac{\rho}{\pi}\right) KP_o \frac{A}{R^2}\right] \exp\left[-2R(\alpha_{sc} + N_A \sigma_{ABS})\right]$$

$$\times \exp\left[-2\sigma_{ABS} \int_0^L N_p(R')dR'\right], \tag{15}$$

where L is the extent of the pollutant cloud. The difference in the power received off resonance and on resonance found by combining Equations 14 and 15:

$$P_r^{off} - P_r^{on} = P_r^{off}[1 - \exp(-2N_A R + N_p L)\sigma_{ABS}]. \tag{16}$$

The difference must be measured relative to the power received at the detector but attenuated by the normally present concentration of gas in the atmosphere given by:

$$P_r^{off} - P_r^{on}|\text{normal} = P_r^{off}[1 - \exp(-2N_A R)\sigma_{ABS}]. \tag{17}$$

For a measurement of the pollutant concentration, the difference between Equations 16 and 17 must equal the noise power given by Equation 7 at a given signal to noise ratio, or:

$$P_r^{off} \exp(-2N_A R\sigma_{ABS})[1 - \exp(-2N_p L\sigma_{ABS})]$$

$$= P_r^{min} = NEP\left(\frac{S}{N}\right)\sqrt{(2\Delta f)}. \tag{18}$$

Substituting for P_r^{off} from Equation 14 and solving for the required transmitted power, P_o gives:

$$P_o = \left[NEP\left(\frac{S}{N}\right)\sqrt{(2\Delta f)}\right]\left(\frac{\pi R^2}{\rho K A^2}\right)\frac{\exp 2(\alpha_{sc} + N_A\sigma_{ABS})R}{1 - \exp(-2N_p L\sigma_{ABS})}.$$

Finally, multiplying the above result by τ and setting $\Delta f \tau = 2$ yields for the required transmitted pulse energy:

$$E_o = \frac{2}{K}\left[NEP\left(\frac{S}{N}\right)\sqrt{\tau}\right]\frac{\pi R^2}{\rho A}\frac{\exp 2(\alpha_{sc} + N_A\sigma_{ABS})R}{1 - \exp(-2N_p L\sigma_{ABS})}, \tag{19}$$

a result first derived by Byer and Garbuny [38]. The required transmitted energy varies as $\sqrt{\tau}$ as expected for a dark current limit detector. It also increases with range due to the Lambertian scattering properties of the topographical reflector with reflectivity ρ into 2π stearadians.

Equation 19 gives the required transmitted energy assuming that the pollutant of interest is also normally present in the atmosphere. The equation simplifies considerably for most pollutants which are normally not present in the atmosphere. For this case, the term $\exp(2N_A\sigma_{ABS}R)$ approaches unity. Finally, if the pollutant cloud itself is optically thin, then the denominator of Equation 19 reduces to $2N_p L\sigma_{ABS}$. In the limit of a very tenuous pollutant cloud the reduced value of $2N_p L\sigma_{ABS}$ leads to a higher required transmitted energy. Thus we expect that there must be an optimum absorption cross-section for a given pollutant density and cloud thickness [38].

As an example of a single ended absorption measurement we consider the detection of CO at $4.7\,\mu m$. We assume an InSb photovoltaic detector with an $NEP = 0.3 \times 10^{-13}\,W\,Hz^{-\frac{1}{2}}$, a 100 ns pulse, a 5 km atmospheric visibility and a $1000\,cm^2$ area telescope receiver with an overall optical efficiency of $K = 0.1$. For an effective target reflectivity $\rho = 0.5$ and $S/N = 1$ on a single pulse, we find from Equation 19 that the required transmitted energy is 20 mJ for a 10 km range. For most ambient pollutant concentration measurements in an urban area a 10 km range is more than adequate so that the advantage of single ended operation may offset the disadvantage of higher required transmitted energy relative to a double ended measurement.

The long path absorption methods have been confirmed by experimental measurements [39]. Using diode lasers, Hinkley and co-workers [4, 29–31] has measured C_2H_4 in automobile exhaust locally and over a parking lot area with a retroreflector return. Recently a diode laser system has been constructed for the monitoring of SO_2 stack emissions. The operation of diode lasers at temperatures near 77 K promises to extend considerably their applications to remote monitoring by absorption.

The topographical absorption method was recently experimentally demonstrated by Henningsen et al. [36]. They used a 20 μJ, 2.1 μm parametric oscillator source to remotely detect CO by topographical backscattering over a 107 m range. Fig. 8.4a shows the CO overtone spectrum taken in the laboratory with the $LiNbO_3$ tunable parametric laser source. Fig. 8.4b shows two rotation lines of CO taken over the 107 m path. The measurement was not made differentially but by a continuous scan over a few minutes time. Even so, atmospheric scintillation did not reduce the signal to noise significantly. The measured CO detection sensitivity, S/N and range confirmed the theoretical results given by Equation 19. Of course more extensive experimental studies need to be made, but the work to date has illustrated the usefulness of the long path absorption methods.

8.2.2.3 Differential absorption

The differential absorption method overcomes the main disadvantage of the long path absorption methods — lack of depth resolution. By monitoring radiation backscattered from distributed Rayleigh and Mie particulate scattering and tuning the transmitted wavelength on and off an absorption line of the pollutant molecule, depth resolved absorption measurements can be made with a depth resolution $c\tau/2$ as shown in Fig. 8.1. The differential absorption method thus combines in an optimum way the depth resolution advantages of the Raman method with the very large absorption cross-sections and resultant high sensitivity of the absorption methods. In this section the required transmitted energy, the signal to noise ratio, and the sensitivity of the differential absorption method is discussed.

The differential absorption method was first suggested by Schotland in 1964 [40] using a searchlight as a light source. The method was recently analysed in detail by Byer and Garbuny [38] and Measures and Pilon [25]. Wright et al. [8] have also considered the application of differential absorption to pollution monitoring. In this section we follow the approach of Byer and Garbuny and first find the required transmitted energy, the signal to noise ratio and then the sensitivity of the measurement versus range.

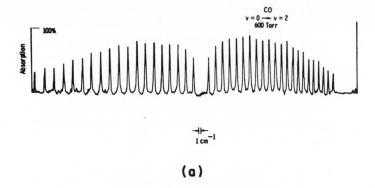

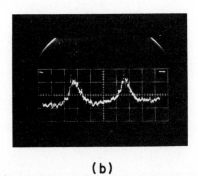

(b)

Fig. 8.4 (a) Vibrational rotational overtone absorption spectrum of CO by a tunable parametric laser. The measured cross-section for CO at atmospheric pressure is $0.75 \times 10^{-21}\,\text{cm}^2$; (b) two rotational lines in CO of $18\,\text{atm}\,\text{cm}^{-1}$ optical density at $107\,\text{m}$ distance. Zero transmission is presented by the small horizontal lines at the beginning and end of the trace (after Henningsen *et al.* [36]).

Before proceeding with the analysis, it is useful to describe schematically pollutant measurement by the differential absorption method. Fig. 8.5 illustrates the returned backscattered power when the transmitted wavelength is tuned on and off the CO absorbing transition. The assumed transmitter and receiver characteristics are shown in the figure. The range derivative of $\ln \left(P_r^{\text{off}}/P_r^{\text{on}} \right)$ equals the CO concentration shown at the bottom of Fig. 8.5. The differential absorption method thus describes a differential in space with the differential tuning on and off the absorption line being understood. Fig. 8.5 illustrates both the depth resolution and the sensitivity of the differential absorption method.

In order to evaluate the required transmitted energy for a differential

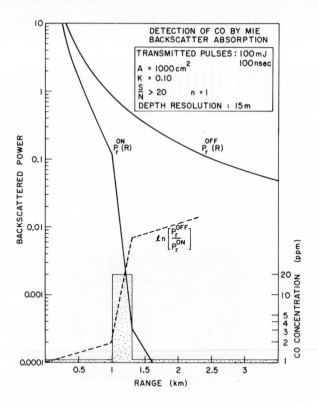

Fig. 8.5 Computer simulation of a depth resolved differential absorption measurement of CO. $P_r^{off}(R)$ and $P_r^{on}(R)$ are the backscattered powers when tuned off and on the CO absorption line. The logarith of the power ratios (dashed line) and the range derivative of $\ln (P_r^{off}/P_r^{on})$ which equals the initially assumed CO concentration is also shown.

absorption measurement we proceed as previously for the topographical absorption method (of Equations 14–19). However, now the effective target reflectivity (ρ/π) is replaced by $\frac{1}{2}c\tau\beta(R)$ with the total backscatter coefficient given by:

$$\beta(R) = \beta_R(R) + \beta_{Mie}(R) \tag{20}$$

where β_R and β_{Mie} are Rayleigh and Mie backscatter coefficients.

Table 8.2 shows that the effective reflectivity due to distributed particular backscattering is approximately four orders of magnitude less than that for topographical targets. Thus the depth resolution advantage obtained by the differential absorption method is acquired at an increased required transmitted energy.

Substituting for ρ/π in Equation 19 we obtain for the required transmitted energy for the differential absorption measurement method:

$$E_o = \frac{4}{K}\left[NEP\left(\frac{S}{N}\right)\frac{1}{\sqrt{\tau}}\right]\frac{1}{c\beta(R)}\frac{R^2}{A}\frac{\exp 2(\alpha_{sc}+N_A\sigma_{ABS})R}{1-\exp(-2N_pL\sigma_{ABS})}. \tag{21}$$

The required transmitted energy now varies as $1/\sqrt{\tau}$ instead of $\sqrt{\tau}$ as for the topographical absorption method because of the bandwidth requirement for depth resolved measurements.

At a given range, the differential absorption method requires about three orders of magnitude more transmitted energy than the topographical absorption method. A 10 mJ pulse energy allows a depth resolved absorption measurement to a 2 km range with a $S/N = 1, n = 1$ and a 16 m depth resolution. The signal to noise ratio improves with increased transmitted energy but only as the square root of the number of averaged pulses. There is also a tradeoff in absorption sensitivity and depth resolution as discussed previously. Obviously, a number of the variables can be adjusted to improve the range and sensitivity of the differential absorption method. However, transmitted energies in the 10 to 100 mJ range are necessary for differential absorption measurements in the i.r. The required energy is reduced somewhat in the visible due to increased scatter and to more sensitive detectors. Examples of SO_2, NO_2 and I_2 detection are given by Measures and Pilon [25] and the required energy for NO_2 detection is discussed by Byer and Garbuny [38].

Schotland [40] has also considered errors involved in the differential absorption measurement due to parameters other than detector signal to noise ratio. These parameters include atmospheric properties, uncertainties in the absorption coefficient and laser frequency, and signal to noise limitations of the returned power measurement. Wright et al. [8] has added digitization noise to the list of important contributions to the overall signal to noise ratio. However, as pointed out by Schotland [40] it is ultimately the returned power signal to noise ratio which limits the range and accuracy of the differential absorption method. Within the past year experimental observations have verified the theroetically predicted parameters of the differential absorption method.

The requirement for 10–100 mJ of tunable output from a laser source has held back the application of the depth resolved absorption method. However, recent advances in laser sources has led to demonstration measurements. In the next section two remote measurement systems are described. It is already apparent from the progress made in the past few years that the differential absorption measurement approach is the preferred remote monitoring method and that it will be an important remote monitoring tool.

8.3 Measurement systems and capability

In this section two operating remote monitoring systems are described. These are research systems that have been under development over the past few years

They represent, however, the state of the art in laser sources, optical components and electronic processing systems. The rapid evolution of tunable laser sources within the next few years will no doubt make the performance of these systems seem limited. However, they do provide a bench mark by which progress in remote monitoring using laser sources can be compared.

8.3.1 Remote monitoring system using a continuously tunable 1.4–4 μm source

The requirement for 10–100 mJ of tunable i.r. laser energy for remote monitoring has led to the development of a continuously tunable parametric laser source.

The tunable source is part of a remote monitoring system which includes: (1) the tunable laser source; (2) the transmitter and telescope receiver optics; (3) a detector and processing electronics. The system was designed with the goal of allowing real time remote monitoring measurements to be made by a single operator working from the mini-computer keyboard. Thus the computer controls the tunable source wavelength, the telescope pointing direction and the data processing and display. The total system has operated as expected and demonstrates that despite its high initial cost, its ease of operation, reliability and versatility make it an economic method of air pollution monitoring.

8.3.1.1 The Nd:YAG pump laser source

The operation of a parametric oscillator requires a diffraction limited pump laser source. We have previously used [41] a TEM_{oo} mode Nd:YAG oscillator operating with a 780 μm spot size and 5–8 mJ of output energy to pump a Nd:YAG amplifier chain. The amplifier chain, which consisted of two 1/4 inch amplifiers and 3/8 inch amplifier and two Faraday rotator isolators for stability, provided over 380 mJ output energy at up to 10 pps. The output energy was limited by the small spatial filling factor of the Gaussian beam. To prevent mode clipping and resulting diffraction problems the beam spot size was chosen at the 1% intensity loss or the $3w_0 = d$ condition, where w_0 is the Gaussian beam spot size and d is the Nd:YAG rod diameter.

To utilize the Nd:YAG rods more efficiently, we have investigated a Nd:YAG unstable resonator oscillator [42]. Our first design, which took into account the focusing properties of the Nd:YAG rod, used a 1/4 inch diameter rod and an output magnification of 3.3. The resulting output coupling was slightly over 80%. The oscillator operated with up to 200 mJ output energy in a diffraction limited beam. Using this source operating at 150 mJ at 10 pps we were able to demonstrate 30 mJ of 0.532 μm output by second harmonic generation in Type

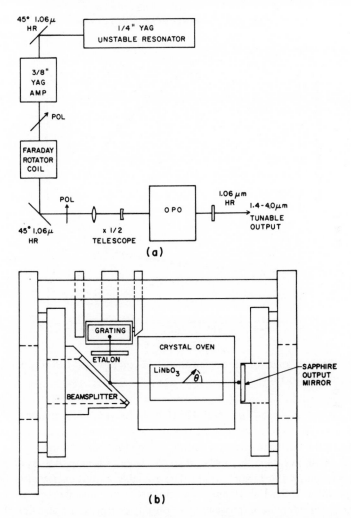

TUNABLE SYSTEM WITH UNSTABLE RESONATOR
AND DOUBLE PASS OPO

Fig. 8.6   (a) Schematic of the Nd:YAG unstable resonator system which
generates 750 mJ per pulse at $1.06\,\mu$m at 10 pps and $\frac{1}{2}$% efficiency.
(b) Schematic of the LiNbO$_3$ parametric tuner design. The $1.06\,\mu$m
pump radiation is incident from left through the dichroic beamsplitter.
A minicomputer controls the stepper motors which scan the grating
and crystal angles for wavelength tuning.

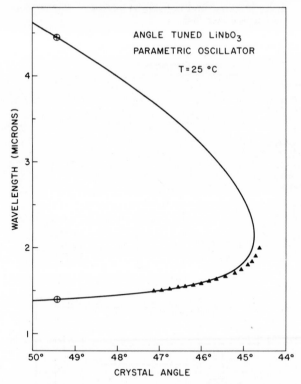

Fig. 8.7   Tuning range for the angle tuned 1.06 μm pumped LiNbO₃ parametric oscillator tuner.

II *KD*P*, 10 mJ of 0.3547 μm and 10 mJ of 0.266 μm. More importantly, the output pumped a LiNbO₃ parametric oscillator which showed a reduced threshold compared to the Gaussian intensity source at the same input energy.

Fig. 8.6a shows the Nd:YAG unstable resonator/amplifier system now employed to pump the LiNbO₃ tuner. The available output energy after the 3/8 inch amplifier is 780 mJ at 10 pps. Thus the use of the unstable resonator oscillator allows considerable improvement in output energy and a simplification of the Nd:YAG pump system. The gain in energy is largely due to the better geometric filling factor provided by the unstable resonator mode. It should be noted that the mode has a plane wave front and in the near field still has the output mirror zero intensity hole in the beam centre. In the far field the mode transforms to a modified Airy disk intensity function which can couple very well to Gaussian profile beams. The above results were obtained in the near field. The Nd:YAG unstable resonator/amplifier system operates with a ½% wall plug efficiency and for up to 1000 h before a flash-lamp change is required. The efficiency and reliability of the Nd:YAG laser source are important parameters in designing a remote monitoring system.

8.3.1.2 LiNbO₃ parametric tuner

The LiNbO$_3$ parametric oscillator operates over the 1.42–4.2 μm tuning range shown in Fig. 8.7 as a singly resonant oscillator (SRO) with the 1.42–2.1 μm signal wave resonated with a 13 cm long optical cavity formed by a grating and 50% reflecting output coupler. The grating provides linewidth control to less than 4 cm^{-1}. In addition the grating provides an absolute frequency reference to within ± 1 cm^{-1} over the tuning range of the oscillator.

The parametric tuner is controlled by an interactive software program using a PDP11E1O minicomputer acting through a CAMAC controller. The software allows the setting of the initial wavelength, scan range, scan rate and repeat scans. It also allows control of the line narrowing tilted etalon for higher resolution spectral scans.

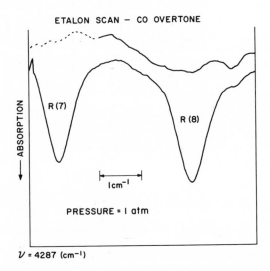

Fig. 8.8 Absorption spectra of the CO overtone $R(7)$ and $R(8)$ rotational lines at 4287 cm^{-1} using the computer controlled tilted etalon scan control.

For 100 mJ of 1.06 μm input energy, the parametric tuner generates a total of 20 mJ output energy in the signal (1.42–2.12 μm) and idler (2.12–4.2 μm) beams. The device is stable and reliable having operated in a research environment for a four month period without degradation of the optical components.

Recently the output energy of the parametric tuner was increased to a total of 85 mJ at 10 pps by using a single pass parametric amplifier. The conversion efficiency from 1.06 μm to tunable output was 30%. It is expected that future improvements in system configuration should lead to output energies in excess of 100 mJ per pulse at linewidths of 0.1 cm^{-1}.

As an example of control of the system for spectroscopy, Fig. 8.8 shows the CO overtone absorption spectra at 1 atm pressure taken with the line narrowing tilted etalon scanned under computer control. The resolution was 0.3 cm^{-1} and the resettability was ± 0.1 cm^{-1}. For remote air pollution monitoring the etalon can be set on and off the absorption peak on every other pulse at repetition rate of up to 10 pps.

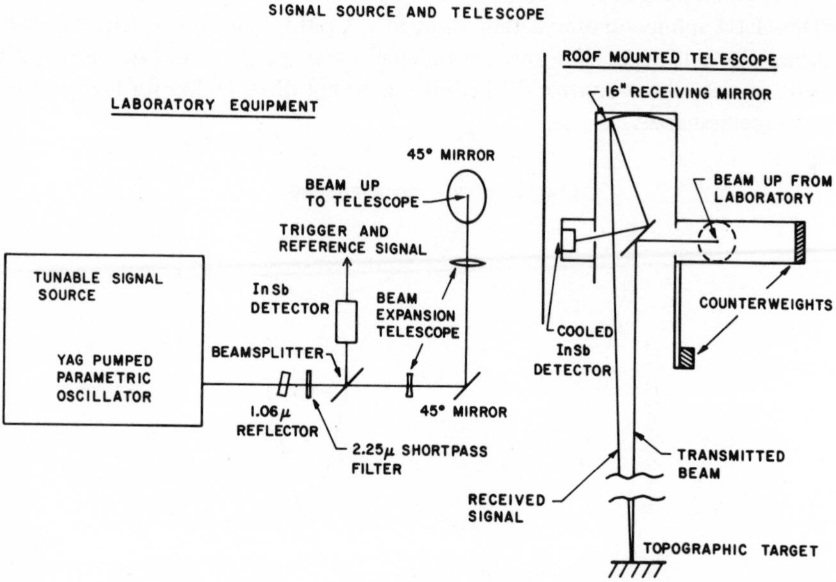

SIGNAL SOURCE AND TELESCOPE

Fig. 8.9 Schematic diagram of the tunable source, monitoring optics and the telescope receiver.

8.3.1.3 Remote monitoring system

The Stanford remote monitoring system consists of a parametric tuner source, a 16 inch ƒ3 diameter receiving telescope. InSb detector and amplifiers, an analogue to digital converter, a fast buffer memory and a PDP11E1O mini-computer. The system operates in a radar fashion using topographical targets or backscatter from atmospheric particulate matter as retroreflectors. The

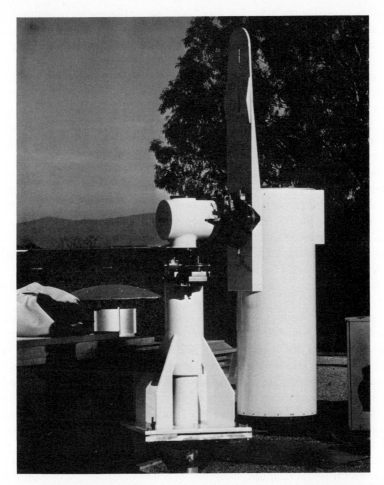

Fig. 8.10 Photograph of the remote monitoring 16 inch $f3$ diameter receiving telescope mounted above the laboratory at Stanford University. Return signals have been detected from the hills seen in the background which are located across the bay at 17 miles distance.

transmitter operates at 10 pps repetition rate allowing near real time remote measurements. A single operator working from the minicomputer can control the system including wavelength selection, telescope direction and data verification displays and recording.

Fig. 8.9 shows a schematic diagram of the optical components of the system. The parametric oscillator supplies 10–30 mJ pulses with a bandwidth of approximately 0.1 wavenumber with the line narrowing etalon. This bandwidth is adequate to resolve the individual rotational lines of the spectra of pollutants. At the output of the parametric oscillator a filter is used to pass either the signal or idler frequency, above and below 2.12 μm respectively. After the filter a detector monitors the transmitted output for the purposes of triggering the received signal recording apparatus and providing an output level reference. In order to minimize the divergence of the transmitted beam, to monitor its range, and eliminate eye safety problems, the parametric oscillator beam is expanded to a 2.5 cm radius with a Galilean telescope using calcium fluoride lenses. Finally, by means of various diagonal mirrors the beam is transmitted from the parametric oscillator to the telescope and into the atmosphere.

The 16 inch $f3$ receiving telescope is operated with the detector at the Newtonian focus. The telescope was specifically designed for remote monitoring and utilizes an alt-azimuth mounting geometry with 45° reflecting mirrors in the x, y axes to reflect the transmitted beam on to an axis collinear with the telescope field of view. Fig. 8.10 shows a photograph of the telescope system. In order to reduce costs, the telescope drive train was designed for ±10 min pointing accuracy so that a belt drive and clutch system could be employed. This enables the telescope to be quickly manually positioned without disengaging gears. Two encoders keep track of the pointing direction at all times.

The telescope focusses the received energy onto a 1 mm² InSb photovoltaic detector which operates at 77 K. The detector size is twice the diffraction limited focal spot size of the telescope. The field of view of the system is 0.41 mrad, which is near the atmospheric limit. Alignment of the optics is, therefore, very critical and a method of optical alignment using a helium neon laser was devised. Basically the procedure involves aligning the components from the source to the detector in sequence. Once accomplished final adjustments are made by monitoring the return signal. The system alignment is stable even though the 750 pound telescope is mounted on an 8 inch diameter aluminium pipe, 30 ft above the laboratory floor. Finally, the telescope was designed to be easily mounted into a van for future field use.

The output of the detector is amplified by a preamplifier and a computer controlled gain switchable amplifier with an overall bandwidth of 1 mHz. As the return signals arrive at the receiver they are amplified by the amplifier and digitized in a 5 mHz 12 bit A to D converter and then stored in a fast buffer memory prior to processing by the computer. In order to accomplish the detection of a particular pollutant the frequency of the transmitted signal may be changed from pulse to pulse. The return signal on the absorption and off the absorption of a particular material may then be examined. At the end of a sequence of pulses the accumulated data is accessed by the PDP11/10 minicomputer to derive the concentration of the atmospheric constituent of interest. Fig. 8.11 shows a schematic of the data processing system. Its primary features

RECEIVER ELECTRONICS AND DATA PROCESSING

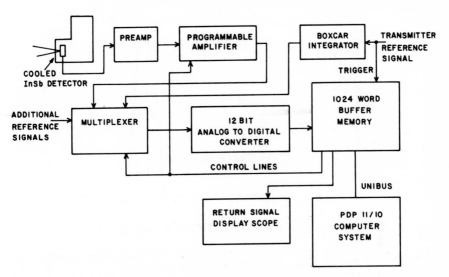

Fig. 8.11 Schematic of the receiver electronics and data processing system.

are a real time display of the received signal via the buffer memory and direct computer control for data storage and processing.

The buffer memory has several features that allow it to be particularly useful in this type of application. The buffer memory serves as a control unit for the digitization process as well as a storage medium. The primary capability of the unit is receiving parallel data from an A/D at up to 10 mHz bit rates. The memory contains 1024-16 bit words of bipolar semiconductor RAM along with appropriate control logic. The number of digitizing operations per laser shot selectable by the user. When the memory is full or half full, depending upon the user's choice, the information it has received may be displayed on an oscilloscope utilizing two built in digital to analog converters. In addition, an interrupt is generated telling the PDP11/10 computer system that it may access the data that has been accumulated. Due to the Unibus type of architecture of the PDP system the buffer memory may be treated as an additional 1 K of memory. Since the output of the A/D is 12 bits, the other 4 bits in each word may be used to contain housekeeping data, such as various amplifier gain settings. In addition, at the conclusion of the digitization of the A/D data, additional memory positions are reserved for more housekeeping data, (i.e. transmitter output level), that may also be digitized by selecting other inputs of the multiplexer in front of the A/D.

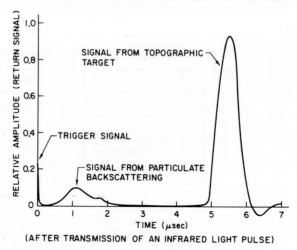

Fig. 8.12   Example of a return signal showing the signal from topographical target at 775 m and from particulate scattering.

Fig. 8.12 shows an example of a return signal. The first peak in the return is due to backscattering from particulate matter in the atmosphere. The second and larger peak is from a stationary target, in this case, Hoover Tower, located at a distance of 775 m from the telescope. For this particular return the transmitter wavelength was $1.06\,\mu m$ and the transmitted power was $12.5\,mJ$ per pulse. The received power from the stationary object return is in the order of $1\,mW$ for a transmitted power of about $1\,MW$.

The system has been used to scan a $400\,cm^{-1}$ atmospheric spectrum near $2.0\,\mu m$ for detection of the 2.01 and $2.06\,\mu m$ CO_2 combination bands. These bands are of interest for remote measurement of atmospheric temperature and density. The scan also showed the edge of the $1.9\,\mu m$ H_2O overtone band which can be used for remote measurement of humidity.

Although the system capability has been demonstrated including topographic scattering to 17 miles and Mie return signals from a 3–4 km range, remote pollutant monitoring has not been pursued due to development efforts on the parametric source. The system is now being re-assembled for dedicated pollution measurement. The versatility of continuous wavelength control and the long-term operational reliability of the laser source and parametric tuner should allow long-term remote monitoring measurements to be made on an hourly and daily basis. We expect that these measurements will further confirm the unique capabilities of remote air pollution measurement with tunable laser sources.

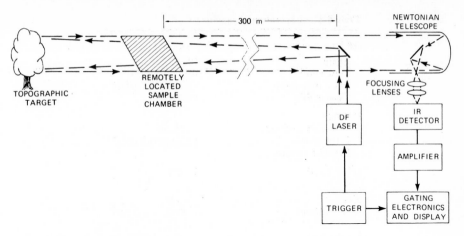

Fig. 8.13   Experimental apparatus.

8.3.2 Remote measurement system using discretely tunable laser source

8.3.2.1 General remarks

Discretely tunable lasers emit laser radiation on many fixed wavelengths. The utility of discretely tunable lasers for remote measurement depends on the spectral coincidence of these fixed wavelengths with absorption features of gases of interest. Because of the high spectral density of both fixed wavelength laser lines and absorption lines of pollutants, many useful spectral overlaps exist. Several gas lasers emit high energy pulses that enable long range measurements to be made. Hence, gas lasers are useful for making range resolved measurements using backscatter from naturally occurring aerosols, and integrated concentration measurements using scatter from topographic targets.

Demonstration measurements were made using the deuterium fluoride (DF) chemical laser and the CO_2 electric discharge laser. The DF laser was used for remote measurement of integrated concentrations of HCl, CH_4, and N_2O. The CO_2 laser was used for measurements of range resolved concentrations of water vapour (H_2O) and has the potential for measuring many other gases.

8.3.2.2 Remote measurement of HCl, CH_4, and N_2O using a DF laser

A diagram of the single ended DF laser system used to monitor HCl, CH_4, and N_2O is shown in Fig. 8.13. This system, reported by Murray $et\ al.$ [45] used a topographic target to provide the backscattered signal and measured the integrated

Table 8.4 Experimental parameters

DF laser transmitter

Energy	100 to 150 mJ pulse^{-1}
Pulsewidth	1.0 μs (FWHM)†
Beam divergence	1.0 mrad (FWHM)†
Typical PRF	1/6 Hz

Receiver

Telescope diameter	31.75 cm
Field of view	3.0 mrad (FWHM)†
HgCdTe detector	
D^* (3.7 μm)	$5(10^9) \dfrac{\text{cm Hz}^{\frac{1}{2}}}{\text{W}}$
Size	1 × 1 mm
Time constant	75 ns (to 1/e point)

† Full width at half maximum.

Table 8.5 Laser wavelengths and absorption coefficients

Gas	DF laser line	Wavelength (μm)	Absorption coefficient (cm^{-1} atm^{-1})
HCl	$P_2(3)$	3.636 239	5.64
CH$_4$	$P_1(9)$	3.715 252	0.047
N$_2$O	$P_3(7)$	3.890 259	1.19

concentration over the path between the lidar and the target. The DF laser beam was transmitted co-axially with the receiver and was pointed through a sample chamber that was positioned 300 m away. The backscattered signal from the topographic target was collected by a Newtonian telescope and focused on to an i.r. detector. The detector signal was amplified and displayed on a chart recorder. Different concentrations of the gases were injected into the sample chamber. The concentrations were measured with both the lidar and an *in situ* monitor and the values compared.

Experimental parameters are shown in Table 8.4. The DF laser operated at between 100 and 150 mJ per pulse, although some variations occurred from line to line and day to day. The HgCdTe detector was designed for use near 10 μm and its detectivity at 3.7 μm was degraded somewhat to 5×10^9 cm Hz$^{\frac{1}{2}}$ W^{-1}. Detectors were available that are twenty times more sensitive in the 3 to 5 μm region. Nonetheless, the HgCdTe detector was used because it was readily available. Even with the use of a sub-optimum detector, the system performance was not limited by detector noise. The laser wavelengths and absorption coefficients that were used are shown in Table 8.5. The wavelengths used for the three gases

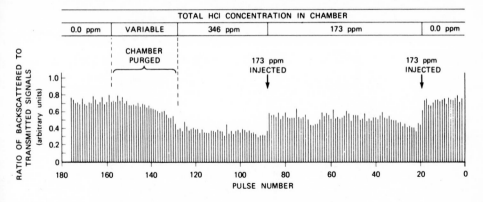

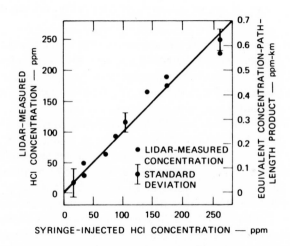

Fig. 8.14 Sample of behaviour of the ratio of backscattered to transmitted signal during system operation on HCl.

Fig. 8.15 Summary of remote measurement tests of HCl in a sample chamber.

were between 3.6 and 3.9 μm. The gaseous absorption coefficients at the available DF laser wavelengths result from spectral coincidences. The HCl absorption coefficient was measured by Bair and Allario [46] and those for CH_4 and N_2O were measured by Spencer *et al.* [47]. The wavelengths were established by Rao [48].

A typical display of raw data for the HCl measurement is shown in Fig. 8.14. The ratio of the backscattered to the transmitted signal is shown as a function of pulse number. Initially, with no HCl in the chamber, the baseline level of the ratio was established. This was followed by two injections of 173 ppm of HCl.

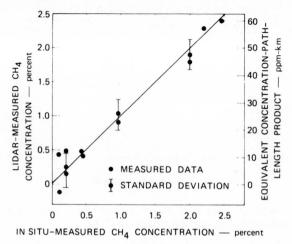

Fig. 8.16   Summary of remote measurement tests of CH_4 in a sample chamber.

Each injection yielded an easily discernible decrease in the plotted ratio. Near pulse number 127, the sample chamber ports were opened and a fan purged the system of HCl. The continuously increasing ratio during the purge verified the decreasing HCl concentration. Ultimately, the ratio achieved a constant value consistent with the initial level with no HCl in the chamber.

A summary of the HCl test for different concentrations is shown in Fig. 8.15. The lidar measured HCl concentration is plotted as a function of the syringe injected value. The solid line represents perfect agreement between the lidar and the syringe values. Each point represents the average of 15 measurements. Each error bar shown represents the S.D. for 15 data points. Generally, good agreement is apparent between the lidar and the syringe values, thus confirming that the absorption coefficient used in the data reduction is accurate. The right vertical axis shows the equivalent product of concentration and path length for the individual measurements. The sensitivity of the system, or minimum detectable concentration, is defined by the concentration that equals the S.D. For HCl, the sensitivity was determined to be 0.05 ppm over a 1 km path, or 50 ppb over 1 km.

A summary of the CH_4 data is shown in Fig. 8.16. Again, good agreement was obtained between the lidar and the *in situ* value, thereby verifying the absorption coefficients used in the data reduction. For CH_4 the system sensitivity was found to be 6 ppm over a 1 km path.

A summary of the N_2O data is shown in Fig. 8.17. The triangles represent data obtained with a juniper tree of a 1 m diameter used as the backscattering target; the circles represent data taken using a plywood board. The two targets

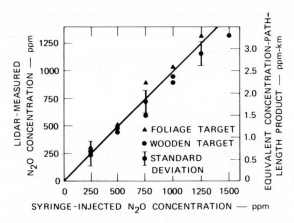

Fig. 8.17 Summary of remote measurement tests of N_2O in a sample chamber.

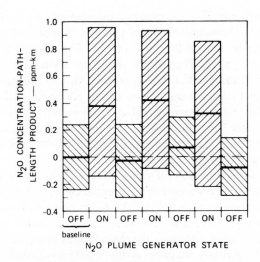

Fig. 8.18 Results of N_2O plume measurement in the open atmosphere near the limit of detectability of the system.

were used to demonstrate system operation on two different materials with different reflectivities. The foliage target was used because it provided a reflectivity typical of field operation. Both targets were placed immediately behind the sample chamber. The lidar results obtained with both targets agree with the *in situ* values. The system sensitivity for N_2O was determined to be 0.24 ppm km.

A plume of N_2O was generated in front of the foliage target to represent

Table 8.6   Measurement error (ppm km) due to interfering species over an 8 km horizontal path*

Target gas	No interference compensation†	Line selection to minimize interferences‡	Interference compensation§
HCl	0.10	0.012	0.0037
CH$_4$	10.0	2.0	0.59
N$_2$O	0.73	0.041	0.012

* Errors calculated from data of [47].

† DF laser lines (strongly absorbed/weakly absorbed) for HCl, CH$_4$, and N$_2$O are, respectively, $P_2(3)/P_1(6)$, $P_1(9)/P_2(5)$, and $P_3(7)/P_1(14)$.

‡ DF laser lines (strongly absorbed/weakly absorbed) for HCl, CH$_4$, and N$_2$O are, respectively, $P_2(3)/P_3(6)$, $P_1(9)/P_1(10)$, and $P_3(7)/P_2(10)$.

§ Compensation is calculated from an estimate of absorption by interfering species. The ratio between the transmission of the strongly absorbed laser lines and that of the weakly absorbed laser lines is assumed to be accurate to within 30% of the actual value.

typical field conditions. The results of that measurement are shown in Fig. 8.18. The product of N$_2$O concentration and path length is plotted as a function of the plume generator state, either 'off' or 'on'. The centre bar represents the mean value of 15 pulses. The extent of the bar above and below the mean represents 1 S.D. of those pulses. The increased S.D. with the generator 'on' indicates that turbulent transport was convecting the plume into and out of the lidar line of sight. The increased average value of the product of concentration and path length is a clear indication that the lidar measured the gas concentration in the plume, even though the concentration was only slightly above the estimated system sensitivity of 0.24 ppm km.

In general, gaseous species in the atmosphere, such as water vapour and HDO, can produce errors in two wavelength differential absorption measurements. The error due to interferences is shown in Table 8.6 in ppm km. These data were calculated using an 8 km horizontal path from the data reported by Spencer *et al.* [47]. The second column represents the measurement error without interference compensation. An arbitrary selection of the weakly absorbed line was used, and this resulted in the measurement error shown in column 2. By carefully selecting the weakly absorbed line so that its transmission by interfering species was matched to that of the strongly absorbed line, the measurement error was reduced to the values shown in column 3. Further interference reduction can be obtained by estimating or measuring the effects of the interfering species at the wavelengths selected. By estimating interference effects to within only 30% of their actual value, the measurement errors shown in column 4 were calculated. The required sophistication of the interference compensation for a given measurement depends on the application and the desired sensitivity. Table 8.6 illustrates the use of two techniques for interference minimization. Further reduction in measurement error can be obtained by measuring the concentration of the interfering species. In principle, a negligible measurement error can be obtained using this technique.

Calculations were made of the performance attainable with a range resolved DF lidar system comprising commerically available components. The signal to noise ratio (SNR) at a given range provides an indication of the expected performance of a range resolved system.

The dominant noise mechanism in an i.r. lidar system designed for maximum range is usually detector noise. The expected SNR can be determined by calculating the signal from Equation 21 and the noise from the detector characteristics. An SNR of 115 at a horizontal range of 10 km with a 100 ns laser pulse width can be obtained for the following parameters:

$$E_0 = 1 \text{ J pulse}^{-1}$$
$$A_r = 1.17 \text{ m}^2$$
$$\lambda = 3.7 \, \mu\text{m}$$
$$K = 0.5$$
$$\beta_{180} = 8 \times 10^{-7} \text{ m}^{-1} \text{sr}^{-1}$$
$$\alpha = 0.0657 \text{ km}^{-1}$$
$$D^* = 10^{11} \text{ cm Hz}^{1/2} \text{ W}^{-1}$$

The backscatter coefficient is characteristic of a low altitude haze and the absorption coefficient is typical of sea level attenuation. The conclusion is that high sensitivity, range resolved measurements are feasible at a 10 km range. Longer range or higher sensitivity can be obtained using multipulse integration.

8.3.2.3 Range resolved measurement of gases using a DIAL system with a CO_2 laser transmitter

A CO_2 lidar system was developed using backscatter from aerosols to provide range resolved measurement of the concentration of gaseous species as reported by Murray et al. [21]. An equipment diagram of the CO_2 lidar system is shown in Fig. 8.19. The laser beam was transmitted co-axially with the receiving telescope. Scattered radiation from naturally occurring aerosols was collected by the telescope and focussed on to a HgCdTe detector. The detector signal was amplified, filtered, digitized, displayed, and stored on magnetic tape for later analysis. A real-time display of the return signal on the absorbed and non-absorbed laser lines was used to monitor system operation.

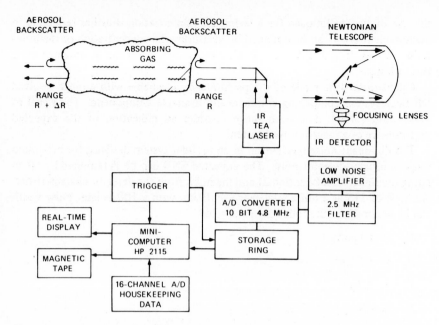

Fig. 8.19 The infrared differential absorption lidar system for range resolved measurement of gaseous species.

Table 8.7 DIAL system parameters

CO_2 *TEA laser*

Transmitted energy	1 J
Lines	$R(12), R(18), R(20)$ on 10 μm band
Pulse width	100 ns (FWHM)
Beam divergence	1.8 mrad (to the 10% point)
Typical PRF	0.5 Hz

Receiver

	Detector 1	Detector 2	Detector 3
Telescope diameter	31.75 cm		
Field of view	3.3 mrad (to the 10% point)		
Hg–Cd–Te detectors	Detector 1	Detector 2	Detector 3
D^*	$2(10^{11}) \dfrac{\text{cm Hz}^{\frac{1}{2}}}{\text{W}}$	$2(10^{10}) \dfrac{\text{cm Hz}^{\frac{1}{2}}}{\text{W}}$	$1.1(10^{10}) \dfrac{\text{cm Hz}^{\frac{1}{2}}}{\text{W}}$
Time constant	3.0 μs	0.98 μs	0.075 μs
Size	1 × 1 mm	1 × 1 mm	1 × 1 mm

The system parameters are shown in Table 8.7. The laser operation on these strong lines was quite stable with 90% of the pulses falling within ± 5% of the average energy. The field of view of the receiver was designed to be nearly twice

Table 8.8 Water-vapour absorption coefficients

Line	$\lambda\,(\mu m)$	$\alpha\,(10^{-4}\,cm^{-1}\,atm^{-1})$
$R(12)$	10.303 458	1.60
$R(18)$	10.260 381	0.935
$R(20)$	10.246 625	8.65

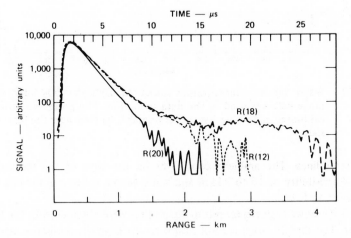

Fig. 8.20 Signal backscattered from atmospheric aerosols for the $R(12)$, $R(18)$ and $R(20)$ lines on the $10\,\mu m$ band of CO_2. The $R(20)$ line is more strongly absorbed by ambient H_2O than is the $R(12)$ or $R(18)$. The $0.98\,\mu s$ detector was used for these results.

the beam divergence of the laser to ensure easy alignment of the system and to minimize the influence of thermal drift during an experiment. Three detector configurations were used during the course of the investigation. At first, a very high sensitivity detector was used because of uncertainty about the magnitude of the backscatter coefficient. This detector had a time constant of $3\,\mu s$, which degraded range resolution. After a large backscattered signal was obtained, the cooled filter was removed, decreasing the time constant to $0.98\,\mu s$ and increasing the detector noise by a factor of 10. To improve range resolution further, a faster detector was obtained having a time constant of $0.075\,\mu s$.

The first target species for the CO_2 lidar system was water vapour, and considerable system optimization was performed on this species. The water vapour absorption coefficients measured by Shumate *et al.* [49] are shown in Table 8.8. These indicate that the $R(20)$ transition is more strongly absorbed than either the $R(12)$ or the $R(18)$. Most of the results were obtained using the $R(20)$ and

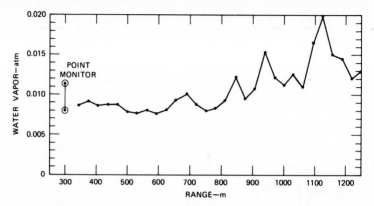

Fig. 8.21 Water vapour concentration measured using the CO_2 lidar. A 60 m range cell was used in the data reduction. The point monitor was a calibrated recording hygrothermograph located near the lidar line of sight.

$R(18)$ laser lines. The ambient conditions on the days of the measurement included a visibility of 16 to 24 km and water vapour content in the neighbourhood of 0.5 to 1.0%.

Fig. 8.20 shows typical lidar return signals that were obtained with the 0.98 μs detector. The sharply rising aerosol return results from the geometric overlap of the transmitter and receiver. Following system convergence, the $1/R^2$ and atmospheric attenuating losses decrease the signals, which drop into the noise between 1.5 and 2.0 km. The differential between the $R(20)$ and the $R(18)$ and the $R(12)$ lines is readily apparent. The rate of change of the differential absorption between these lines is proportional to the concentration of the absorbing species.

The water vapour concentration was calculated using the data of Fig. 8.20. The resulting water vapour profile is shown in Fig. 8.21 as a function of range. The point monitor was a recording hygrothermograph located on the roof of a building at a range of 300 m near the lidar line of sight. Good agreement was obtained between the point monitor and the lidar values for short ranges. The lidar determined concentration appeared to undergo random fluctuations at less than 1 km, whereas the return signal was relatively free of noise out to a range of 1.5 km. This is indicative of the propagation of noise effects from the return signal through the DIAL equation. This demonstrates the distinction between the differential signal to noise ratio and the SNR of the return signal.

Since the above results were obtained two major changes have significantly improved system performance. The installation of a fast detector with a 75 ns time constant improved the range resolution and better receiver realignment

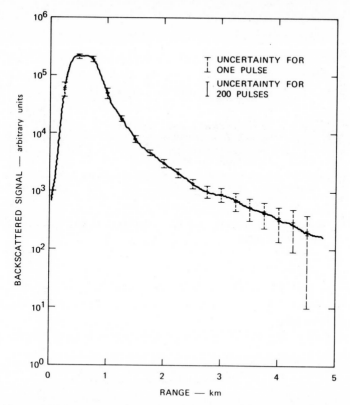

Fig. 8.22 Typical return signal summed over 200 pulses showing uncertainties as a function of range.

increased the system range. The results shown in Fig. 8.22 were obtained with these modifications. The signal did not drop into the noise until a range of a factor of 2 over the results obtained earlier. The error bars shown in Fig. 8.22 represent the uncertainties for a single pulse and also for the 200 pulses that were summed to obtain the curve. The single pulse uncertainty is the standard deviation of the return signals at each range. The multipulse uncertainty is the single pulse uncertainty divided by the square root of the number of pulses. The increasing uncertainty as a function of range indicates the decreasing SNR.

The concentration derived from the new data is shown in Fig. 8.23. The uncertainties associated with the return signal are propagated through the data reduction process. The resulting uncertainties in concentration were calculated using the standard theory [50] and are shown in Fig. 8.23. In this case, the increasing uncertainty in the multipulse measurement equalled the measured

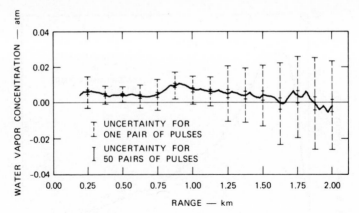

Fig. 8.23 Range resolved water vapour concentration and uncertainties as a function of range.

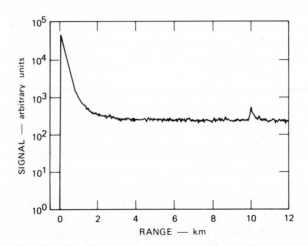

Fig. 8.24 Backscattered signal including atmospheric return and scatter from the coastal foothills at a range of 10 km.

concentration near 1.5 km. This range represented the maximum range for which the measured values have significance. The negative concentration values have no significance in reality and are displayed to illustrate how the measurement degrades with increasing range.

A range resolved DIAL system can simultaneously be used to determine the integrated species concentration using topographical reflections. Fig. 8.24 is an example of a return signal on the $P(14)$ line of the $10\,\mu m$ band of CO_2 with the

Table 8.9 Calculated CO_2 lidar system sensitivities

Species		Absorption coefficient (cm^{-1} atm^{-1})	Sensitivity (ppb km)
Name	Formula		
Ammonia	NH_3	120.0	0.42
Benzene	C_6H_6	2.3	22.0
1,3 butadiene		3.45	15.0
Carbon dioxide	CO_2	$1.8\ (10^{-3})$	28 000.0
Ethylene	C_2H_4	33.0	1.5
Freon 113	$C_2Cl_3F_3$	19.2	2.6
Freon 11	CCl_3F	31.0	1.6
Freon 12	CCl_2F_2	92.0	0.55
Methanol	CH_3OH	19.4	2.6
Ozone	O_3	12.7	4.0
Perchloroethylene	C_2Cl_4	28.5	1.8
Sulphur dioxide	SO_2	6.73	7.5
Sulphur hexafluoride	SF_6	800.0	0.063
Trichloroethylene	C_2HCl_3	14.0	3.6
Vinyl chloride	C_2H_3Cl	6.79	7.4
Water vapour	H_2O	$8.36\ (10^{-4})$	60 000.0

lidar system pointed west toward the coastal foothills near Menlo Park, California. A well defined topographical reflection primarily from foliage at 10 km was obtained. In such a configuration, the range resolved concentration of a pollutant can be obtained to a range limited by the SNR. Simultaneously, the integrated concentration of the pollutant over a longer range between the lidar system and an arbitrary topographical target can be obtained. Wide applicability of such combination range resolved and integrated concentration lidar systems is foreseen for use in source monitoring and ambient air quality monitoring.

In a prediction of the range resolved performance of the CO_2 lidar system, the following parameters were used:

$$E_o = 15\ \text{J pulse}^{-1}$$

$$A_r = 1.17\ \text{m}^2$$

$$\lambda = 10.0\ \mu\text{m}$$

$$K = 0.5$$

$$\beta_{180} = 1.2 \times 10^{-7}\ \text{m}^{-1}\ \text{sr}^{-1}$$

$$\alpha = 0.0927\ \text{km}^{-1}$$

$$D^* = 2(10^{10})\ \text{cm Hz}^{1/2}\ \text{W}^{-1}.$$

The backscatter coefficient is characteristic of a low altitude haze and is somewhat less than it is in the 3 to 4 μm region. The detector characteristics are those consistent with available detectors. The calculation of SNR using Equation 21 yields a value of 50 at a range of 10 km for a single pulse. Thus, high sensitivity, range resolved measurements are feasible at 10 km. Longer range or higher sensitivity can be obtained using multipulse integration.

The sensitivity of the CO_2 lidar system to different gases has been calculated based on an SNR of 200. The sensitivity values for 16 representative compounds are shown in Table 8.9. The absorption coefficients vary widely. Those for CO_2 and water vapour are relatively small. This is necessary to permit long range transmission of the beam. The absorption coefficient of many other gases is high, yielding a sensitivity of more than 23 ppm in a km. Range resolved multigas monitoring of these gases appears feasible at a 10 km range with commercially available components.

From Table 8.9 it is evident that line tunable i.r. laser sources have adequate coincidences to allow measurements of a number of important molecular species. The other advantages of these sources of high pulse energy, commercial availability and relatively low cost make them promising for future remote monitoring applications.

8.4 Conclusion

In this chapter we have presented a brief overview of laser remote monitoring. Two operating systems were described to illustrate the present state of the art of this rapidly developing field.

From recent demonstration measurements it is now evident that remote monitoring using tunable laser sources will indeed become a unique monitoring tool. The method allows single ended, sensitive, depth resolved absorption measurements to be made in a real time radar fashion. Furthermore, by wavelength tuning the method is capable of selectively monitoring any pollutant molecule with an optical absorption transition. Although, in a research and development state at this moment, recent demonstrations show that in the future remote laser monitoring systems will prove to be a cost effective general monitoring tool for both ambient and source pollutants.

Acknowledgements

At Stanford Research Institute numerous staff members have made essential contributions to the development of discretely tuned i.r. lidar technology. The experiments were performed by J.E. van der Laan, J.G. Hawley and D.D. Lee. The development of an on line data display and the data reduction were performed by M.F. Williams. Valuable contributions to planning the experiments were made by E.K. Proctor, R.S. Vickers, R.D. Hake, M.L. Wright and W.B. Grant.

The work at Stanford Research Institute was supported by a grant from the U.S. National Science Foundation programme on Research Applied to National Needs and by a contract from the U.S. Geological Survey.

At Stanford University the system was developed with the help of Dr R.L. Herbst, R. Baumgartner and D. Wolfe with support from the Electric Power Research Insitute and NSF/RANN.

We wish to express our appreciation to Mary Farley for timely preparation of the manuscript.

References

[1] Byer, R.L. (1975). *Opt. Quant. Elect.* **7**, 147–77.
[2] Wang, C.P. (1974). *Acta Astronautica* **1**, 105–23.
[3] Derr, V.E. (1972). *Remote Sensing of the Troposphere*. C55.602 T75 U.S. Government Printing Office: Washington, D.C. NOAA.
[4] Hinkley, E.D. (1976). *Laser Monitoring of the Atmosphere.* Springer-Verlag: Berlin.
[5] Bowman, M.R., Gibson, A.J. and Sandford, M.C.W. (1969). *Nature* **221**, 456; see also Gibson, A.J. (1969). *J. Sci. Instr.* **2**, 802.
[6] Rothe, K.W., Brinkman, U. and Walther, H. (1974). *Appl. Phys.* **3**, 115–19.
[7] Zuev, V.E. (1976). 'Laser-Light Transmission Through the Atmosphere' in *Laser Monitoring of the Atmosphere*, (ed. Hinkley, E.D.) Springer-Verlag: Berlin.
[8] Wright, M.L., Proctor, E.K., Geasioek, L.S. and Liston, E.M. (1975). *A Preliminary Study of Air Pollution Measurement by Active Remote Sensing Techniques*, Stanford Research Institute project (1966).
[9] Collis, R.T.H. and Russel, P.B. (1976). 'Lidar Measurement of Particles and Gases by Elastic Backscattering and Differential Absorption', in *Laser Monitoring of the Atmosphere*, (ed. Hinkley, E.D.) Springer-Verlag: Berlin, Germany.
[10] Collis, R.T.H. (1970). *Appl. Optics,* **9**, 1782–8.
[11] Allen, R.J. and Evans, W.E. (1972). *Rev. Sci. Instr.* **43**, 1422–32.
[12] Viezee, W., Collis, R.T.H. and Lawrence, J.D. (1973). *J. Appl. Meteorology* **12**, 140–8.
[13] Schotland, R.M., Sassen, K. and Stone, R. (1971). *J. Appl. Meteorology* **10**, 1011–17.
[14] Chemesha, B.R. and Nakamura, Y. (1972). *Nature* **237**, 328–9.
[15] Kent, G.S. and Wright, R.W.H. (1970). *J. Atm. Terrestrial Phys.* **32**, 917–43.
[16] Collis, R.T.H. and Uthe, E.E. (1972). *Opto-electronics* **4**, 87–99.
[17] Hall, F.F. Jr. (1974). 'Laser Systems for Monitoring the Environment', in *Laser Applications*, (ed. Monte Ross) Academic Press: New York.
[18] Beran, D.W. and Hall, F.F. Jr. (1974), *Bull. Am. Meteorological Soc.* **55**, 1097–105.
[19] Mason, J.B. (1975). *Appl. Optics,* **14**, 76–8.
[20] Cooney, J. and Pina, M. (1976). *Appl. Optics,* **15**, 602–3.

[21] Murray, E.R., Hake, R.D. Jr., van der Laan, J.E. and Hawley, J.G. (1976). *Appl. Phys. Letts.* **28**, 542–3.

[22] Clifford, S.F., Ochs, G.R. and Want, T-i, (1975). *Appl. Optics,* **14**, 2844–50.

[23] Grant, W.B. and Hake, R.D. Jr. (1975). *J. Appl. Phys.* **46**, 3019–23; see also, Svanberg, S. (Goteberg, Sweden, private communication).

[24] Grant, W.B., Hake, R.D., Liston, E.M., Robbins, R.C. and Proctor, E.K. Jr. (1974). *Appl. Phys. Letts.* **24**, 550–2.

[25] Measures, R.M. and Pilon, G. (1972). *Opto-electronics* **4**, 141–53.

[26] Wang, C.C. and Davis, L.L. Jr. (1974). *Phys. Rev. Letts.* **32**, 349–51.

[27] Wang, C.C., Davis, L.L. Jr., Wu, C.H., Japar, S., Niki, H. and Weinstock, B. (1975). *Science* **189**, 797–800.

[28] Wang, C.C., Davis, L.I. Jr., Wu, C.H. and Japar, S. (1975). *Appl. Phys. Letts.* **28**, 14–16.

[29] Hinkley, E.D. (1976). *Opt. Quant. Elect.* **8**, 155–67.

[30] Hinkley, E.D. and Kelley, P.L. (1971). *Science* **171**, 635–9.

[31] Hinkley, E.D., Ku, R.T. and Kelley, P.L. (1976). 'Techniques for Detection of Molecular Pollutants by Absorption of Laser Radiation', in *Laser Monitoring of the Atmosphere*, (ed. Hinkley, E.D.) Springer-Verlag: Berlin, Germany.

[32] Menzies, R.T. (1972). *Opto-electronics* **4**, 179.

[33] Menzies, R.T. (1973). *Appl. Phys. Letts.* **22**, 592–3.

[34] Menzies, R.T. (1976). 'Laser Heterodyne Detection Techniques', in *Laser Monitoring of the Atmosphere*, (ed. Hinkely, E.D.) Springer-Verlag: Berlin, Germany.

[35] Peyton, B.J., DiNardo, A.J., Cohen, S.C., McElroy, J.N. and Coates, R.J. (1975). *IEEE J. Quant. Elect.* **QE-11**.

[36] Henningsen, T., Garbuny, M. and Byer, R.L. (1974). *Appl. Phys. Letts.* **24**, 242–4.

[37] Kildal, H. and Byer, R.L. (1971). *Proc. IEEE,* **59**, 1644–63.

[38] Byer, R.L. and Garbuny, M. (1973). *Appl. Optics,* **12**, 1496–505.

[39] Snowman, L.R. (1972). Technical Report R72ELS-15, General Electric Electronic Laboratory: Syracuse, New York.

[40] Schotland, R.M. (1974). *J. Appl. Meteorology,* **13**, 71–7.

[41] Byer, R.L., Herbst, R.L. and Fleming, R.N. (1975). 'A Broadly Tunable IR Source', in *Laser Spectroscopy,* (ed. Haroche, S., Pebey-Peyroula, J.C., Hansch, T.W. and Harris, S.E.) pp. 207–26. Springer-Verlag: Berlin, Germany.

[42] Herbst, R.L., Komine, H. and Byer, R.L. (1977). *Optics Commun.* **21**, 5–7.

[43] Herbst, R.L., Fleming, R.N. and Byer, R.L. (1974). *Appl. Phys. Letts.* **25**, 520–3.

[44] Baumgartner, R., Wolfe, D., Warshaw, S. and Byer, R.L. (1976). 'Remote Air Pollution Monitoring System Using a Tunable 1.4–4.0 μm Infrared Source', presented at the Conference on Laser Engineering and Optical Systems, May 1976, San Diego, California.

[45] Murray, E.R., van der Laan, J.E. and Hawley, J.G. (1976). *Appl. Opt.* **15**, 3140–8.

[46] Bair, C.H. and Allario, F. (1976). *Appl. Opt.* **16**, 97–100.

[47] Spencer, D.J., Denault, G.C. and Takimoto, H.H. (1974). *Appl. Opt.* **13**, 2855–68.

[48] Rao, K.N. (1976). *Physical Chemistry*, (ed. Ramsay, D.A.) Butterworth: U.K.

[49] Shumate, M.S., Menzies, R.T., Margolis, J.S. and Rosengren, L.G. (1976). *Appl. Opt.* **15**, 2480–8.

[50] Beers, Y. (1962). *Introduction to the Theory of Errors*, Addison-Wesley: Massachusetts.

CHAPTER NINE

Planning and execution of an air pollution study

9.1 Introduction

Pollution monitoring is an expensive business and it should not be undertaken lightly. In a world of limited resources, any monitoring programme will probably have taken priority over some other socially useful exercise. It will, therefore, be assumed in the following discussion that data are being acquired because:

(a) Some undesirable effect has been noted or;

(b) There is reason to suppose that if certain levels of contamination of the air or precipitation from the air are reached some harmful effect will occur; and

(c) There is some possibility of remedial actions being taken, should this be found desirable.

Fig. 9.1 illustrates the features and functions of any air pollution monitoring exercise which has been designed in accordance with the above assumptions.

The first row of boxes represents the path along which the effluent travels from its source(s) to the receptor(s) where the harm may possibly be done (the source to sink history). The second row shows the various types of monitoring corresponding to each of these stages (i.e. the monitoring network). Some of these stages may be absent in some instances (see Section 9.3 below).

If the information acquired by the monitoring network is to be of any use, then it must be transmitted (row 3), probably with some processing en route, either in 'real time' or after sufficient data had been collected (row 4) to those responsible for interpreting it in terms of what plant control and/or receptor protection measures are necessary (row 5).

Additional information would normally be available in the form of historical analogues and/or mathematical/physical models to help in any decision making. Experience might also show that the scale of monitoring could be varied according to plant-operating or meteorological conditions and so some monitoring

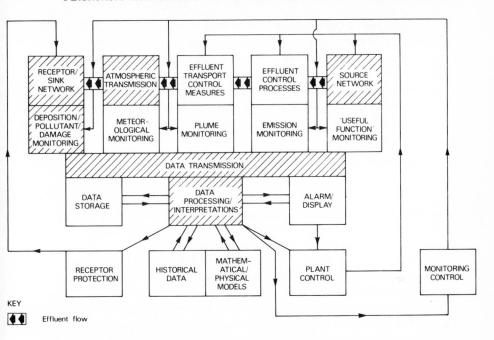

Fig. 9.1   Features and functions of a useful air pollution monitoring exercise.
The hatched sections will always be present.

control procedure might be desirable. If it is established that no problems exist,
then the monitoring could be terminated.

The way in which the data from any monitoring exercise are handled will, of
course, depend very much on the objectives of the survey. These, in turn,
together with any prior knowledge of the likely source to receptor history,
should have determined the form of the monitoring network, the quantity of
data to be collected and the rate of acquisition of data, within overall limits set
by the money, manpower, instrumentation and resources available. It is, therefore, necessary that any discussion of data handling and analysis of results should
include likely source to receptor histories, monitoring networks and also objectives.

In some states and countries, legislation has already been enacted or is being
prepared, to lay down MACs (maximum allowable concentrations) or some
equivalent terms for some pollutants for various types of exposure. Where this is
the case, it might be argued that the sole purpose of a monitoring programme is
to ensure that the law is being complied with and that the historical data, available to those responsible for control measures, are reflected in the legislation.
This view could be correct if the laws have been carefully framed, but where the

law includes such details as the dispersion models to be used in calculating GLCs (ground level concentrations) and the acceptable patterns of monitoring sites, the legislation has probably been made too inflexible to be efficient. This may well have been done for reasons of bureaucratic convenience.

In the following discussion it will be assumed that a flexible (i.e. the best practicable means) approach is possible, that each situation is to be evaluated on its own merits and that no Procrustean regulations exist.

9.2 Objectives of the monitoring programme

9.2.1 General

Assuming that a monitoring programme is contemplated for cases (a) or (b) of Section 9.1 above, in case (a) the object of monitoring exercise will be:
(i) To identify the pollutant (if this is not already known);
(ii) To identify the source of the pollutant (if this is not already known);
(iii) To assess the effectiveness of any control measures taken.
In case (b), the object will be:
(i) To determine existing dosage levels of the pollutant, so that an objective assessment may be made as to whether control or receptor protection measures are necessary now or in the future;
(ii) If source control measures are taken, to assess their effectiveness.

9.2.2 Pollutant identification

When damage has been observed, the only certain way to identify the pollutant responsible is to establish a direct causal relation between the pollutant and the damage. This is best done by identifying the pollutant at the site of the damage, e.g. asbestos fibres in carcinomas [1], or by reproducing the damage in controlled experiments, which will also establish dosage thresholds for damage.

An alternative is to establish statistical correlations between damage and dosage. This is much less satisfactory, because of the many correlations existing between pollution dosage and many other human activities (e.g. pollution and urban living) likely to cause 'damage'. The initiation of high-cost pollution control programmes on the basis of correlations alone may well result in a waste of resources.

9.2.3 Source identification

Having recognized that it is desirable to reduce the dosage levels for a particular pollutant or group of pollutants, it will be necessary to establish the major source(s). The assumption that sources are responsible in ratio to their total emissions is another pitfall which could lead to serious misapplication of resources, if expensive control measures were imposed irrespective of source type. The dosage levels observed from a given source will be a complicated function of emission rate, the location of the source with respect to the receptor, the topographical features of the area, the prevailing meteorological conditions and the height of the plume above the receptor (see Section 9.3.5).

Procedures for source identification are discussed in Sections 9.8 and 9.9.

9.2.4 Economic assessment of 'damage' versus control

If it has been established that certain sources cause damage, before control measures are taken, a proper assessment of the relative costs of the damage and control procedures needs to be made. The costs of the control procedure should include the effects of any change in the emission due to the control process and any effects on ground or water of its products.

It may, therefore, be necessary to monitor changes in minor, but possibly more toxic, components of the emission resulting from operation of the control equipment, as well as the major pollutant for which the control process was initiated.

Other side effects of the control process which will be important are changes in the physical nature of the emission (e.g. reduction in temperature or addition of liquid droplets) which could adversely affect the plume rise and so reduce the dilution of the residual effluent at ground level, in addition to making the plume more visible.

9.2.5 'On-line' plant control

If plant operation leads to acceptable dosages under most weather conditions, it may be necessary to operate some form of control under meteorological conditions which result in abnormally low dilution at the receptor. An example of such a control measure would be the substitution of low pollutant fuel, too expensive for continuous use.

The monitoring system providing the information leading to initiation of such control might consist of:

(i) Meterological instruments, e.g. to measure wind direction and mixing depth to ensure correct source identification;

(ii) Fixed dosage monitoring at sensitive or representative receptor points;

(iii) A mobile survey unit; or

(iv) Combinations of two or all of these systems.

Alternatively, if control equipment is installed, monitoring will be necessary to detect its failure. This would normally involve effluent monitoring, but some back up from field monitoring would help in deciding whether shut down of the plant is necessary following control equipment failure.

9.2.6 Control of future developments

Provided acceptable dosages from pollutants have been established and existing levels are found to be satisfactory, a monitoring programme might be designed to assess changes in the general level of pollution likely to result from future developments. The objective here would be to assist with decisions as to the need for restrictions on such development in certain areas and/or the introduction of control measures at some time in the future. The validation of plume rise and dispersion models for tall stack emissions, so that these models could be used to determine suitable stack heights for future large installations, is an example of an objective in this category.

9.2.7 Receptor protection

If control measures fail or are too expensive, or if there is an accident , receptor protection may be the only effective way of dealing with pollution. This would be initiated as a result of plant function monitoring, effluent monitoring or field monitoring.

Receptor protection may also be necessary as a result of long-term exposure to low levels of pollution. These may be quite harmless in most respects but result in damage in a particular, peculiarly susceptible, location or process. An example is sulphur deposition (e.g. from a smelter) over land used for grazing cattle, resulting in an eventual deficiency of selenium in their diet. In this case additions of small quantities of selenium to the feed may be all that is required to correct the problem [2].

9.2.8 Detection of long-term trends

It has only recently been established (e.g. [3, 4]) that the atmosphere as it exists today is largely a mixture of organic effluents and very different from the inorganic atmospheres to be expected in lifeless worlds in a physically similar situation to our own. There is no reason to suppose that the atmosphere is naturally in some state of delicate equilibrium, beneficial to all existing creatures and likely to be upset only by the 'unnatural' activities of man. The composition of the atmosphere is continually changing, and these changes have led to the replacement of many plant and animal species. Man's days are, therefore, probably numbered if he does not interfere with the atmosphere. If he does, then he may be reversing a trend that would have led to his destruction or he may be accelerating or even initiating such a trend. In any case, long-term monitoring of atmospheric constituents at 'background' stations appears to be prudent. Some interesting facts have already come to light as a result of such monitoring, e.g. in recent decades the long-term trend of reducing CO_2 levels due to growth of vegetation has been reversed, possibly owing to emissions from fossil fuel burning [5].

Changes in the dissolved material in precipitation may have important effects on soils, e.g. control of particulate emissions could lead to increased acidity in rain as a side-effect. In general the requirements for such monitoring will be the opposite to those for source identification or source study networks: here it is the 'background' that is to be measured, so the sites should be chosen for minimal effect from individual sources.

9.2.9 Monitoring control

Because adequate monitoring is usually expensive, one object of monitoring should be to establish or to confirm that certain pollutants are not likely to be harmful at the levels found and consequently that further monitoring is unnecessary. Alternatively, it may be established that pollutant monitoring will be necessary only in certain meteorological conditions, so continuous meteorological monitoring, supported by a pollutant monitoring facility for use as required, may be sufficient.

9.3 Effluent history from source to receptor

9.3.1 General

Effluent control processes and/or effluent transport control measures (Sections 9.3.3 and 9.3.4) may or may not be present in any particular instance, but in any useful monitoring exercise there will always be a source network, and atmospheric transmission to a receptor/sink network.

The dosage received by a receptor will depend on the rate of emission of the effluent into the atmosphere, the effluent transport control measures taken, the dilution and attenuation during the atmospheric transmission and its own ability to take up or reject the pollutant (Section 9.3.6). In general it will be possible to avoid adverse effects by source control, effluent control, effluent transport control or receptor protection.

9.3.2 Source network

The source network may consist of a single tall chimney or a large number of small sources distributed over a large area (e.g. motor vehicles or houses in a city). In the case of secondary pollutants the source (or sources) is the plume (or plumes) some distance from the source(s) of the primary pollutants.

Providing the sources of the primary pollutants are known, information on their characteristics and locations should be included in the historical data available to assist the data analysis and decision-making processes.

9.3.3 Effluent control processes

A control process may be defined as the modification, for other than economic reasons, of the effluent produced as a by-product of the primary function of a plant or process. The object of this modification will be the improvement of the environment.

It is important that, in assessing the true value of the improvement, a full assessment of the environmental impact of the control processes be made, e.g. a sulphur-removal process will not, in general, affect the total addition of sulphur to the environment, unless the control processes produce a useful product and the production of this useful product by existing methods in consequently reduced. The installation of a control process means that monitoring the useful function will no longer be sufficient to determine the emission, unless detailed information on the control process is available and the process itself is very reliable.

9.3.4 Effluent transport control procedures

The plant should be constructed and operated in such a way that the maximum possible dilution of plume material before it returned to ground level is ensured. These considerations, which might be termed 'effluent transmission control procedures' include:

(i) Avoiding downwash (flagging) into the lee of the stack by maintaining a high efflux velocity.

(ii) Avoiding downdraughts in the lee of adjacent buildings or topographical features by careful site selection and building a tall stack.

(iii) Avoiding negative or low buoyancy of the effluent gases by emitting these at a high temperature.

(iv) Ensuring maximum plume rise by (i) to (iii) above and by emitting from as few stacks as possible.

The detailed discussion of how the objects of the transmission control procedures can be achieved and how the plume rise should be calculated has been pursued at length in the literature and will not be repeated in detail here (see, e.g. [6, 7]).

Wind tunnel modelling may be necessary to ensure proper stack design in some cases [8]. In general the control will be acheived by site selection, and the design of the plant, but some 'on-line' flexibility, such as a damper system, to obtain reasonable efflux velocities at low load without unnecessarily high effluent velocities on full load, is possible.

9.3.5 Atmospheric transport

The natural defence of living organisms against toxic effluents is the great capacity of the atmosphere to dilute and in many cases subsequently to rid itself of these materials. The mechanism which ensures the dilution of effluent gases necessary to enable life to exist is atmospheric turbulence. This turbulence is produced in two ways. Firstly, by the stirring of the wind caused by the drag of the surface and objects projecting from it and secondly by convection currents rising from the surface when the surface is warmer than the air.

At a particular location for a given wind direction, the depth of the turbulent layer (the mixing layer) is determined by the strength of the wind and the degree of surface heating. It follows that, for low-level sources near a surface receptor, dilution is at a minimum when the surface is colder than the air (i.e. there is no convection) and there is little or no wind.

This is not true for emission from tall chimneys or from sources which have a large initial buoyancy. In conditions of reduced turbulence near the ground, the mixing layer is very shallow and plumes from these sources will rise above it into the 'free atmosphere' where turbulence is almost always very low. Effluent from such sources will not be detected at the ground until conditions change and the turbulence originating at the surface reaches the height at which their plumes finally level out. The greatest ground level concentrations for high sources will, therefore, occur in conditions of either strong winds or strong convective activity. Even in these conditions, proper stack design can ensure dilution of the stack gases by factors of the order of 10^4 at ground level providing the gases are buoyant. This may be compared with reductions in source strength by factors of 2 to 200 achieved by most control processes.

The design of an effective monitoring network and the realistic interpretation of the data from it will, therefore, depend primarily on a detailed understanding of the atmospheric transmission processes. For this reason, Section 9.4 will outline the more important aspects of these processes.

9.3.6 The receptor/sink network

In addition to dilution by atmospheric turbulence, as it travels from source to receptor, the pollutant may be attenuated by deposition on the surface or modified by reactions with other atmospheric gases or aerosols.

Distant receptors may, therefore, be protected from the effluent by this attenuation process as well as by the atmospheric dispersion. Conversely, successful effluent transmission control procedures, by eliminating high ground level concentrations and high deposition rates near the source, may result in some increase in the very low concentrations (which will be small anyway), at very long distances where the pollution has become well mixed through the boundary layer.

Receptors may be regarded as active if they ingest air or passive if the air passes over or through them under the action of the wind. In the former case, if the ventilation rate is independent of wind speed, the dosage they receive will be a product of pollutant concentration, ventilation rate and exposure time and the efficiency with which the receptor retains the pollutant. In the latter case, it will be the product of the ground level concentration, exposed surface area, and a 'deposition velocity'. The deposition velocity will be a function of the wind speed and turbulence, the physical characteristics of the receptor and the chemical or biological resistance of the receptor to the pollutant [9]. In the case of aerosols the problem is complicated by the inertia of the particles; large particles tend to impinge on projections into the air stream, while smaller particles follow the streamlines round the projection. Impaction on the surface does not necessarily mean retention by it, as the particle may bounce off or be re-entrained by the wind (saltation). Particle fall velocity will also be important in determining the dosage received by horizontal surfaces if the particles are large enough.

Therefore, it will be inferred that a proper assessment of relations between air concentrations and damage may require a great deal of supplementary information about the properties of the receptor and the pollutant.

Chemical reactions of a pollutant with other atmospheric gases or aerosols are also complicated processes. Solar radiation, relative humidity, temperature and the presence of suitable catalysts or materials forming intermediate products can all be important.

9.4 Outline of the more important features of the atmospheric transport

9.4.1 General properties of mixing layers

In Section 9.3.5 the importance of the atmospheric transport in determining effects at the surface was emphasized. A full discussion of transport processes would, of course, take a whole book. The prime purpose of this section is, therefore, to bring the more important features of these processes to the attention of the reader and to direct him towards more detailed descriptions.

If we consider a type of surface to begin at some well-defined topographic or topological discontinuity, e.g. the coast, the boundary between hills and plains, the edge of an extensive urban area, then an approach to an equilibrium boundary layer will be reached only after a time of travel greater than the ratio of the momentum of the air stream to the body forces acting on it. As these forces include the effect of the rotation of the earth, this ratio will be of the order of the inverse of the Coriolis parameter, which in middle latitudes is 10^4 s. Thus, for fetches less than 10^4 times the wind speed (in m s^{-1}), the boundary layer should be considered as a growing or an 'internal' boundary layer. A discussion of the growth of internal boundary layers is given, for example in [10].

In addition to the fetch, the properties of the mixing layer over a given type of surface will depend upon the following external factors:

(a) The properties of the incident air stream (i.e. its condition as it arrives over the windward edge of the type of surface being considered).

(b) The roughness characteristics of the surface considered.

(c) The free stream or surface geostrophic wind speed (U_G).

(d) Some property determining the magnitude and direction of the heat flux between the air stream and the surface considered.

It is difficult to express (d) as an external parameter as all the relevant properties (e.g. potential temperature difference between the air at some level and the surface, surface heat flux, Richardson number at some level, Monin—Obukhov length at some level) will all, to some extent, be determined by factors (a)–(c). It is customary to estimate one of the above and use this as the independent variable. Definitions of Richardson number and Monin—Obukhov length are given, for example in [11].

Where horizontal temperature (i.e. density) gradients exist, there is a variation in the pressure gradient and hence in the geostrophic or free stream wind velocity with height, in addition to the shear due to surface friction. In this case there may be appreciable turbulence exchange at heights well above the top of the 'barotropic' mixing layer [12].

9.4.2 Meteorological factors determining air stream characteristics

Features (a), (c) and (d) of Section 9.4.1 above will, to a large extent, be determined by a combination of factors depending on the large-scale features of the pressure system affecting the site at the time and on diurnal heating or cooling of the surface. Fig. 9.2 gives vertical time sections over land showing the properties of the mixing layers generated by each of the principal air stream types. Fig. 9.3 shows the locations of the air streams in relation to typical features of a middle latitude synoptic weather map. Each of these air stream types will now be discussed briefly.

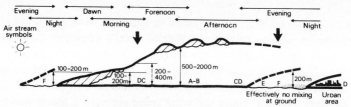

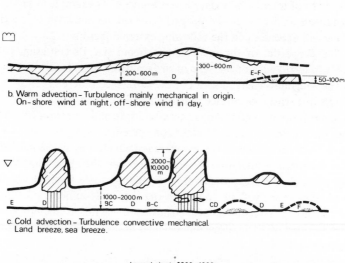

a. Settled Antiajclonic – Turbulence mainly convective in origin.

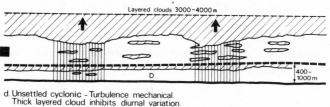

b. Warm advection – Turbulence mainly mechanical in origin.
On-shore wind at night, off-shore wind in day.

c. Cold advection – Turbulence convective mechanical.
Land breeze, sea breeze.

d. Unsettled cyclonic – Turbulence mechanical.
Thick layered cloud inhibits diurnal variation.

KEY

	Fog or Cloud		Well defined ⎫ Boundary of
	Precipitation		— — III defined ⎭ turbulent layer
			↑↓ Upward or downward motion
			Ground fog

Fig. 9.2   Time sections of the lower atmosphere in each of the four principal air stream types of middle latitudes. The time axis is calibrated in general terms (morning, afternoon etc.) as the actual times will vary with location and time of year. The heights are indicated at various points along the time scale as these again will be functions of location, time of year etc. The corresponding 'Pasquill' categories are indicated by capital letters just above the surface.

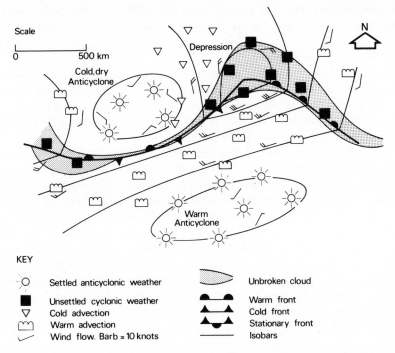

KEY

$\overset{\cdot}{\underset{\cdot}{\bigcirc}}$	Settled anticyclonic weather	⬛➤	Unbroken cloud
⬛	Unsettled cyclonic weather	▲▲	Warm front
▽	Cold advection	▲▲	Cold front
⌒⌒	Warm advection	▲▲	Stationary front
⌒	Wind flow. Barb = 10 knots	——	Isobars

Fig. 9.3 Locations of principal air stream types in relation to typical features of a middle latitude (Northern Hemisphere) surface, synoptic weather map.

9.4.3 Settled anticyclonic (Fig. 9.2a)

The principal dynamic features of these air streams are:

(i) Large-scale divergence (i.e. flow outwards from the high-pressure centres) across the isobars and corresponding subsidence (descending motion) in the lower half of the troposphere.

(ii) The descending air above the mixing layer has a stable vertical temperature gradient and low humidity.

(iii) The wind speeds are light or moderate.

(iv) Turbulence in the mixing layer is typically convective in origin.

The typical diurnal history of the mixing layer is shown in Fig. 9.2a.

Under clear skies at night a surface radiation inversion develops (over rural areas) leading to very low levels of turbulence (Pasquill category F [11, p. 367] and consequently extremely restricted turbulent diffusion, so there is no effective mixing layer. Later in the night, if the air near the surface is sufficiently moist, fog will form. If the fog is deep enough, the maximum radiation cooling will be transferred from the surface to the top of the fog layer. Convective turbulence will then develop in the fog layer and any low level emissions of pollutants will become well mixed through the fog layer. The top of the mixing layer is now the top of the fog.

As the sun rises, the surface will be warmed sufficiently for the fog to evaporate near the ground, low stratus cloud will develop and eventually disperse. Further surface heating then results in the depth of the boundary layer continuing to increase and again, if there is sufficient moisture present, fair weather cumulus cloud will form later in the forenoon.

The convective themals will penetrate some distance into the stable capping layer of dry subsiding air and some of this air will be fed into the top of the mixing layer, helping the warming process (see [13] for a quantitative discussion). As the intensity of solar radiation begins to decline after noon, the supply of heat will eventually be insufficient to maintain the warming process and shortly after the maximum temperature is reached the air near the surface will begin to stabilize. The large scale turbulence at higher levels will then dissipate, as the supply of further energy from the surface is cut off, and the diurnal cycle is then complete. This is the classic air pollution situation for low level emissions. Modifications due to the presence of surface features and the importance of effective emission height on pollutant behaviour in these conditions will be discussed in Sections 9.4.8 to 10.

9.4.4 Warm advection (Fig. 9.2b)

Warm advection occurs when the underlying surface is cooler than the surface upwind. Typical examples over north—west Europe are the south—westerly wind in the warm sector of a depression or the southerly flow ahead of warm fronts (Fig. 9.3).

The principal features of this type of air stream are:

(i) A deep, stably stratified flow with high humidity.

(ii) Moderate or strong winds.

(iii) Little organized vertical motion, mainly weak subsidence.

(iv) Turbulence in the mixing layer, mainly mechanical; some weak convection over land during the day.

The diurnal history is illustrated in Fig. 9.2b.

Overnight there will be extensive low cloud; this may well reach the surface as sea or hill fog. Diurnal heating may burn off the cloud, or at least cause it to break, but the diurnal range in mixing layer height will be much smaller than in an anticyclonic situation. At a sufficient distance from the coast, if the cloud has been dissipated by surface heating and the wind speed is light enough, a surface radiation inversion and fog patches may develop in the evening. However, in maritime locations low stratus cloud will usually spread over the area from the direction of the coast later in the evening, causing the surface air to warm again and a mechanically stirred mixing layer to re-form.

9.4.5 Cold advection (Fig. 9.2c)

These streams occur typically on the western sides of low pressure areas (Fig. 9.3) and also locally as a result of sea breezes and shallow katabatic winds (see Sections 9.4.8 and 9.4.9). The potential temperatures at heights up to several thousand metres are lower than those at the surface, except intermittently during the night and in subsiding air between clouds.

Other features of these streams are:

(a) Winds moderate or occasionally strong.

(b) Convective cloud often giving showers.

(c) Clear subsiding air between the clouds may give locally limited mixing depths corresponding roughly to the cloud base.

(d) Turbulence — partly mechanical, but the mixing depth is controlled largely by the convective instability.

(e) Little net vertical motion.

The diurnal history is shown in Fig. 9.2c. Winds and shower activity may be insufficient to prevent some stabilization of the surface air after sunset, but any fog patches will be shallow and clear if convective clouds move overhead probably at times during the night, but in any case soon after sunrise. The radiational cooling under clear skies may be very rapid because of the low humidity.

9.4.6 Unsettled cyclonic (Fig. 9.2d)

These streams show fairly rapid changes in wind direction and generally overcast skies, often with rain. A general upward vertical motion and the baroclinic nature of the flow often result in an ill-defined top to the mixing layer. However, in some situations the presence of a warm moist air mass above may lead to a sharp temperature inversion restricting vertical mixing. There will be a component of flow across the isobars towards the low pressure. Turbulence near the surface is mainly mechanical in origin and there is little diurnal variation.

9.4.7 Frequency of occurrence of different situations

The frequency with which the different types of air streams will occur at a given site will depend upon its geographical location. They will often be associated

with a particular wind direction. For the U.K., a rough breakdown would be:
   Settled anticyclonic, 30% (variable, often NE to E wind);
   Warm advection, 50% (SE to W wind);
   Cold advection, 10% (NW to N wind);
   Cyclonic, 10% (variable).

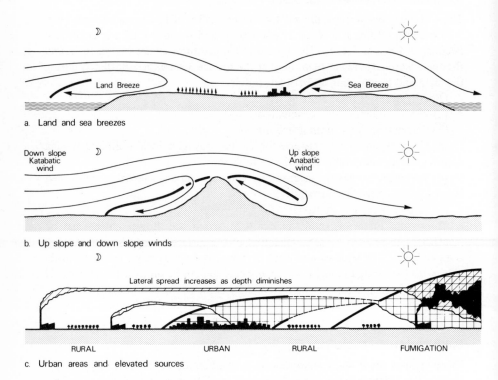

a. Land and sea breezes

b. Up slope and down slope winds

c. Urban areas and elevated sources

Fig. 9.4   Effect of surface features: (a) coasts; (b) high ground; and (c) urban areas and source height on pollutant dispersion.

9.4.8 Land and sea breezes

When the pressure gradient is slack, i.e. mainly in settled, anticyclonic conditions, the temperature difference between land and water leads to flow from the cooler towards the warm surface. At some height (usually $< 1000\,\text{m}$) and at some distance (up to $\sim 80\,\text{km}$) from the shore, the normal flow will be maintained [14]. Thus this situation, illustrated in Fig. 9.4a, leads to a limited mixing depth situation and also, in circumstances in which the land or sea breeze opposes the general flow, to the possibility of recirculation of pollution.

9.4.9 Upslope and downslope winds

Local winds also develop when air cooled by radiation at night over upland regions runs downhill to displace warmer air at lower levels (Fig. 9.4b). These katabatic winds are usually turbulent and in narrow valleys or fjords can be quite strong. Mixing within the stream will be quite efficient but there will be little exchange between the cold air stream and the warmer air above it. On the other hand, when low ground is covered with fog in the early morning, and the upper slopes clear, the differential heating may cause upslope (anabatic) winds to develop (Fig. 9.4b). The top of such circulations is generally limited by the snow line.

9.4.10 Urban areas and elevated sources

An urban area enjoys its own microclimate and the reduced levels of turbulence associated with radiation inversions in rural areas are seldom encountered in built-up areas [15]. Temperatures are also a few degrees higher than they are in rural areas in light winds [16]. The consequence of these effects is that the urban area develops its own internal boundary layer and that low-level emissions which are trapped within this layer become well mixed within it (Fig. 9.4c).

In settled anticyclonic conditions at night the warmer urban air rises above the cooler rural air on the lee side of the city, and an 'urban plume' drifts downwind. As the convective boundary layer develops over the rural area during the morning, the urban plume will eventually be brought down to the ground (right-hand side of Fig. 9.4c), leading to a rapid increase in ground level pollution. As the mixing depth increases, the concentration will fall from this peak fumigation value [17].

Fig. 9.4c also illustrates some features of the behaviour of plumes from elevated sources. These will generally level out at some height depending on atmospheric conditions, stack height and heat emission rate. The initial dilution caused by turbulence due to relative motion will be much greater than that experienced by inert tracers and will generally be comparable to that from a point source at 10 to 20 km upwind in 'Pasquill F' [18] conditions by the time the plume has travelled a few hundred metres. If the air stream at plume level is stably stratified, the depth of the plume will diminish (Fig. 9.4c) under the action of gravity forces as it drifts further downwind, resulting in additional lateral spread, with the result that the plume width is often greater than in a neutral boundary layer [18]. This again differs from the behaviour of inert plumes, which usually show a much reduced lateral spread [19].

As the plumes from elevated sources drift over urban areas at night, they may be fed into the urban boundary layer, or if the source is large enough, they may pass over it (middle of Fig. 9.4c). It is interesting to note that there could be some descending air flow on the approach to the urban plume owing to the presence of a surface wind over the urban area and consequent downwind displacement of the streamlines. The opposite effect would occur during the day when wind speeds over the urban area would in general be lower because of the greater surface roughness there.

9.5 The monitoring network

9.5.1 General

It is possible that a monitoring network would contain no pollutant sensors at all but would consist solely of meteorological instruments, backed up by information on the current functioning of the source and experience in the form of historical data or dispersion models. However, in general the monitoring network will record some or all or the variables listed in Section 9.5.2 to 8 below.

9.5.2 Function monitoring

The effluent is a by-product of some useful or desirable function (e.g. manufacturing, power supply, transport). One way to monitor the rate of emission of effluent is to monitor the useful function. The relation between the rate of emission of effluent and the rate of performance of the useful function will first have to be established. This need not necessarily involve measurements of the effluent. For example, if the effluent is produced by burning fuel and the products of combustion are known, the rate of emission could be calculated from knowledge of the rate of fuel usage and the amount of each component of the fuel retained in the (unemitted) ash.

The units of function monitoring will be quantities such as megawatts for electricity generating stations, cars per kilometre or cars per second for roads, kilograms of fuel burnt per second for heating systems, etc.

9.5.3 Emission monitoring

Continuous, accurate monitoring of the emissions from any plant is a difficult undertaking because of the difficulty of maintaining the instrumentation in good working order under adverse conditions. Although trials of limited duration are usually sufficient to relate emissions and function monitoring where no control equipment is installed, continuous monitoring of a controlled pollutant is much more necessary to ensure that the control equipment is working properly. The physical changes [temperature, liquid carry over (drift) etc] produced by the control processes should also be monitored to ensure that the transport control procedures (Section 9.3.4) remain effective in ensuring maximum dilution of the remaining pollutants (and possibly newly introduced pollutants resulting from the control process). One difficulty with emission monitoring is that the feature usually recorded is the concentration (or possibly flux) of effluent at some point or points in the ducting or stack, while the total emission is required. Conversion of one measurement to another requires an accurate knowledge of the total volume emission of gas from the plant and, if the pollutant and flow rate (in the case of flux measurements) are not uniform across the duct, of the distribution and flow, requiring multipoint sampling. Leakage of air into the duct near the sampling point can seriously affect the results.

A second point, sometimes overlooked, is that if the emission is hot, then if the (gaseous) emission is expressed as a volume it must be corrected to atmospheric temperature when GLC is determined. For this reason it is better to express emission as $kg\,s^{-1}$ rather than $vol.\,vol.^{-1}$ of flue gas or $m^3\,s^{-1}$.

9.5.4 Plume monitoring

Plume monitoring involves monitoring the effluent between the emission point(s) and its arrival at the receptor network. Remote sensing equipment (lidar, cloud searchlights, correlation spectrometers, laser spectroscopy, etc.), airborne sensors (using aircraft, including kites, balloons and model aircraft) and instrumented towers have all been used. All these techniques tend to be expensive and/or difficult to operate over extended periods, so their main function has tended to be the validation or development of plume rise and dispersion models and/or pollutant—reaction models.

It is possible to use remote sensing techniques to monitor emissions as an alternative, or a supplement, to plant monitoring. Closed-circuit television monitoring of emissions from the stack has been used to detect excessive dust emission. Plume photography is perhaps the least expensive method of monitoring the near-field general properties of emissions.

9.5.5 Meterological monitoring

The most relevant meteorological parameters are wind speed, direction, vertical temperature gradient, where relevant the depth of the mixing layer, and lateral and vertical components of atmospheric turbulence. If practicable, these should be measured over the whole plume depth using tower-borne instruments, a balloon or aircraft. In practice, only near-surface observations are normally available. If dosage monitoring is impracticable on a universal basis, previous experience (including dispersion models) will enable the most disadvantagous meteorological conditions to be defined. If some source or emission control is deemed necessary in those conditions, meteorological monitoring may be a more practical way of initiating the control measures than dosage monitoring, because of the very large number of monitoring sites required to cover the whole area round the source. However, from large sources with very tall stacks, the highest ground level concentrations are observed in unstable meteorological conditions with light or moderate winds and a limited mixing height; the plume levelling off at the top of the mixing layer but being wholly retained within it. In such conditions, a slight change in the wind speed and/or mixing layer height can result in the plume escaping from the mixing layer, giving zero or low ground-level concentrations (see, for example, [20]). A mobile sampling unit (Section 9.6.8), fitted with remote sampling equipment in addition to point monitors, which is able to establish whether or not high GLCs were in fact occurring when the meteorological monitoring indicated they were likely, could save a large proportion (maybe 60 to 75%) of unneccessary control measures in such conditions.

9.5.6 Damage monitoring

In a way damage monitoring—using the word damage in the sense of any interference with normal processes—is effected at all times by the general public. They will complain if a pollutant offends their senses; if they think it is affecting their property; or if they can detect some physiological effect.

The purpose of any damage monitoring programme should, therefore, be: to quantify any qualitative monitoring by the general public; and to detect damage of more subtle forms, not immediately apparent to the public at large.

The quantifying of damage is necessary, if the correlations between damage and dosage, necessary to put any economic assessment of control processes on a rational basis, are to be established. It is, of course, easier to suggest that damage should be quantified than to specify how this should be done, especially in the case of damage to health or amenity. However, we continually have to cost benefits and services in practice in all aspects of life, e.g. with regard to costs of medical treatment, so there is some basis of experience to help with this difficult task.

9.5.7 Dose monitoring

Damage will be related to dose rather than ground level concentration and it may be desirable to use an instrument which simulates the exposure of the receptor, if it is not possible to measure the dose the receptor receives. An example of such a collector is the directional deposit gauge [21], which attempts to simulate the soiling of vertical surfaces or the passage through windows and doors of airborne dust.

The measurement of GLC alone may lead to difficulties in relating damage to a particular source. As an example, if the deposition velocity of a pollutant on a surface were proportional to the wind speed, and the pollutant concentration were proportional to source strength but inversely proportional to the wind speed, one might find little correlation between damage and concentration. This could mean that fluctuations in concentration were due primarily to variable wind speed rather than variable emission rates. It would be wrong to infer that the pollutant was not responsible for the damage. It is, therefore, important to ensure that all variables that are likely to effect the dose received by sensitive receptors, as well as ground level concentration, are monitored.

9.5.8 Monitoring ground level concentration

The results of pollution measurement are usually expressed as concentration, either as mass mass^{-1} of air, vol. vol.$^{-1}$ of air or mass vol^{-1} of air. The latter has, unfortunately, lately become the most widely used. I say unfortunately because mass vol.$^{-1}$ of air is not conserved when the pressure or volume of the air changes, and this is especially misleading when one is comparing samples taken at different heights. The use of these units has arisen because sampling rates are usually expressed as volumes per unit of time while the response of the receptor usually depends on the mass of pollutant sampled.

The conversion to concentration requires a knowledge of the volume sampling rate and the time taken to produce the response. As atmospheric concentrations fluctuate in time and space, the peak concentration reported during a given sampling period will depend in a rather complicated way on the volume sampling rate and the response time of the instrument. If the sampling instrument is mounted in a vehicle, the movement of the vehicle will also effect some smoothing. The highest peak readings recorded during any sampling period will, therefore, be given by the instruments which effect minimum smoothing i.e. those with a fast response and, in the case of vehicle mounted samplers, a slow speed of traverse of the plume.

The requirements for high sensitivity are that as large a mass as possible of the pollutant should be collected, implying, for a given sensor, as large a volume as possible should be sampled, i.e. the opposite requirement for fast response. The response time of an instrument itself may be complicated by factors such as turbulent mixing of the sampled air within the instrument and adsorption and desorption of the pollutant on surfaces within the instrument. If the instrument records continuously then readings could be taken far more frequently than the performance of the instrument merits, leading to unnecessary expense in data processing and storage.

Whether or not the measurement is expressed as concentration or dosage (concentration × sampling time), it is essential to include the sampling time as additional information. The minimum useful sampling time is about three times the response time of the instrument. The design of pollution monitoring systems will be discussed in Section 9.6.

9.5.9 Optical effects of pollutants

The optical effects of emission constitute one form of damage to the environment. The visible appearance of plumes may cause offence near the source, while some reduction in sunlight and visibility may occur at long distances, even on a global scale.

Measurements to assess optical effects may be made either by remote sensors or by point samplers such as the nephelometer [22]. Such measurements are often regarded as pollution monitoring rather than damage monitoring. However, the optical effects of pollutants in the atmosphere are complicated by the gas to aerosol conversion, the shape, size, refractive index and dielectric constant of the particles, all of which may change with the relative humidity if the particles are hygroscopic. The scattering will be directional; both scattering and absorption will be dependent on the light wavelength for a given particle size. Polarization may also be important.

Although good correlations between nephelometer readings of the light scattering coefficient and mass of suspended particulate have been demonstrated [23], these relations are necessarily of local application and such readings are, therefore, to be regarded primarily as an indicator of one of the effects of pollution.

Although most gases are colourless, some like NO_2 absorb visible light. A brown haze may therefore result either from scattering and absorption by particles of sub-micron size and/or from the presence of NO_2.

9.6 The design of pollution monitoring systems

9.6.1 General

Having decided that pollution monitoring is necessary, one is faced with the prospect of reconciling: the monitoring requirements; the funds available; the staff available; and the instrumentation available.

One must decide how many samplers to purchase, whether they are to be installed in fixed patterns, in a vehicle or vehicles, or to be transported from one pattern to another in different conditions or at different stages in the monitoring programme. The duration of sampling should also be anticipated and the minimum averaging period (see Section 9.6.2) specified.

9.6.2 Choice of minimum averaging period

Assuming that there is a choice of samplers, one of the most important factors determining the final selection will be the shortest averaging period that is thought to be relevant to the problem under investigation. In general, a faster response instrument is more difficult to operate, more expensive and produces more data to handle than a corresponding slow response instrument. On the other hand, if the peak concentrations over periods of a few seconds or minutes are of paramount importance, daily or hourly average readings may have little relevance. It is small consolation to the condemned man to tell him that the daily average HCN concentration in the gas chamber is well below the MAC (maximum allowable concentration). The averaging period should therefore be included in the definition of each MAC. If MACs have been defined, then the minimum averaging period included in their definitions could determine the minimum averaging period for a monitoring programme.

It is difficult to lay down any general guidelines, but the following give some indications of requirements for different applications.

Order of minimum averaging period	Type of survey
10 s	Mobile sensors, acute respiratory effects; studies of 'puffs'
3 min	Useful for studying acute health effects if faster response not available
1 h	Time average concentrations; dispersion studies, diurnal changes, discrete source studies; damage to plants
24 h	Effects of weather systems; chronic health effects; area source studies
1 month	Seasonal and annual variation; long-term effects from global sources.

A faster response instrument than that specified will be more than adequate for the task indicated, provided the sensitivity and zero stability are adequate (see Section 9.6.3).

9.6.3 Choice of instruments

Having fixed the minimum sampling period one may be faced with a choice of instrument. If possible, discussion with groups already monitoring in the same field is desirable, but progress in instrument development is so rapid that this is often not possible.

The following factors must all be considered carefully:

472 · AIR POLLUTION ANALYSIS

(a) Is the response time short enough?

(b) Is the response specific to the pollutant considered; if not, what inter-ference can be expected from other pollutants?

(c) Cost of the instruments.

(d) Estimated running cost.

(e) Quality and number of staff required to operate the network and provide first line maintenance.

(f) Service available from manufacturers in the event of serious malfunction (whether replacement instruments are available if some have to be taken out of service).

(g) Power requirements.

(h) Is the output suitable for the anticipated data transmission recording system?

(i) Size and weight.

(j) Housing requirements.

(k) Zero stability and zero checking facility.

(l) Sensitivity, calibration stability and calibration check (preferably at more than one point).

(m) Has the instrument an established record of trouble-free running in the field?

(n) Are the weather conditions to which the proposed network is to be exposed likely to create difficulties not experienced in previous use?

Continuously recording instruments are generally not the best for monitoring low background concentrations of pollutants. In this case it is better to pass a large volume of air through some suitable collector (filter, scrubber, etc.) and then determine the total quantity of pollutant collected by a suitable analysis technique. It is important that the pollutant is fixed on collection, so that it does not evaporate or disappear in some other way later in the sampling period.

The efficiency of filters may change as they become clogged with collected material — a factor which may lead to the underestimation of particulate matter collected in clean air [24].

9.6.4 Choice of mobile, fixed, transportable or combined sampling system

The development of compact, fast response instruments with modest power requirements has made the choice of a mobile sampling facility increasingly attractive in recent years. If the facility also includes remote sensing equipment, such as the correlation spectrometer, differentiation between effects from large, high level sources and adjacent low level sources becomes much easier. Data transmission difficulties are avoided and far fewer sampling instruments are needed – a great advantage since the cost of the samplers is probably at least an order of magnitude greater than that of older, slow response instruments.

The disadvantages of mobile sampling is that, unless aircraft are used, sampling is restricted by the available road network. Interference from pollution produced by vehicles or other adjacent small sources (or even the sampling vehicle itself) may be difficult to quantify if point samplers alone are employed. The influence of individual small sources upwind on ground level concentration measurements may be very important, e.g. if SO_2 is being sampled, readings from a single house 50 m upwind, a hamlet of 100 houses 1 km upwind and town of 10^4 houses 10 km upwind could all be comparable if coal or some other sulphur-containing fuel was being used. Sheltering effects of hedges, copses etc. can reduce readings by a factor of 2 or more for pollutants which are taken up by vegetation. Undulating country produces variability, reduced ventilation in valleys may trap pollution from local sources and shelter samplers from the effects of distant sources. Increased turbulence on ridges and on lee slopes may bring down pollution from elevated plumes which has been passing over lower ground nearer the source.

These effects will tend to produce random fluctuations in the readings of GLC from a mobile sampler and systematic differences between samplers in a fixed network. The success of a fixed sampling system will therefore be very dependent on the care with which the sites are chosen in regard to the prime purpose of the monitoring programme.

A transportable system has the advantage that sites can be changed if they are found to be unsuitable, or the network can be varied to suit changing conditions: e.g. it can be set out each day in what is anticipated to be the downwind direction. The power requirements, size and weight of the instruments available would be important factors in such surveys. Data reduction could also present problems if other than average values over the period of exposure were required, as individual records (autographic or possibly magnetic tape from cassette recorders) would have to be synchronized and processed after the event. On-line use of the data would be difficult.

A mixed system, with a mobile facility backed up by carefully selected continuous recording at a few selected sites, is an attractive option if sufficient funds and staff are available.

9.6.5 How many pollutants should be monitored?

If all sources produced pollutants in the same ratios, if all subsequent depletion processes were the same and if there were no chemical reactions in the atmosphere, monitoring the concentrations of one pollutant would be sufficient to quantify the behaviour of all the others. This situation may be approached fairly closely in the vicinity of a single source, provided there is no appreciable background, but in any comprehensive area or distant source survey, full understanding of the dispersion and modification processes may require monitoring of several secondary as well as primary pollutants. Once again, it is difficult to lay down any general guidelines.

Measurements of a second pollutant may provide useful information on the origins of the first. For example, if smoke is emitted only from low level sources and SO_2 from both high and low levels, then occurrence of high SO_2 GLCs only in association with high smoke readings would be a strong indication that most of the observed SO_2 came from low level sources as well.

9.6.6 Height and exposure of samplers

In general, there will be a change of pollutant concentration with height. Near individual sources this will be due to the location of the source. For example, with ground level sources, the concentration will decrease with height; with elevated sources, the concentration will increase with height until the axis of the plume is reached. If the pollutant is absorbed by the surface or projections from it, there will be an increase in concentration with height due to the depletion of the pollutant in the surface layers, once the gradients due to initial source location have been sufficiently smoothed out. In general, the more turbulent the flow, the less gradient there will be near the ground, and conversely in stagnant conditions or very smooth flow, pollution from distant sources will tend to disappear near the surface. This is one of the reasons for the spatial variability in pollution already mentioned in Section 9.6.4.

It is, therefore, desirable to site sampling inlets at a reasonably constant height not too close to the surface (say 2 m), to avoid obviously sheltered locations where absorption can begin above the sampling height (e.g. orchards, near hedges, woods, high walls, etc.) especially if the pollutant is expected to be taken up at their surfaces. In undulating country, it should be recognized that location may be an important factor and that full effects of a distant source are unlikely to be experienced in valleys and possibly on low ground upstream of a ridge. On the other hand, low level emissions may be trapped and give locally high concentrations at sheltered locations.

When monitoring within area sources such as a city, the important consideration is the spatial variability of pollution within the area. For a given emission density, as the area becomes larger and the general level of pollution rises, the effect of nearby individual small sources becomes relatively less important compared with the combined effect of all the distant sources, but may still be predominant in unfavourable locations. If we choose ideal sites, with no local sources (e.g. the centres of parks) these will not be typical. Pollution levels may in fact be unrealistically low compared with inhabited areas, as the development of stable layers over the relatively cool open spaces may effectively cut them off from the sources at times. In these conditions, GLCs may fall dramatically if the clear area is a sink for the pollutant(s) considered. It would, therefore, seem to be better to choose sites with an average density of development for the part of the city which the site is intended to represent.

If information of the vertical profile of pollutants is required, it may be possible to site samplers on suitable towers. An alternative is to locate the intakes on the tower and the sampler at ground level [17]. Careful assessment of the likely losses, time delay and time smoothing effect of the connecting tube is necessary with this arrangement.

9.6.7 Lay out and spacing of instruments in fixed surveys

9.6.7.1 General

A wind direction measured at one height at one location [25] over a relatively short period ($\leqslant 1$ h) is generally an imprecise indicator of the direction of a plume from a particular source in the vicinity of the wind recorder. Marked changes of wind with height (sea-breezes etc. (see Section 9.4)) aside, it is probably safe to conclude that the best estimate of plume direction deduced from such a wind direction measurement would fix the plume as being with a 45° sector centred about that direction, except in convective conditions with light winds, where the uncertainty could be greater. The only way one could be reasonably sure of measuring a concentration near the axis (peak value for the distance considered) from such a plume would be to have a fairly dense network of instruments as discussed in Section 9.6.7.2 below.

On the other hand, if one wishes to arrive at the frequency distributions in, for example, different wind directions, or at different distances from the source, much of the information from such a dense network would be redundant, so different spacing criteria are necessary (Section 9.6.7.3).

The number of sites that are operated in a network within an area source depends largely on the resources available for, and the objects of, the survey (Section 9.6.7.4).

9.6.7.2 Discrete source surveys—verification of dispersion models

The main requirement in the verification of dispersion models is accurate location of the plume axis. As wind direction measurement does not generally enable one to do this (Section 9.6.7.1) we have to arrange the instruments in arcs (of at least three instruments) in such a way that the average reading of the instrument giving the highest reading is not less than some large fraction (say 0.9) of the axial value (C_A).

Assuming a Gaussian distribution of material a cross wind (i.e. concentration C at a point θ radians off the plume axis is given by:

$$C = C_A \exp - \theta^2/(2\theta_y^2)$$

where θ_y is the standard deviation of the cross wind distribution in radians), the above criterion means that

$$\frac{2}{\theta_s} \int_0^{\theta_s/2} \exp - (\theta^2/(2\theta_y^2)) \, d\theta > 0.9$$

i.e.

$$\int_0^{t=\theta_s/(2\theta_y)} \exp - (t^2/2) \, dt > 0.9(\theta_s)/2\theta_y)) \tag{9.1}$$

where θ_s is the angular spacing between samplers. Equation 9.1 implies

$$\theta_s/(2\theta_y) < 0.8.$$

Taking 0.08 radians as a typical value of θ_y [26] for a one hour sampling period, we have

$$\theta_s < 0.128 \text{ radians}$$

i.e.

$$\theta_s < \text{about } 8°,$$

so the maximum angular spacing of samplers should be about $7\frac{1}{2}°$. This separation also permits estimation of θ_y as well as C_A.

The distance of maximum concentration from a single emitter will be determined by the height of the plume axis and the dispersive properties of the atmosphere. If the vertical spread of the plume increases as some fractional power $(1/n)$ of the distance downwind, then the distance of maximum concentration will increase as (plume height)^n. For tall chimneys the average distance of the maximum appears to be about $H^2/12$, where H is the plume height at the average wind speed [26].

The variation of GLC with distance is relatively flat in the vicinity of the concentration maximum, so the distance does not appear to be very critical.

Minimum requirements: arcs of at least three instruments at three distances or at two distances supported by single instruments nearer to and further from the source than the arcs and if possible between them. One of the arcs should be at the distance where the maximum GLC is expected to occur most frequently, calculated as indicated above or by some other method. See [27] for details of calculations of distance of maximum GLC.

9.6.7.3 Discrete source surveys — statistics of incidence of various levels of pollution

In this case the principal object would be to see how the frequency of occurrence of various levels of pollution from the discrete source(s) being investigated varies with direction and/or distance from the source.

In order to assess the variation with direction, the sites would normally be set out in a ring or concentric rings centred at the source. The spacing requirement is the opposite to that of Section 9.6.7.2 above: what is now required is that, when one site is in the plume, the adjacent sites are normally unaffected. If we regard unaffected as recording $< 10\%$ of the average recorded at the in plume sampler, again assuming a Gaussian distribution we have

$$0.1 \, (2) \int_0^{\theta_s/2} \exp - \theta^2/2\theta_y^2 \, d\theta > \int_{\theta_s/2}^{3\theta_s/2} \exp - \theta^2/2\theta_y^2 \, d\theta$$

as the criterion in this case.

Taking $\theta_s = 0.08$ radians, this criterion is satisfied if

$$\theta_s/2\theta_y > 0.22,$$

i.e. the separation of the sites should be at least $13°$; $22\frac{1}{2}°$ appears to be adequate for most locations [25].

If resources are sufficient for only one ring, this should be set out as close as possible to the estimated distance of maximum ground level concentration (Section 9.6.7.2). It would be an advantage, in this case, to have additional sites at around half and at twice the ring radius in the direction of the wind which is expected to give the most frequent and/or highest ground level concentrations.

If important topographical features are present, sites should be chosen where the maximum effects from these features are expected, in addition to, or possibly in place of, the regular pattern.

9.6.7.4 Area surveys (i.e. sites within the source area)

It is more difficult to specify the number of spacing of sites in an area survey. In general it will depend on the resources available for, and objects of, the monitoring exercise. The correct order of density seems to be something like $\sqrt{N}$ where N is the population in tens of thousands for cities over 250 000 (i.e. a minimum of five sites) or R, where R is the radius in kilometres, where the situation is not complicated by topography. In the latter case, instruments should be sited in areas which are expected to be especially difficult.

A number of objective methods for calculating the spacing of sites have appeared in the literature but they all presuppose some knowledge of the spatial variability of pollution and the correlations between readings at different sites, neither of which would be known until some measurements had been made.

In general, the centre of GLC will be displaced from the centre of emission by an amount which depends on the effective source height of the emissions. Dilution factors will be much greater for high level emissions than for low level emissions and the meteorological conditions resulting in the minimum atmospheric dispersion will also be different (Section 9.4). It is, therefore, important to site samplers near the centre of emission and downwind in the prevailing direction for stable, light wind conditions (for low level emissions) and downwind in the prevailing directions for strong winds and for convective conditions to observe maximum effects for high or medium level emissions.

9.6.7.5 Distant source surveys

Most of the requirements for siting instruments for this type of survey were mentioned in Section 9.6.6. To recapitulate:

(a) Effects of local, low-level sources should be avoided;

(b) Sheltering effects by absorbing features (hedges, copses etc.) should be avoided;

(c) Sheltering effects resulting from reduced atmospheric dispersion (siting in valleys, etc.) should be recognized and considered in relation to the overall objectives of the survey.

9.6.7.6 Global effects surveys

If a pollutant is to have a global effect, it is likely to have a long life in the atmosphere and therefore it is unlikely that sheltering effects will be important. The major consideration will, therefore, be the choice of sites in which fluctuations caused by individual sources are a minimum, i.e. as remote as possible from human activity.

9.6.7.7 Multipurpose surveys

Although a survey may have been designed for a specific purpose, e.g. monitoring a point source, it may also provide much useful information on other features of the pollution in the locality. It may be possible to maximize this subsidiary function at little additional expense by consideration of such possibilities at the planning stage. However, one should avoid the pitfall of trying to use information from existing surveys for some purpose other than that for which it was intended, if there are serious deficiencies in the data for that purpose. For example, although it may be reasonable to estimate the likely distribution of 24 h readings from a dispersion model predicting hourly average concentration distributions, it would be unwise to attempt to modify the dispersion model in light of any discrepancies which arose between observed and predicted 24 h readings. Comparison between a wide range of observed and predicted 1 h readings would be the only reliable way to do this.

9.6.8 Mobile monitoring

9.6.8.1 Surface systems

The location of the vehicle at any particular time must be recorded if the measurements made by a mobile system mounted in a motor vehicle are to be meaningful. The simplest way of doing this is simply to mark autographic recordings of the monitored pollutant(s) when recognizable features (intersections, etc.) are passed. The time the vehicle was at any point can then be deduced from the route map by interpolation. However, if the object of the survey is to produce measurements, at, for example, direction intervals of $1°$ referred to an origin at some prominent source, the reduction of such records can be a time-consuming business. Automatic recording of distance travelled and bearing, using a gyro or corrected magnetic compass, greatly facilitates such analysis. Data logging and if possible some computing facility should be available in the vehicle to support the navigational equipment.

Systematic traversing of the plume at one or more distances down wind is probably the most fruitful procedure in single source monitoring. However, if the standard deviation of the cross wind distribution of plume material is $5°$ (about 0.08 radians), a traverse which takes the vehicle out of the measurable plume at both sides will probably be at least $5 \times 5°$, i.e. 4 km at 10 km downwind. Allowing for orientation of the road not being ideal and the turn around at the end of each traverse, one is hard pressed to approach a total of ten traverses an hour. Four or five is a more realistic average figure, especially if a change of distance is involved between traverses.

Meandering of the plume tends to distort the pattern and so each traverse gives only an approximation to the instantaneous cross wind spread of the plume, therefore several hours sampling will be required to define the plume envelope in a way that could be related to patterns observed from fixed networks.

9.6.8.2 Airborne systems

Aircraft are generally expensive to operate but have the advantage of greater freedom of movement, three-dimensional capability and higher speed of traverse. Location now involves height as well and it is not possible to use the dead reckoning method of position-fixing because of the effect of wind drift. On-line presentation of processed data to the observer is more difficult than in a motor vehicle; especially if a light aircraft is involved.

Aircraft surveys are therefore more applicable to specialist investigations, especially those involving chemical reactions in the atmosphere, rather than for obtaining reliable information on dispersion characteristics of sources.

9.7 Data handling

9.7.1 Data transmission

If the information from the monitoring network is to be used on a real time basis then transmission to some control point will be essential. Even if the data are not to be used on-line for control purposes, data logging and inspection of the performance of the monitoring network are facilitated by such transmission. In general, each sensor will transmit at some predetermined interval and it is desirable that the value of the variable transmitted be averaged over the period between transmissions. Radio or telephone lines may be used.

In the absence of any control facility, the signal will still have to be transmitted from the sensor to some suitable recording device located nearby and in the case of electrical outputs it is important that the connecting cable be protected from any interference.

9.7.2 Data storage

For on-site recording, autographic or magnetic tape cassette recording is the most suitable. Accurate time marking is essential if the data are to be correlated with variables recorded at other sites. For the central data logging facility, magnetic tape or punched paper tape are suitable as the immediate output. If the data are fed into a computer, either subsequently or on-line, then magnetic disc storage of either the raw or processed data is often preferred.

In choosing the interval between readings for digital recording the following factors are important: the response time of the instrument; the expected minimum time for an effect from the pollutant on the receptors of interest to be noticed; and limitations to data storage and expenditure on computing time.

It is desirable to run autographic records in parallel with digital recording because information is not completely lost if the digital or transmission system fails and because of the ease with which one can assess the main features of the data (and often malfunctioning of the sensor) by quick visual inspection of anaolgue records. Multipoint recorders may be used in conjunction with the data logger, or on-site if more than one variable is being recorded there.

9.7.3 On-line alarm/display systems

The alarm would indicate to the plant control engineer when a situation requiring remedial action has arisen. It may be audible, visible or both. The network data may be displayed on a map, either digitally or by dials, or by a print-out. The print-out may be produced at intervals automatically or by interrogation. If the central logging system includes a computer, the display could include averages, peaks, standard deviations etc., over longer periods (e.g. over 1 h if the data were recorded every 1 to 5 min). In this case the alarm might operate when a given time average value was exceeded or a level was exceeded for more than a given number of scans.

Presentations for subsequent assimilation will be dealt with under 'Analysis of results' (Section 9.8).

9.7.4 'On-line' recognition of defective readings

It is obviously undesirable to initiate expensive control measures on the basis of high readings of pollutant concentration which subsequently are shown to have been erroneous. However, the elimination of erroneous data is often a difficult business because one is constantly seeking the abnormal, and elimination of extreme readings as suspect could defeat the object of the exercise. Familiarity with the instrumentation will normally expose fairly well-defined failure patterns and the bulk of defective readings can usually be eliminated by setting up suitable criteria to cater for these. The elimination process is greatly aided if there is some redundancy in the network, and/or built in calibration facilities and zero checks.

9.8 Analysis of results

9.8.1 General

In addition to the data available from the monitoring network, there will be background information which is of relevance (historical data, for example (Section 9.8.2)) and possibly concurrent monitoring of relevance to the investigation by other organizations (e.g. meteorological data by the national weather service). To some extent the choice of an area for specialized monitoring may be influenced by the existence of useful parallel monitoring or historical data.

Correction of defective readings on-line was mentioned in Section 9.7.4; retrospective correction is discussed in Section 9.8.3. The presentation of valid data is discussed in general terms in Section 9.8.4 and physical models in Section 9.8.5. Examples follow in Section 9.9.

9.8.2 Availability of historical data

9.8.2.1 Source inventories and characteristics

It is possible that some emission monitoring (or function monitoring) will permit hourly estimates to be made of emissions from some or all of the sources affecting the sampling area. However, it is more likely that only general historical information, such as total sales of fuel to householders, traffic density in selected census periods etc., will be available. Even so, it should be possible to build up a reasonably accurate estimate of emission from the various types of sources at different times of the day and year and in different weather conditions. In addition to estimates of the emission from each source, an estimate of the effective height of release should be made. This will vary with meteorological conditions and to some extent with rate of emission. In many cases it is likely that the plumes from small sources with short chimneys will be subject to interference from the wakes of adjacent buildings, so the effective height in moderate or strong winds may be less than the actual height of release (see, for example Section 5.5 of [4] and also [28]).

The location of sources with respect to any marked topographical features (Section 9.8.2.3) is also important (see, for example, pp. 30–3 of [29]).

9.8.2.2 Climatological data

The following information is likely to be of use:

(i) Frequency of occurrence of different wind speeds at different times of day in different direction ranges (if available over range of effective source heights) including local winds such as sea breezes, etc. (Section 9.4).

(ii) Frequency of occurrence of different mixing depths [in association with (i) if possible].

(iii) Strength or extent of stable layers.

(iv) Rainfall and humidity.

(v) Sunshine and cloud cover.

(vi) Atmospheric pressure.

9.8.2.3 Topographical information

The existence of any marked topographical features which prevent the spreading out of pollution, disrupt and prevent the proper rise of plumes or reduce the effective height of plumes passing overhead should be noted. In a well-designed survey they should have been taken into account when the location of sampling points was decided (Section 9.6). Important features include changes of surface characteristics at urban boundaries, woods and forests, coasts, rivers, etc.; and hills, mountains, valleys etc.

9.8.2.4 Experience of similar source networks or other information (e.g. epidemiological) or damage/dosage relations for the pollutants under investigation

Such information as is available from earlier monitoring in approximately similar locations with similar sources would provide valuable assistance, both in planning the network and in deciding if the readings, when obtained, are reasonable In the absence of proven dispersion or of other physical models, developing analogue prediction models with the aid of any additional available data may be the most practicable course (see also Section 9.8.4.6).

Where dosage/emission relations have been established, dosage/damage data from earlier field or laboratory studies will help determine or may have already established whether the observed pollutant levels are a cause for concern.

9.8.3 Elimination of erroneous readings

Introduction of errors when samples are taken from field monitoring points to a central laboratory — often with a time lapse of several weeks — has been discussed by Paterson [29], who used a correlation technique for deciding whether or not observations were reliable. Continuously recording instruments are normally not subject to common errors, except as a result of bad design or calibration technique.

Where there has been some change in zero or sensitivity between calibrations, it is usually possible to decide at which point the observations became unreliable if there is some redundancy in the network, e.g. if the readings are normally correlated with those from adjacent sites and the correlation pattern changes markedly. If more than one pollutant is being monitored, correlations between pollutants may become anomalous. However, some faults, e.g. volume sampling rate errors, may be common to all pollutants.

Calibration checks should, if possible, involve at least one intermediate point in addition to checking zero and full-scale readings, otherwise changes in response characteristics (e.g. linearity) may not be detected.

9.8.4 Statistics

9.8.4.1 General

The display of the collected results is probably the most important part of any monitoring exercise, if it is to be of any general use. It must also be borne in mind that many of those who may need to make decisions based on such information are not experts in chemical analysis, meteorology or statistics so the presentation of the results in an easily assimilable manner is very important.

9.8.4.2 Mean values over specified averaging periods

Because the concentration of pollutants fluctuates with time, the highest concentration measured during a given sampling period decreases as the length of the averaging period increases.

The mean of all the three-minute periods in an hour will, of course, be the hourly mean, but we might, for example, compare the highest three-minute mean during an hour with the hourly mean value and express this ratio of the highest mean over the shortest averaging period divided by the mean over the whole sampling period as a peak/mean ratio.

In general the peak/mean ratio decreases as the mean increases, because a high mean value usually implies that pollutant plume directions have remained steady during the sampling period, so there is less variability in short-period average concentrations, and also that the sampling point is near the centre rather than near the edge of the pollutant cloud distribution over the sampling period.

It is, therefore, important to qualify information on peak/mean ratios by such statements as 'at the point of maximum (hourly average) concentration' or 'for occasions when the hourly average GLC exceeded 1 pphm', if they are to be interpreted and applied correctly.

9.8.4.3 Frequency distributions

The frequencies with which given concentrations, for given averaging periods, are exceeded at each sampler, together with the absolute maximum concentration, are probably the most important statistical data provided by the survey. Separate frequency distributions may be prepared for different periods of the day or year (e.g. night/day, summer/winter), for different wind directions or weather situations etc., if sufficient data are available. Data from several sites may be combined in some or all of these distributions.

Other useful information is the distribution of periods for which a given concentration is exceeded during a day, year or similar interval. The data are sometimes plotted on log probability paper, which has the advantage that if the frequency distribution is log normal, the cumulative frequency distribution appears as a straight line, the 50% value is the geometric mean and the slope of the line gives the geometric standard deviation.

9.8.4.4 Diurnal or annual variations

The readings over specific periods of the day, if readings have continued over several days, or specific parts of the year, if readings have continued for several years, may be averaged to investigate diurnal or annual variations. Some measure of the scatter of the individual readings about the mean values should also be obtained to enable one to decide whether any cyclic variations which appear are significant. It may be convenient to subdivide the data according to wind speed and/or cloud cover if diurnal effects are being investigated.

9.8.4.5 Mean values in different weather situations and/or wind directions

Both the emission and the dispersion of the effluent material before it returns to ground level may be affected by weather conditions. Subdivision of data by time of day or year (Section 9.8.4.4) will, to some extent, take account of this, but further investigation of the effect of temperature, vertical temperature gradient, wind direction and/or speed may be desirable. Any marked effects should be apparent from mean values but the scatter of the individual readings from which the means were deduced is also important in assessing significance.

A convenient method of displaying wind direction effectiveness is the 'pollution rose' (see Section 9.9).

9.8.4.6 Correlations and regression analysis

Correlations between concentrations of the same pollutant and any of the following measured over the same or different time intervals may be of interest: meteorological variables; other pollutant concentrations at the same site; damage; the same or other pollutant concentrations at other sites.

Correlations between successive measurements of one pollutant at the same site at different time intervals (auto-correlations) are also useful (e.g. in deciding the distance of the principal sources affecting the site). If multiple regression techniques are used it must be remembered that relations with no physical basis are not likely to be applicable elsewhere unless data from other locations (Section 9.8.2.4) has been included in their development. If possible the relations developed should have their basis in some simple physical model. It must also be remembered that relations developed by statistical techniques only are insufficient to establish causal relations between variables especially where dosage/damage relations are concerned. The existence of a third, common, unmonitored factor should always be suspected if it is difficult to establish the causal relation independently.

9.8.5 Physical models

9.8.5.1 Dispersion models

In a dispersion model, the ground level concentration pattern is calculated for each source by assigning an effective height to the plume (see for example

[6, 7]) and then using one of a number of techniques to estimate the spread of the effluent material as it travels downwind. The overall pattern is then derived by summing the contributions of the individual sources.

Small sources of similar type, e.g. domestic, commercial or vehicles, may be treated as area or line sources, with the total estimated or observed emission divided by the estimated area or length of the source to give the source strength.

Broadly speaking, dispersion models fall into two types: (a) diffusivity or 'K-theory' models; and (b) Gaussian plume models.

In type (a), a diffusivity K and a three-dimensional wind velocity vector $\mathbf{U}$, which will in general vary with position, are attributed to each part of the pollutant's path between source and receptor areas and the distribution of material calculated by solving the diffusion equations with appropriate boundary conditions.

These models are often simplified by treating $\mathbf{U}$ as one-dimensional and $\mathbf{U}$ and K as functions of height only.

In type (b), the distribution of material is assumed to be normally distributed about the effective height (H) of the plume so that the concentration at points (x, y, z) with the source of strength Q as origin is given by

$$C(x, y, z) = \frac{Q}{2\pi \mathbf{U} \sigma_z \sigma_y} \exp \frac{-(H-z)^2}{2\sigma_z^2} \exp \frac{-y^2}{2\sigma_y^2}.$$

Additional terms, representing the effect of reflection of the plume at the ground or at the top of the mixing layer, may be included. σ_z and σ_y, the vertical and lateral standard deviations of the concentration distributions, increase with distance downwind and a number of methods have been proposed for estimating them either from measurements of atmospheric turbulence or from the general properties of the mixing layer.

In both 'K-type' and Gaussian models, the effects of surface deposition of material and reactions between pollutants may be included in the calculations if these are thought to be important. Other effects, such as the additional spreading of the plume caused by its rise (Section 9.4.10), may also be taken into account [18]. Neither model is completely satisfactory, because 'K-theory' applies only when successive movements of the dispersing material are uncorrelated, which is certainly not the case, while the Gaussian model applies only when the turbulence is stationary and homogeneous.

As a compromise, Moore [7, 20] has developed a modified Gaussian model in which, among other innovations, σ_z is assumed to vary with source height. This approach gave good agreement between observed and calculated ground level concentrations downwind of power station chimneys.

9.8.5.2 Box and cell models

In a box model, the effects of individual sources are ignored, and the air stream (often termed an airshed) under investigation is treated as if it were a single box or a number of boxes (cells). The depth of the boxes is normally the maximum depth of the mixing layer or internal boundary layer [e.g. the urban boundary layer, (Fig. 9.3)] during the period of each time step.

The surface area of each box may correspond to some reasonably homogenous part of the area or may be fixed arbitrarily, e.g. as the area round each pollutant sampling site or a square of given side length. Unless special precautions are taken to reduce spurious mathematical diffusion effects, the sides of the boxes should be kept parallel and normal to the local wind direction in multiple box or cell models.

For each of a number of time steps, whose duration is normally between one hour and one day, depending on the degree of spatial and temporal variability which the model is designed to represent, the emissions within each box are assumed to be uniformly mixed throughout it. Pollutant removal or conversion processes could be taken into account if this were deemed necessary. The inflow and outflow of pollutant from each box are calculated from estimated wind speeds and directions. In its simplest form, this type of model ignores high pollution produced by nearby low level sources and the zone of zero pollution around high level sources. A detailed discussion of box and cell models is given in [11], beginning on p. 385.

9.8.5.3 Calculation of trajectories

When the time of travel of a pollutant is many hours, the wind speed and direction may change appreciably between the time it leaves the source and the time it arrives at the receptor.

In this case it will be necessary to construct a trajectory, e.g. to use estimates of wind speed and direction, to work backwards from the place and time an 'incident' was observed, or forwards from the time and place of a release of pollutant, to find the probable path of the pollutant. One difficulty is that air at different levels is often travelling in different directions and may be rising or descending. Danielsen [30] has demonstrated that real isentropic trajectories, particularly in the vicinity of a rain-producing weather system, may be very different from those estimated from two-dimensional (isobaric) flow patterns. This difficulty, which has not been resolved, may be especially important in deciding the origin of dissolved material in precipitation.

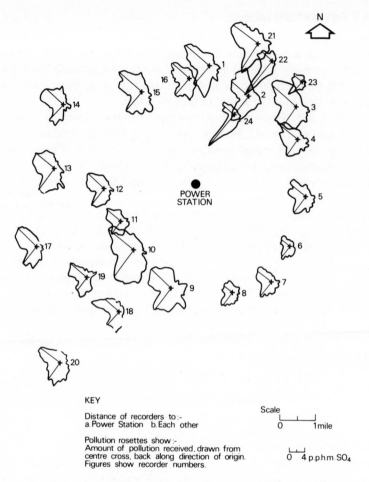

KEY

Distance of recorders to:-
a. Power Station b. Each other

Pollution rosettes show :-
Amount of pollution received, drawn from
centre cross, back along direction of origin.
Figures show recorder numbers.

Scale

0   1mile

0   4 p.p.h.m. SO₄

Fig. 9.5 Variation of average SO$_2$ concentration with wind direction at each recorder site at High Marnham.

9.9 Examples of monitoring networks and data presentations

9.9.1 General

The principal features of some examples of the various types of surveys enumerated in Sections 9.6.7 and 9.6.8 are given in Sections 9.9.2 to 6. Multi-purpose surveys (Section 9.6.7.7) are not included separately, but where a network has been used to provide information on subjects other than the major purpose of the investigation, these are mentioned. The discussions are restricted to major programmes mostly occupying periods of a year or more, but are not intended to be comprehensive.

9.9.2 Discrete source surveys

9.9.2.1 National air pollution control administration studies of SO_2 at Keystone, Homer City and Conemaugh power plants [31]

Type of measurements: 'instantaneous' three-dimensional SO_2 recordings of plume cross-sections by helicopter at 4 to 4.8, 10 and 16 km downwind. Six to ten 30 min average SO_2 bubblers on ground beneath plume. Supporting meteorological measurements: wet and dry bulb temperatures and pressure by helicopter and radiosonde, pilot balloon measurements of wind speed and direction to above plume level.

Presentation of data: Tabulations of original readings in three large volumes of over 1000 pages in all. Autographic charts were used for continuous monitoring and a number of supplementary investigations were made concurrently with the main investigations (e.g. plume photography, isotope ratios).

9.9.2.2 C.E.G.B. studies of SO_2 around High Marnham and West Burton power stations [25, 32, 17]

Continuous ground-level recording (autographic) of SO_2 (minimum averaging period: 3 min) with patterns of up to 22 recorders shown in Fig. 9.5 (High Marnham) and Fig. 9.6 (West Burton). These patterns conform to the spacing criteria for 'statistical' point source surveys given in Section 9.6.7.3.

Figs. 9.5 and 9.6 also illustrate one of the forms of data display pollution roses (Section 9.8.4.5). The variability of average concentrations at instruments in similar locations, even with careful siting, is noticeable in Figs. 9.5 and 9.6. This is probably due to local sheltering (and possibly source) effects. The high readings with pollution coming from the north-west to south-west direction illustrate the dominance of distant sources over the large power station emissions.

Methods of identifying the power station contribution were: (i) recognition

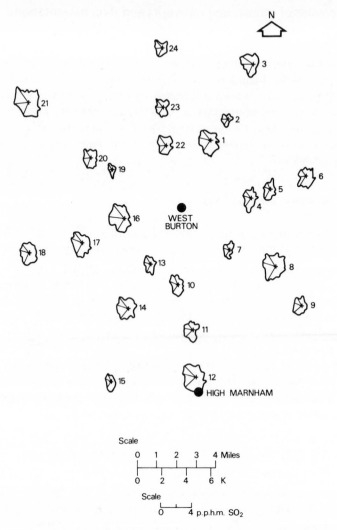

Fig. 9.6 Variation of average SO_2 concentration with wind direction at each recorder site (Summer 1966).

of a characteristic peaky trace above the relatively steady background contribution; and (ii) comparing the results from adjacent recorders, subtending an angle of about 90° at the source, and subtracting the mean reading of the outer two as background from the readings of the inner three as plume. Plume readings (⩾ 1 pphm), however, rarely showed on more than one meter at a time [25].

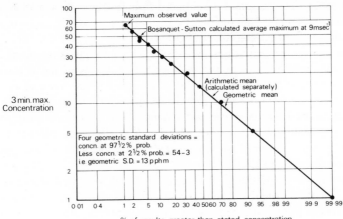

3 min maxima

	No. of recorder-hours in which given 3 min max. was recorded. Recorders							
	12	11, 15	10, 14	7, 8, 9, 13.	4, 5 16, 17	1, 6, 18, 19, 20, 22	2, 23	3, 21 24
	Distance from source							
Concn. (pphm)	$\frac{1}{2}$ 0.8	3 5	$5\frac{1}{2}$ 9	7 11	10 16	$12\frac{1}{2}$ 20	14 23	17 (miles) 28 (km)
0	173	322	258	419	452	635	409	285
1–4	27	10	19	62	61	91	22	22
5–9	46	36	44	115	118	67	16	16
10–14	13	35	25	82	36	18	5	1
15–19	8	17	9	30	7	9	0	1
20–24	6	26	7	9	6	5	2	1
25–29	8	8	6	6	2	1		
30–34	6	8	3	1	1			
35–39	5	4	0		0			
40–44	2	3	1		1			
45–49	1	2						
50–54	1	0						
55–59	2	1						
60–64	0	0						
65–69	2	2						
70, 71, 86, 90	1 each							
	131	152	114	305	232	191	45	41

Frequency of occurrence of given concentration ranges W 65/6 and S 66.

Fig. 9.7 Tabular display of 3 min maxima of SO_2 from High Marnham at different distances. Upper diagram shows distribution at 5 km plotted on log probability paper.

Wind speed and direction were obtained from instruments on a 45 m tower at the power station site and supplementary data on lapse rate and 450 m wind were estimated by the Meteorological Office. Frequencies of occurrence of 3 min or hourly average concentrations were displayed graphically as functions of wind speed. Other presentations of the data were frequencies of the occurrence of various concentration ranges for different averaging times. Some of these were subdivided by temperature lapse conditions, with further subdivisions by wind speed ranges and between winter and summer. Data display was mostly tabular, but log–probability plots were also used (Fig. 9.7). Occasions with exceptionally high GLCs were described in some detail. Variation with distance from the source was discussed in [32].

Concentration distributions were compared with values calculated from dispersion equations, and background concentration distributions, mostly from distant sources in this rural area, were also tabulated. So, although the principal objective of the survey was acquisition of statistics of GLCs around power stations, it could also be regarded as multi-purpose survey (Section 9.6.7.7) because of the important applications to dispersion models and distant sources.

9.9.2.3 C.E.G.B. Tilbury–Northfleet plume rise and dispersion experiment [33]

The instruments used to sample SO_2 were the same as those described in Section 9.9.2.2, this time set out with the primary purpose of testing dispersion models (Fig. 9.8). The angular spacing of $7\frac{1}{2}°$ on the arcs conforms to the requirements of Section 9.6.7.2. In support of the ground level measurements, the plume height was recorded by cloud-searchlight [33] and lidar [34] on several hundred occasions. Meteorological data were provided by instruments on a 184 m tower and at the power station site.

The experiment began in 1963 and recording was by autographic charts at each site, supplemented by magnetic tape recorders in the final year (1968). The chart data was digitized on semi-automatic chart reading equipment for each 12 h period when the 114 m wind direction from the meteorological tower indicated that the plume from either power station should have been over the sampling network. Wind speed and direction were processed in the same way, readings being taken every three min. The computer then produced maps of the hourly average concentration at each site, the peak reading for the hour and, using the computed trajectories from the two stacks, maps of SO_2 attributed to the Northfleet and Tilbury plumes. Fig. 9.8 is an example of such a map showing the total contribution at each site for the hour (i.e. without subdivision as to source).

With the plume heights fixed by observations at the sites, the analysis was

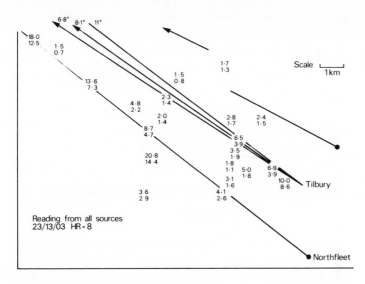

KEY

Date time group, bottom left-hand corner

03	Time of beginning of period (0330)
13	Day
23	Month (11th month of second year; November 1964)
HR=8	8th hour of period, i.e. 10·30 - 11·30

Upper reading	Maximum during hour
Lower reading	Average during hour (p.p.h.m.)
	N.B. Decimal point of maximum reading gives meter locations on diagrammatic print-out

Line 11°	Direction of 114 m wind
8·1°	Direction from SO_2 distribution in near arc
6·8°	Direction from SO_2 distribution in second arc
	The bearing is measured in degrees off the meter axis, positive readings indicate that the axis of the plume lies east of the meter pattern axis

Fig. 9.8   Surface pollution pattern (hourly mean values, period 10.30 to 11.30 h; 13 November 1964) [41].

directed towards testing and developing dispersion formulae. Fig. 9.9 shows a comparison between mean values of the one h average maximum ground level concentrations and the values calculated from the prediction models developed. The data are presented as functions of wind speed for the major stability sub-divisions 'unstable', 'slightly stable' and 'stable'. The model is described in [20] and other references given in that paper.

Scatter of the individual hourly readings about the mean for the hour in the different stability groups as functions of the wind speed are illustrated in Fig. 9.10.

In 1969 this study was transferred from the S.E. Region to the 2000 MW power station at Eggborough in the N.E. Region [35], where the sampling

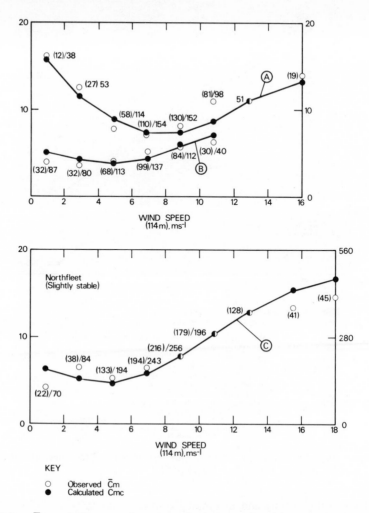

KEY

○ Observed $\bar{C}m$
● Calculated Cmc

Fig. 9.9 $\bar{C}m$ and Cmc as a function of wind speed in (a) unstable; (b) stable; and (c) slightly stable meteorological conditions.

network has been designed both to provide statistics of SO_2 concentration incidence and to test dispersion models (Fig. 9.11).

Data storage is on paper tape produced at a central data logger, where signals from each SO_2 recorder and the supporting meteorological data from a television mast almost 400 m tall are received every three minutes. The tapes are processed each day by computer, which again plots maps of peak and mean

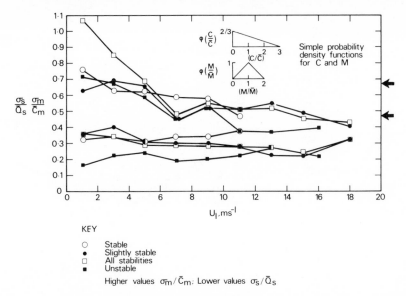

KEY

○ Stable
● Slightly stable
□ All stabilities
■ Unstable

Higher values $\sigma_{\bar{m}}/\bar{C}_{m}$; Lower values $\sigma_{\bar{s}}/\bar{Q}_{s}$

Fig. 9.10 Scatter of value of observed maximum GLC (σ_m/C_m) and SO_2 emission ($\sigma_s/\bar{Q}_s$) as a function of wind speed for various stability groups (Northfleet plume).

hourly concentrations at each site, together with the maximum and minimum reading and number of valid readings received each hour.

9.9.3 Area surveys

The B.P. Reading survey [36] was a detailed study designed to test dispersion and empirical models for a town of 120 000 inhabitants, spread over an area of about 100 km^2. There were forty monitoring sites for SO_2 (Fig. 9.12) – far more than would normally be installed, in a routine survey (Section 9.6.7.4). Concentrations were averaged over 6 h. Supporting meteorological data [36 (II)] were made at a height of about 13 m from a fixed and a transportable mast and from a balloon sonde.

Detailed source inventories were prepared and estimates of emission from low level sources were made from historical data (Section 9.8.2.1). Six hourly fuel consumption figures were available for some of the larger sources. Data were displayed as contour maps (Fig. 9.12 is an example) for all wind directions and for each direction octant. Empirical expressions gave better correlations with the

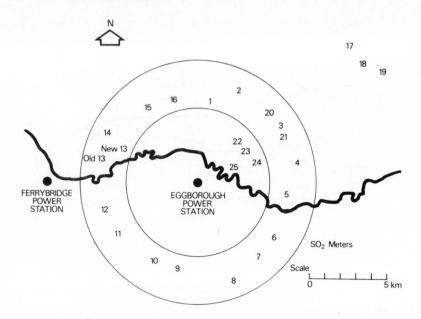

Fig. 9.11 Eggborough SO$_2$ survey, layout of SO$_2$ recorders.

data than physically based dispersion equations. However, Pasquill [11] demonstrated that the physical models gave a good representation of the overall average value and frequency distribution. There were large discrepancies between individual observed and predicted values due, Pasquill suggests, to the practical uncertainties in specifying meteorological conditions precisely at all points and times.

An interesting report giving guidance on network design in urban areas based on sampling requirements and information available from the results of past surveys (e.g. at Nashville and Sheffield) is given by Blokker [37].

9.9.4 Distant sources and global effects

One of the latest examples of a network set up to look at the effects of distant sources is the 'Long Range Transport of Air Pollutants' LRTAP, an O.E.C.D. project, measuring chemical components in precipitation and SO$_2$ and particulate sulphate in air. Fig. 9.13 shows the location of some of the sampling stations and some interesting historical data relevant to the project, the 1972 emissions of SO$_2$ in 127 km squares over Europe. The map appears in a paper [38] in which decay rates for SO$_2$ and transformation rates of SO$_2$ into particulate sulphate are

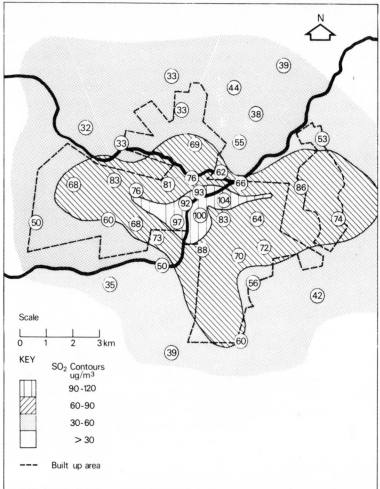

Fig. 9.12 SO$_2$ averages for the whole year (all wind directions) at Reading.

derived from the data. The sampling period for the measurements was 24 h. Part of the work involved trajectory calculations and reference is given in the paper to the method used to calculate them.

An alternative approach to the same problem is described in a paper by Smith and Jeffrey [39]. Vertical cross-sections of SO$_2$ and sulphate were measured by a sampling aircraft flying approximately normal to the wind direction over the North Sea. SO$_2$ emission was calculated for 20 km side squares over the U.K. and trajectories from source to sampling area were calculated, with estimates of the uncertainty in the calculated paths.

Measurements of atmospheric dust concentrations over the North and South Atlantic using a mesh technique are described in [40]. These authors found the concentrations of Saharan dust decreased by an order of magnitude over a distance of about 3000 km but that the dust was still present in quantities of around $10 \mu g\,m^{-3}$ over the West Indies, three orders of magnitude higher than concentrations of dust over the western approaches to the British Isles, despite the large industrial emissions in north-eastern U.S.A.

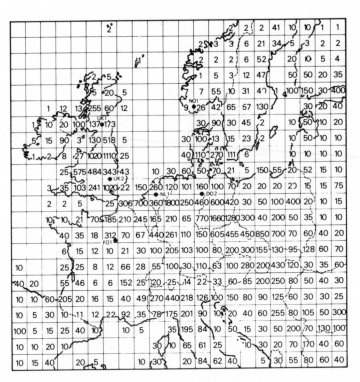

Fig. 9.13 European SO_2-emission 1972. Unit: 10^3 metric tonnes y^{-1}. (OECD-project 'Long Range Transport of Air Pollutants' [LRTAP] preliminary report). Better information has now been received for many countries. For instance, total 1973 emissions of SO_2 for the U.K. were about 5.7 million metric tonnes, which is 12.3% less than the 6.5 million tonnes indicated in Fig. 9.1. Similar uncertainties may of course apply to the estimated emissions of other countries.

Estimated values of SO$_2$ decay and transformation rates, etc.

Samp. Station	Date of 1st day (1973)	No. of days	k_0 10^{-5} s^{-1}	k_1 10^{-6} s^{-1}	h_0 10^3 m	R_{SO_2}	R_{SO_4}	$\bar{a}_d$ µg m^{-3}	$\bar{q}_d$ µg m^{-3}	S_a µg m^{-3}	S_q µg m^{-3}	$\bar{b}_d$ µg m^{-3}	S_b µg m^{-3}
DO2	1 April	62	1.3	0.78	2.2	0.514	0.466	14.4	13.1	12.6	8.4	2.5	2.0
FO1	1 April	62	2.2	1.9	0.57	0.537	0.538	24.9	25.9	26.8	12.9	5.9	8.4
NO1	1 Dec.	59	1.4	1.2	1.2	0.527	0.551	6.4	5.1	7.5	5.0	3.0	3.2
NO1	15 Dec.	58	0.72	1.6	1.8	0.469	0.602	8.4	7.0	10.1	6.1	4.5	3.7
NL1	1 June	61	2.5	4.8	2.0	0.602	0.532	10.2	11.0	10.0	4.4	10.5	9.4
NL1	1 Aug.	60	1.6	2.2	1.7	0.722	0.370	17.3	19.5	16.8	8.5	11.0	9.7
NL1	1 Oct.	61	1.2	3.1	1.0	0.564	0.673	25.5	27.8	26.9	14.1	11.1	10.8
UK1	1 April	60	3.6	1.2	0.95	0.455	0.469	28.6	28.2	18.9	10.9	6.2	4.1
UK1	1 Oct.	61	3.9	1.5	0.64	0.437	0.389	36.5	41.1	18.7	11.5	9.7	7.2
UK2	1 Aug.	62	2.7	3.0	1.0	0.678	0.688	8.2	8.1	10.1	6.7	5.5	6.8
UK2	1 Oct.	53	3.6	2.3	0.65	0.527	0.551	12.8	11.8	13.1	8.8	5.0	5.8

Symbols (mean value over period designated by a bar): k_0 estimated SO$_2$ decay rate; k_t estimated transformation rate SO$_2$–SO$_4$; h_0 estimated mixing height parameter; R_{SO_2} empirical correlation coefficient between observed and computed SO$_2$ concentrations; R_{SO_4} empirical correlation coefficient between observed and computed SO$_4$ concentrations; $\bar{a}_d$ mean value of observed SO$_2$ concentrations; $\bar{q}_d$ mean value of computed SO$_2$ concentrations; S_a empirical S.D. of observed SO$_2$ concentrations; S_q empirical S.D. of computed SO$_2$ concentrations; $\bar{b}_d$ mean value of observed SO$_4$ concentrations; S_b empirical S.D. of observed SO$_4$ concentrations.

Acknowledgements

Figs. 9.6–8, 9.12 and 9.13 are reproduced by permission of Pergamon Press. Figs. 9.9 and 9.10 are reproduced by courtesy of the Council of The Institution of Mechanical Engineers from 'A Simple Boundary Layer Model for Predicting Time Mean Ground-level Concentrations of Material Emitted from Tall Chimneys' by Dr Moore. Figs. 9.5 and 9.11 are reproduced from internal C.E.G.B. reports; similar diagrams appeared in the Journal of the Institute of Fuel and the Tribune de Cebedeau respectively, and are reproduced by courtesy of the publishers.

References

[1] W.H.O. International Agency for Research on Cancer (1973). *Ann. Occ. Hygiene* **16**, 9–18.
[2] Summers, P.W. (1975). Private communication.
[3] Junge, C. (1972). *Q. Jl. R. met. Soc.* **98**, 711–29.
[4] Lovelock, J.E. (1972). *Atmos. Environ.* **6**, 579–80.
[5] Munn, R.E. and Bolin, B. (1971). *Atmos. Environ.* **5**, 363–402.
[6] Briggs, G. (1969). Plume rise, U.S.A.E.C. Division of Technical Information.
[7] Moore, D.J. (1975). *Atmos. Environ.* **8**, 441–58.
[8] Robins, A.G. (1975). *Proc. Instn. Mech. Engrs.* **189**, 44–54.
[9] Chamberlain, A.C. (1966). *Proc. R. Soc. A* **290**, 236–65.
[10] Pasquill, F. (1972). *Q. Jl. R. Met. Soc.* **98**, 469–94.
[11] Pasquill, F. (1974). *Atmospheric Diffusion* (2nd edition), Ellis Horwood, Chichester.
[12] Wipperman, F. (1973). *The planetary boundary layer of the atmosphere*, Deutscher Wetterdienst, Offenbach.
[13] Carson, D.J. (1973). *Q. Jl. met. Soc.* **99**, 450–67.
[14] Slade, D.H. (ed.) (1968). *Meteorology and atomic energy*, U.S.A.E.C.
[15] Bringfelt, B., Hjorth, T. and Ring, S. (1974). *Atmos. Environ.* **8**, 131–48.
[16] W.H.O. (1970). Urban climates, Technical Note No. 108.
[17] Martin, A. and Barber, F.R. (1973). *Atmos. Environ.* **7**, 17–38.
[18] Moore, D.J. (1973). *Farad. Symp. Chem. Soc.* **7**, 222–8.
[19] Singer, I.A. and Smith, M.E. (1966). *Air Water Pollution Int. J.* **10**, 125–36.
[20] Moore, D.J. (1975). *Proc. Instn Mech. Engrs* **189**, 33–43.
[21] Lucas, D.H. and Moore, D.J. (1964). *Air Water Pollution Int. J.* **8**, 441–54.
[22] Charlson, R.J., Horvath, H. and Pueschel, R.F. (1967). *Atmos. Environ.* **1**, 469–78.
[23] Pilat, M.J. and Ensor, D.S. (1971). *Atmos. Environ.* **5**, 209–16.
[24] Biles, B. and Ellison, J. McK. (1975). *Atmos. Environ.* **9**, 1030–2.
[25] Martin, A. and Barber, F.R. (1966). *J. Inst. Fuel* **39**, 294–307.
[26] Moore, D.J. (1969). *Phil. Trans.* **265**, 245–60.
[27] Moore, D.J. (1974). 2nd IUTAM Symp. 1973. *Adv. in Geophys.* **18B**, 201–21.

[28] Smith, M.E., (ed) (1968). *Recommended guide for the prediction of the distance of airborne effluents*, A.S.M.E.
[29] Paterson, M.P. (1975). The atmospheric transport of natural and man-made substances, Ph.D. Thesis, University of London.
[30] Danielsen, E.F. (1974). *Adv in Geophys.* **18B**, 73–94.
[31] Schiermeur, F.A. and Niemeyer, L.E. (1970). *Large Power Plant Effluent Study* (Lappes), Vols. 1 and 2, U.S. Department of Health Education and Welfare (1970); Vol. 3, U.S.E.P.A. (1972).
[32] Martin, A. and Barber, F.R. (1967). *Atmos. Environ.* **1**, 655–78.
[33] Lucas, D.H., James, K.W. and Davis, I. (1967). *Atmos. Environ.* **1**, 353–66.
[34] Hamilton, P.M. (1969). *Phil. Trans.* **A265**, 153–72.
[35] James, K.W. (1973). *Tribune de Cebedeau*, **361**, 1–7.
[36] Marsh, K.J. (1967). (and M.D. Foster) *Atmos. Environ.* **1**, 527–50; (with K.A. Bishop and M.D. Foster) *Atmos. Environ.* **1**, 551–60; (with V.R. Withers) (1969) *Atmos. Environ.* **3**, 281–302.
[37] Blokker, P.C. (1973). Major aspects in air pollution monitoring in urban and industrial areas, Stichting Concawe, the Hague, Report No. 7/73.
[38] Eliasson, A. and Saltbones, J. (1975). *Atmos. Environ.* **9**, 431–6.
[39] Smith, F.B. and Jeffrey, G.H. (1975). *Atmos. Environ.* **9**, 643–60.
[40] Parkin, D.W., Phillips, D.R., Sullivan, R.A.L. and Johnson, L.R. (1972). *Q. Jl. R. Met. Soc.* **98**, 798–808.
[41] Moore, D.J. (1967). *Atmos. Environ.* **1**, 389–410.

Index